AF754547

GENETICS OF ADAPTATION

Georgia Genetics Review III

VOLUME 3

Genetics of Adaptation

Edited by
RODNEY MAURICIO

Reprinted from *Genetica*, Volume 123(1–2) 2005

Springer

A C.I.P. Catalogue record for this book is available from the Library of Congress.

ISBN 1-4020-3476-8

Published by Springer,
P.O. Box 17, 3300 AA Dordrecht, The Netherlands.

Printed on acid-free paper

All Rights Reserved
© 2005 Springer
No part of this work may be reproduced, stored in a retrieval system, or transmitted in any form or by any means, electronic, mechanical, photocopying, microfilming, recording or otherwise, without written permission from the Publisher, with the execption of any material supplied specifically for the purpose of being entered and executed on a computer system, for exclusive use by the purchaser of the work.

Printed in the Netherlands

CONTENTS

Genetica (2005) 123: 1–2 © Springer 2005

Natura non facit saltum

One of the most enduring controversies in evolutionary biology is the genetic basis of adaptation. Darwin emphasized 'many slight differences' as the ultimate source of variation to be acted upon by natural selection. In the early part of the 20th century, the 'Biometrical School' emphasized the importance of gradual transformation in the evolution of adaptive traits. Opposed to this view were the 'Mendelian geneticists,' who emphasized the importance of 'macromutations' in evolution. In his landmark 1930 book, *The Genetical Theory of Natural Selection*, R.A. Fisher seemingly resolved this controversy, demonstrating that mutations in genes of very small effect were responsible for adaptive evolution. As H.A. Orr and J.A. Coyne stated in their 1992 paper (Am. Nat. 140: 725–742): 'the neo-Darwinian view has ... triumphed, and the genetic basis of adaptation now receives little attention. Indeed, the question is considered so dead that few know the evidence responsible for its demise.'

Orr and Coyne reexamined the evidence for this neo-Darwinian view and found, surprisingly, that both the theoretical and empirical basis for it were weak. Orr and Coyne encouraged evolutionary biologists to reexamine this neglected question: what is the genetic basis of adaptive evolution? The answer to this question, said Orr and Coyne, could come only from 'genetic analysis of adaptive differences between natural populations or species.'

The study of the genetics of adaptation is an emerging field of inquiry that is central to the study of organic evolution. Ultimately, an understanding of adaptive evolution will require detailed knowledge of the genetic changes that accompany evolutionary change. The genetic basis of phenotypic variation for traits involved in adaptive responses is often complex. This complexity arises from segregation of alleles at multiple interacting loci (Quantitative Trait Loci, or QTL), whose effects are sensitive to the environment. Thus, an understanding of the genetic basis of adaptation must begin with an analysis of what QTL affect variation in the adaptive trait within and between populations (or species), and what are the effects and gene frequencies of alleles at each QTL.

Beyond the molecular and statistical components, the study of the genetics of adaptation also requires an understanding of the role these characters play as adaptively important traits. In other words, placing the genetics in a realistic ecological context must be a main goal of this research agenda. Although a comprehensive dissection of complex traits is most feasible today using model organisms, the promise of the genomic revolution is that we will soon be able to extend these approaches to any organism where compelling evolutionary or ecological questions remain.

In 2001, nearly 10 years after the publication of Orr and Coyne's call to action, I organized a symposium on the genetics of adaptation on the campus of the University of Georgia in Athens to assess the progress the field had made over the past decade. This meeting brought together over 50 scientists from as far away as Alaska, Germany and Finland to discuss the advances in both molecular genetic and statistical techniques that have allowed for considerable progress to be made in this field. This meeting was generously supported by a grant from the University of Georgia's 'State of the Art Conferences' program administered by the Office of the Senior Vice President for Academic Affairs and Provost. The papers presented in this volume are the tangible product of that third annual Georgia Genetics Symposium.

Almost all the speakers invited to the symposium have contributed papers to this volume. In addition, several poster presenters were invited to contribute a paper. All contributions to this volume were peer-reviewed by at least 2 external reviewers. In addition, many of the manuscripts were reviewed by graduate students in one of my graduate classes in evolutionary genetics. I am grateful to all the reviewers for giving their time.

The contributors were selected to represent a diversity of study systems and approaches. Orr and Phillips were each invited to give an overview of the field and their contributions appropriately start the volume. The next two papers are based on a theoretical approach. In particular, Zeng provides an overview of the statistical issues involved in QTL mapping and provides a preview of two critical extensions of QTL mapping: accounting for correlations among traits and mapping of eQTLs from microarray data. Several examples of the study of the genetics of adaptation within plant populations follows, including a detailed analysis of epistasis by Juenger et al. Two examples of the study of the genetics of adaptation within animal populations are reported in papers by Nachman and Jones.

The volume continues with a transition between within-population studies and studies with a broader focus on the genetics of species differences. The application of the study of the genetics of adaptation is a critical component of 21st century agricultural research and the paper by Boerma and Walker demonstrate the power of that approach. Two additional contributions by Paterson and Ross-Ibarra continue along this vein by considering the role of the study of the genetics of adaptation in crop evolution. Finally, I conclude the volume with what I hope is a provocative paper that challenges ecologists and genomicists – two important contributors to present and future studies of adaptation – to integrate their respective disciplines.

Finally, I am most appreciative of the patience the contributors showed as this volume was compiled. Some authors required more time and persuasion than others – which resulted in the considerable delay of the more prompt authors' contributions. I can only beg their forgiveness and hope that the final product was worth the wait. I believe that it is. As you will read, many of the papers in this volume are first-rate and contain some creative approaches to the study of the genetics of adaptation. The volume is filled with provocative ideas and suggested future directions. My sense is that many doctoral dissertations could find their birth by the careful reading of this volume.

RODNEY MAURICIO
University of Georgia
Athens, Georgia, USA

Genetica (2005) 123: 3–13 © Springer 2005

Theories of adaptation: what they do and don't say

H. Allen Orr
Department of Biology, University of Rochester, Rochester, NY 14627, USA (Phone: +1-585-275-3838; Fax: +1-585-275-2070; E-mail: aorr@mail.rochester.edu)

Received 24 May 2002 Accepted 14 May 2003

Key words: adaptation, adaptive landscape, experimental evolution, Fisher's model, mutational landscape, QTL analysis

Abstract

Theoretical work on adaptation has lagged behind experimental. But two classes of adaptation model have been partly explored. One is phenotypic and the other DNA sequence based. I briefly consider an example of each – Fisher's geometric model and Gillespie's mutational landscape model, respectively – reviewing recent results. Despite their fundamental differences, these models give rise to several strikingly similar results. I consider possible reasons for this congruence. I also emphasize what predictions do and, as important, do not follow from these models.

Introduction

After a delay of many decades – a delay due largely to the reign of the neutral theory – adaptation has begun to receive serious attention. As usual, the reason has more to do with experimental than theoretical progress. At least three kinds of empirical study have renewed interest in adaptation and, in particular, in the genetics of adaptation.

The first is quantitative trait locus (QTL) analyses. In most of these studies, the character difference analyzed is of obvious adaptive significance (e.g., floral differences affecting pollinator attraction in the monkeyflower *Mimulus* [Bradshaw et al., 1998]) and the results plainly provide information on the genetics of adaptation. In other cases, the character difference may be of less obvious adaptive significance but the QTL results themselves suggest that the character diverged under natural selection, i.e., a disproportionate share of 'plus' factors reside in the high line suggesting a history of directional natural selection (Orr, 1998b; Zeng et al., 2000). The second kind of experimental study is molecular population genetic. The discovery of codon bias made it clear that, despite much talk of neutrality, natural selection acts with astonishing subtlety and ubiquity. This conclusion has been supported by more recent work estimating the proportion of amino acid substitutions driven by adaptive evolution. Smith and Eyre-Walker (2002), for instance, recently concluded that about 45% of all amino acid substitutions between *Drosophila simulans* and D. *yakuba* are adaptive. The third kind of experimental study involves microbial experimental evolution. While QTL and molecular population genetic work often involve natural differences between taxa, experimental evolution involves a degree of human intervention. Microbes are typically placed in novel laboratory conditions (e.g., high temperature) and the increase in fitness that occurs during adaptation is tracked through time. Despite this artificiality, these experiments provide extremely high resolution information on the genetics of adaptation, especially when combined with whole genome sequencing. Work in DNA bacteriophage, for example, suggests that 80–90% of nucleotide changes seen during such experiments are adaptive (Wichman et al., 1999), with a surprising number of changes occurring in parallel, i.e., across independently evolving lines (Wichman et al., 1999; Bull et al., 1997).

This empirical work collectively leaves little doubt that adaptive evolution is common – far more common than many would have been guessed two decades ago. Unfortunately, though, theoretical work on adaptation has continued to lag behind its experimental counterpart and population genetic theory remains largely concerned with neutral or deleterious alleles. Though the reasons for this are partly clear – the neutral theory provides an important null hypothesis and it is easier mathematically to study neutral or deleterious alleles – one begins to get the feeling that population geneticists have been laboring over the wrong thing. This neglect of adaptation likely contributes to the common feeling among working evolutionists that population genetic theory has little to say about their day-to-day research: a theory that slights adaptation is unlikely to be of much use to most evolutionists. Fortunately, a few potential starts to a mature theory of adaptation have now been made (Gillespie, 1984, 2002; Gerrish & Lenski, 1998; Orr, 1998a, 2000; Gerrish, 2001).

Here I briefly review these efforts. These theories can be broken into two classes, those that are phenotype based and those that are DNA sequence based. I consider an example of each: Fisher's geometric model, in which adaptation occurs in a continuous phenotypic space, and Gillespie's mutational landscape model, in which adaptation occurs in a discrete DNA sequence space. I discuss recent results from each model. I also emphasize places where these fundamentally different models yield surprisingly similar results. Finally, I briefly consider possible connections between the models. Throughout, my approach will be non-mathematical and unrigorous. Hopefully, such an informal tour will be of some use to experimentalists who, though interested in adaptation, have neither the time nor background needed to wade through a technical literature.

My goal in the present paper is also partly negative. I emphasize not only what these models allow us to say about adaptation but what they do *not* allow us to say. I take this opportunity, in other words, to clear up several misconceptions about predictions that do and do not follow from these models.

Fisher's geometric model

Population genetic models take such a familiar form that it is easy to overlook a respect in which they are odd. These models begin with selection coefficients but say nothing whatever about where these coefficients come from. It is vaguely assumed of course that selection coefficients emerge from the phenotypic effects of mutations on one or more characters but the mapping from phenotype onto fitness is never made explicit. Although this shortcut suffices for many evolutionary questions, it leaves us in an awkward position when thinking about adaptation. If we want to know, for instance, if mutations of large phenotypic effect are less likely to be favorable than those of small effect, we obviously need a model that allows mutations to have different *phenotypic* sizes, not just different selection coefficients. We need, in other words, a model that systematically maps phenotypic effects onto fitness effects. The simplest such model was introduced by Fisher (1930) in his book *The Genetical Theory of Natural Selection*.

Fisher's so-called geometric model captures the fact that organisms must fit their environment in many ways. They must hunt the right prey, avoid the right predators, resist the right diseases, detoxify the right compounds, and so on. Fisher argued that this problem of conforming to many constraints could be captured by a simple geometric model. In particular, we can imagine that each character in an organism is represented by one axis in a coordinate system. If there are n characters, we have n axes and thus an n-dimensional phenotypic space. Some combination of trait values at these n characters represents the best combination of values in the present environment. For convenience, we can place this (local) optimum at the origin of our n-dimensional coordinate system. Figure 1 shows a simple example of Fisher's model for an organism that is comprised of just two characters ($n = 2$). Because of a recent change in the environment, the population has been thrown off the optimum O and now resides at position A. For simplicity, Fisher's model assumes that fitness falls off from the optimum at the same rate in all directions.

The object of adaptation is to return to the optimum. The problem – and this is the key problem confronting Darwinian evolution – is that the population must attempt this return to the optimum by using *random* mutations, i.e., those

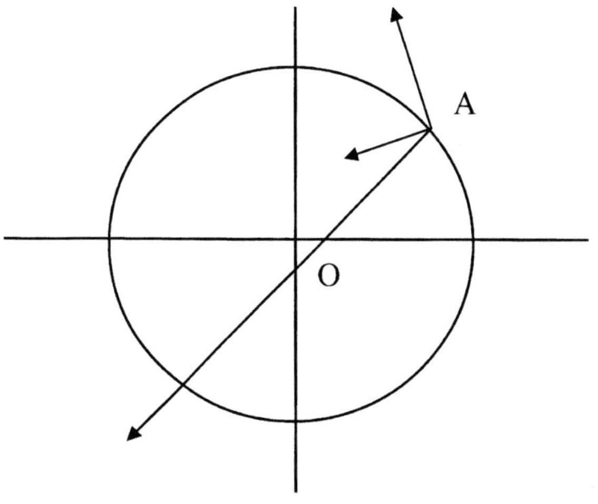

Figure 1. Fisher's geometric model of adaptation for an organism that is comprised of $n = 2$ characters (the x and y axes). The optimal combination of trait values sits at the origin, O. The population presently sits at position A. Several random mutations (vectors) are shown. Those mutations that land within the circle (and so are closer to the optimum) are favorable; those that land outside the circle (and so are farther from the optimum) are deleterious. Note that different mutations can occur in different phenotypic sizes.

that have random direction in phenotypic space. Several random mutations are shown in Figure 1. Obviously those mutations that happen to land nearer to the optimum (and so fall within the circle shown in Figure 1) are favorable, while those that land farther from the optimum (and so fall outside the circle) are deleterious.

The critical point of Fisher's model is that mutations can come in different phenotypic sizes. Mutations are vectors and some vectors might have bigger magnitudes, r, than others, as shown in Figure 1. In general, we can imagine mutation involving a distribution, $m(r)$, of mutations having different sizes. In Fisher's model, the fitness effect of a mutation thus emerges from its size and direction in phenotypic space. Fisher's model provides, in other words, a statistical mapping of phenotypic effect onto fitness effect – a mapping that emerges naturally from the challenge of conforming to many constraints.

An adaptive substitution in Fisher's model (as in reality) involves a two step process. If a mutation is to contribute to adaptation, it must first be favorable. Second, it must also escape accidental loss when rare. Most favorable mutations do not make it, reflecting the known low probability of fixation of $2s$ for a unique beneficial mutation.

I now turn to key results obtained in Fisher's model. Because these have been well reviewed (Barton, 1998; Orr, 1999; Barton & Keightley, 2002) my treatment is brief.

What Fisher's geometric model says

Fisher (1930) used his model to answer one of the simplest possible questions about adaptation: Are phenotypically small or large mutations more likely to be favorable? He showed that the answer is small. Indeed mutations having infinitesimally small phenotypic effects ($r \to 0$) have a 50:50 chance of being favorable, while mutations of larger effect suffer a rapidly declining chance of being favorable. Fisher famously interpreted this to mean that small mutations are the stuff of adaptation.

As is now well known, Kimura (1983) showed that Fisher's interpretation was confused. Although small random mutations *are* more likely to be favorable, they are also more likely to be accidentally lost by genetic drift when rare. Taking both factors into account, Kimura concluded that mutations of *intermediate* size are the most likely to play a role in adaptation.

But Kimura's distribution is also not what it first seems. The reason is that Kimura neglected the fact that adaptation typically involves *multiple* substitutions that gradually approach an optimum. Kimura's distribution only corresponds to that for the first step of such an adaptive walk. When we allow for a gradual approach to an optimum involving many substitutions, we get a different answer from Kimura: the distribution of factors fixed during adaptation is nearly exponential, where we assume that the optimum stays put during the bout of adaptation we study and ignore mutations of very small effect (Orr, 1998; see also Figure 2). This result is surprisingly robust (Orr, 1998a, 1999), arising more or less independently of the shape of the fitness function (e.g., Gaussian, quadratic, linear), the form of the distribution of mutational effects (so long as small mutations are more common than large and the typical mutation is small relative to the distance to the optimum), and dimensionality of the organism (so long as $n > 10$ or so). Thus if Fisher's simple picture of adaptation tells us anything about adaptation, it tells us that the expected distribution

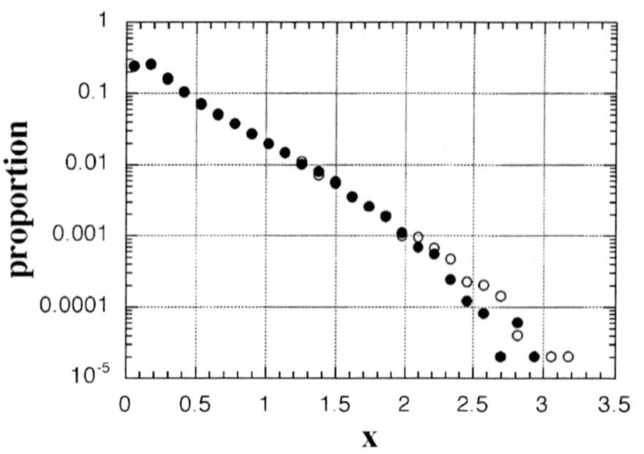

Figure 2. The distribution of factors fixed during adaptive walks to the optimum in Fisher's model. The distribution is approximately exponential (straight line on a semi-log plot). Open circles refer to computer simulations performed at $n = 25$ dimensions; filled circles at $n = 50$ dimensions. Fifty thousand substitutions over many realizations of adaptive walks were recorded in each case. As in Kimura (1983), a uniform distribution of mutational effects was provided to natural selection.

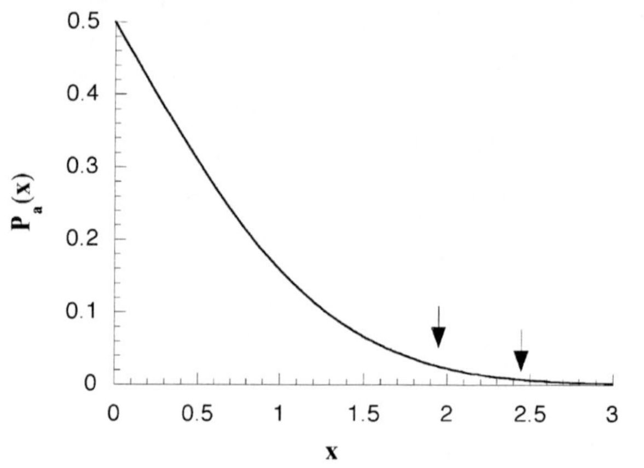

Figure 3. The expected sizes of the largest factors fixed in Fisher's model. The curve is Fisher's famous probability that a random mutation of a given size will be favorable. The left arrow gives the expected size of the leading factor at $n = 100$ dimensions and the right at $n = 500$ dimensions. As in Figure 1, the distribution of mutational effects was uniform. The approximate size of the largest factor fixed is from Orr (1998, equation 17).

of effect sizes is nearly exponential – not that given by Fisher or Kimura.

Two other results characterize adaptive walks to fixed optima. The first is that the mean phenotypic sizes of the factors fixed at substitutions $k = 1, 2, 3,\ldots$ fall off as an approximate geometric sequence (Orr, 1998a, Eq. 11). Early substitutions thus tend to be larger than later. The second result is that the expected size of the largest factor fixed during an adaptive walk is larger than either Fisher or Kimura implied (Orr, 1998a, Figure 7 and Eq. 17). With Fisher, the reason is obvious: he ignored the accidental loss of small mutations. With Kimura, the reason may not be first obvious, but is equally simple: the expected maximum of a series of draws at $k = 1, 2, 3,\ldots$ must be larger than the expected value of a single draw at $k = 1$ (given by Kimura). The biological point is perhaps best made in Figure 3, which shows Fisher's probability that a mutation of a given size will be favorable as well as the expected size of the largest factor fixed at $n = 100$ and $n = 500$ given a uniform distribution of mutational effects. By the time $n = 500$, the largest factor fixed (the 'leading factor') is large enough that a random mutation of this size suffers a tiny 0.0067 chance of being favorable. This contrasts to a 0.5 probability for the infinitesimally small mutations that Fisher believed underlaid adaptation. The leading factor fixed during adaptation does not therefore correspond to a Fisherian infinitesimal one. Instead, it corresponds to a mutation that, according to Fisher's calculation, suffered an absurdly small chance of being favorable.

What Fisher's geometric model doesn't say

There are several conclusions that cannot be drawn from Fisher's model, or at least from studies of it that have been performed so far. Some are straightforward while others are subtle.

The most obvious limitation is that we cannot say anything about adaptation from standing genetic variation. All studies of Fisher's model to date, including those of Kimura (1983), Orr (1998a, 1999, 2000), Hartl and Taubes (1998), Poon and Otto (2000), Barton (2001), and Welch and Waxman (2003), consider evolution from new mutations. Although this represents an obvious theoretical limitation, it is unclear to what extent it represents a biological limitation. Despite a long quantitative genetic tradition that emphasizes the significance of standing variation, we have no idea if most long term evolution (yielding fixed species differences) has much to do with such variation (especially as a substantial portion of standing phenotypic variation, at least for *Drosophila* bristles, reflects transposable element insertion polymorphisms, which do not appear to often contribute to species evolution [Long et al., 2000]). It is entirely possible that a good deal of long term evolution involves the fixation of new mutations.

The second limitation is that we cannot say much about adaptation when the environment changes on a fast time scale and the population chases a moving optimum. All studies of Fisher's model so far have focused on the simple case of a *single* bout of adaptation: the environment shifts and we study the population's approach to the new optimum. The consequences of a moving optimum seem clear in one case only. If the optimum moves away from the population at the same rate that the population moves to the optimum, it is as though the population is forever taking a first step and the distribution of factors fixed must collapse to that given by Kimura (weighted by the distribution of mutational effects), not an exponential. For similar reasons, we have no reason to believe that adaptation to a moving optimum will generally involve an exponential distribution of factors. This represents an obvious problem for future work.

The remaining limitations are more subtle. The most important – and misunderstood – is that we can say nothing about the *absolute* sizes of the favorable mutations fixed during adaptation in any actual case, as emphasized by Orr (2001). We can say that these factors are larger than Kimura predicted and far larger than Fisher predicted but we cannot speak of absolute effects. Precisely the same limitation applies to Fisher's and Kimura's own analyses. The problem is not that we cannot write down equations for these quantities; we can. The problem is that these solutions depend on parameters that are, in any actual case, unknown. One is the dimensionality, n, of the organism. The absolute size of the first factor fixed or the largest factor fixed depends on n, reflecting the fact that large mutations have a greater chance of fixation in simple (few dimensions) than complex (many dimensions) organisms (see Orr, 1998a: Eq 11, 17 and Figure 7). Thus, the first or largest factor fixed might be large in a simple (or highly modular) organism but small in a complex (or less modular) one. The second unknown quantity is the distribution of mutational effects, $m(r)$, provided to natural selection. The absolute size of the factors fixed during adaptation obviously depends on this distribution. To see this, consider a trivial case in which the optimum is 100 units away but the organism produces mutations only of size 0–0.001 units. The sizes of the first factor and the largest factor fixed will clearly be small relative to the distance to the optimum but for reasons having nothing to do with Fisher's argument. Adaptation cannot fix what mutation does not make.

We also cannot say anything about the distance to the optimum traveled by the first or largest factor fixed in any actual case. The reason once again is that the answer depends on dimensionality. Roughly speaking, fixed factors travel $\sim r/\sqrt{n}$ of the way to the optimum (Orr, 2000, Eq. 5; Barton & Keightley, 2002). Factors of a given size thus travel further in simple than complex organisms. This represents one of the 'costs of complexity' emphasized in Orr (2000).

Despite all this, several important results emerge from Fisher's model that *do* hold over nearly all n and across a variety of distributions of mutational effects: (1) the distribution of factors fixed is nearly exponential; (2) early substitutions have larger effects than later, with mean effects falling off as an approximate geometric sequence; and (3) the leading factor fixed is larger than predicted by Fisher or Kimura.

While the above limitations are empirical – the answers depend on unknown parameters – at least two features of Fisher's model may be inherently unrealistic. The first is that while characters are scaled so that fitness falls off at the same rate over all characters, Fisher's model also assumes that mutational effects are random over these orthogonal, scaled characters. This is not necessarily true, as Fisher himself noted (Fisher 2000, p. 302). Strictly speaking, then, Fisher's model considers evolution in an idealized organism in which mutation is isotropic over the space defined by a set of independent, selectively equivalent characters. Some of the above results – e.g., Fisher's probability that a mutation is favorable, Kimura's distribution of factors fixed at step one, and the exponential distribution of effects fixed throughout a walk – might fail in more complicated models that allow non-isotropic mutational effects. But it is important to bear in mind that Fisher's model merely tries to capture the essence of Darwinian adaptation. And this essence is that organisms must adapt by using mutations that are random with respect to an organism's needs. Fisher captured this sense of randomness in a particularly natural way: by letting mutations have random direction in phenotypic space. Fisher's model thus tells us what to expect in the simplest mathematical caricature of Darwinian adaptation. (But see Orr (2000) and Barton & Keightley (2002) for the idea that actual organisms may have an 'effective dimensionality,' n_e.)

The second artificial feature of Fisher's model is that it features no necessary last substitution. The reason is that Fisher's model considers a continuous phenotypic space in which a population can always go further to the optimum. The result is that adaptation invariably appears complicated: adaptive walks involve many steps and the typical factor has a small effect. But real adaptation in real organisms occurs in a discrete space of DNA sequences. One consequence is that there *is* a DNA sequence that is (locally) best and, once reaching this sequence, adaptation is complete, at least for this bout of adaptation. There *is* therefore a necessary last substitution at the DNA level.

This concern opens up a new set of questions that cannot be answered in Fisher's model: How many substitutions occur before the population reaches a local optimum? What proportion of the overall increase in fitness that occurs during an adaptive walk is due to the first substitution? How much is due to the largest substitution? To answer these and other questions, we require a model of adaptation that is explicitly DNA sequence based. I consider such a model below.

Gillespie's mutational landscape model

Models of adaptation in sequence space were first introduced by Maynard Smith (1962, 1970). Although he considered evolution in a space of protein sequences, most theorists now consider evolution in a space of DNA sequences. Several such models have been introduced (reviewed in Gillespie, 2002). Here I describe one, Gillespie's 'mutational landscape' model (Gillespie, 1984, 1991).

The mutational landscape model follows adaptation at a gene or small genome. The region of interest is L base pairs long. The model assumes that adaptation is due to point mutations and that mutation is weak ($Nu << 1$, where N is population size and u is per nucleotide mutation rate) and selection strong ($Ns > 1$, where s is a selection coefficient). (It is important to note that selection is strong only relative to population size: s might well be small in absolute terms.) Under these conditions, a population is essentially fixed for a wild-type sequence at any point in time. We imagine that the present wild-type was, until recently, the fittest allele available. But following an environmental change, the wild-type has slipped in fitness and at least one favorable mutation is now possible.

The population's challenge is to evolve from the present, less than ideal, allele to the fittest one available. It does so by mutating the wild-type. This process generates a large number of different sequences. One of Maynard Smith's (1962, 1970) and Gillespie's key insights was that we need not consider all of these sequences. Instead, we need only consider those $m = 3L$ alternative sequences that can be reached by mutation at a single base (each of the L sites can mutate to three different nucleotides, hence $m = 3L$). The point is that double, triple, etc. mutants are so rare that they can be safely ignored (Gillespie, 1984). In effect, then, natural selection at the DNA level has a short horizon, seeing only one mutational step into sequence space. This short horizon represents an important constraint on adaptation that has no analog in Fisher's model.

At this point, the mutational landscape model makes an important assumption: although the wild-type is no longer the fittest allele available, it is nonetheless of high fitness. This makes good biological sense. Because environments are autocorrelated through time, a wild-type might well slip in fitness, but it is unlikely to plummet. More specifically, the mutational landscape model assumes that the wild-type allele has fitness rank i, where i is small: of the $m + 1$ relevant alleles (m single-step mutants plus wild-type), the wild-type is the ith best. This means that, of the m single-step mutations, a small number ($i - 1$) is favorable and all the rest are deleterious (see Figure 4).

Figure 4. The fitness ranks of alternative sequences in the mutational landscape model. The present wild-type has rank i. The small number of alleles to the right of i ($j = 1, 2, \ldots, i - 1$) are favorable, while the many alleles to the left are deleterious.

Although each of these $i - 1$ favorable alleles suffers a chance of accidental loss each time it appears, mutation is recurrent and one allele will ultimately be fixed. At that time, onestep in adaptation is complete and the process repeats. The new wild-type now produces its own suite of m one-step mutant sequences, one or more of which might be favorable. If so, a new wild-type is again fixed. This process continues until the population arrives at a sequence that is fitter than *all* its one-step mutational neighbors. Adaptation has, at that point, reached a local optimum and is complete.

Although the above process is simple, we have avoided a key issue: How do we assign fitnesses to alternative alleles? This is one of the trickiest issues in all models of adaptation, particularly given the nearly complete absence of relevant data. But Kimura (1983) and Gillespie (1983, 1984, 1991) suggested a way out: we can randomly assign the fitnesses of alleles from some probability distribution. Although this may not at first sound satisfying – we will, after all, still have to make some assumptions about this distribution – Gillespie's (1983, 1984, 1991) key insight was that these assumptions turn out to be far weaker than one might guess. Indeed, the choice of fitness distribution is almost irrelevant. The reason is fairly profound and is worth understanding.

The key fact is that the wild-type allele has high fitness. This allows us to import a body of probability theory known as extreme value theory, which describes the properties of the largest several values drawn from a distribution. Remarkably, extreme value theory shows that these properties are *independent* of the exact distribution that one is drawing from. It does not matter, in other words, if the distribution of allelic fitnesses is normal, or gamma, or exponential, or log-normal, or Weibull, etc. – the fittest few alleles behave in the same way regardless. (The only exceptions involve exotic distributions like the Cauchy – which has no mean – or those that are truncated on the right. Formally, a distribution must belong to the 'Gumbel type,' which includes most ordinary distributions. See Gumbel, 1958; Leadbetter, et al., 1980; also see the Appendix in Orr, 2003). Extreme value theory's independence from the distribution drawn from is reminiscent of the Central Limit Theorem and the two results are similarly robust (Leadbetter et al., 1980). Extreme value theory thus allows a deep and important simplification in the study of adaptation: we can draw conclusions about adaptive evolution that do not depend on the arbitrary choice of fitness distribution.

For present purposes, the most important result from extreme value theory involves the differences in fitness between the best allele (fitness rank $j = 1$), the next-best allele (fitness rank $j = 2$), and so on (see Figure 4). These fitness spacings show particularly simple behavior (Gillespie, 1991; Orr, 2003a,b), behavior that lets us answer many questions about adaptation. I review some recent results below.

What the mutational landscape model says

One of the simplest questions we can ask about adaptation at the DNA level is: What is the distribution of fitness effects among beneficial mutations? Because extreme value theory tells us the fitness spacings between any high fitness wild-type and those rare mutant alleles that have even higher fitness, we can calculate the expected distribution of fitness effects (ΔW) among beneficial mutations. This distribution has two surprising properties (Orr, 2003a): (1) it is always exponential; and (2) it always has the *same* mean no matter what the fitness rank, i, of the current wild-type allele. Natural selection is thus presented with the same expected distribution of fitness effects among new beneficial mutations, whether the current wild-type is the second-best allele, the third-best, and so on. This invariance property is unexpected and counterintuitive.

Natural selection will ultimately fix one of the few beneficial mutations available, a choice that depends on the probabilities of fixation of the various mutations. Perhaps the most important question we can ask about this event – the unit event in adaptation – is: How far does a population 'jump' when natural selection fixes a favorable mutation? This question comes in two flavors. One involves fitness rank and the other magnitude of fitness increase. Both turn out to have simple answers.

First fitness rank. If a population is fixed for the ith fittest allele, does the population typically jump to the fittest available mutation ($j = 1$), to the next-fittest ($j = 2$), or to a mutation that is only slightly better than wild-type ($j = i - 1$) at the next substitution (Figure 4)? Better yet, what's the *mean* fitness rank of the favorable mutant jumped to? The answer is that the population will on average jump to the

$$E[j] = \frac{i+2}{4} \quad (1)$$

best allele (Orr, 2002). Remarkably, this result depends *only* on the present fitness rank i and is independent of everything else, including the distribution of allelic fitnesses. Adaptation by natural selection is thus characterized by a simple rule that maps present fitness rank onto future fitness rank. This jump in rank is also large. If a population is presently fixed for, say, the $i = 20$th best allele, it will typically jump to about the fifth best allele at the next substitution. Adaptation does not therefore incrementally inch from a wild-type to slighter better mutants. It instead leapfrogs such mutants, immediately arriving at a much better one.

We can also find the size of the mean fitness jump that occurs when a favorable mutant allele is substituted (Orr, 2002). Although this calculation is much harder than the above one, the answer turns out to be just as simple. It is

$$E[\Delta w] = \frac{2(i-1)E[\Delta_1]}{i}. \quad (2)$$

$E[\Delta_1]$ is the mean fitness gap between the fittest ($j = 1$) and next-fittest ($j = 2$) mutant alleles, a quantity that *does* depend on the form of the distribution of allelic fitnesses (normal, exponential, etc.). But the biologically important point is that Eq. 2 is nearly insensitive to starting wild-type fitness rank, i. For non-trivial i, the mean fitness jump is just $E[\Delta W] \approx 2E[\Delta_1]$. (This approximation is easily derived given that the distribution of fitness effects among beneficial mutations is exponential. Weighting this exponential by the probability of fixation, $2s$, one finds that among fixed favorable alleles is gamma distributed with a mean of $2E[\Delta_1]$.)

While Eq. 2 is simple, it doesn't quite tell us what we'd most like to know. It lets us predict the mean 'size' of a given substitution but the answer depends on quantities that are generally unknown (e.g., $E[\Delta_1]$). It is hard therefore to see how this prediction could be tested. Fortunately, though, we can ask related questions that are more easily tested. These new questions hinge on the fact that adaptation of a DNA sequence involves a last substitution. We can thus ask: (1) What proportion of the overall increase in fitness that occurs during a bout of adaptation is due to the *first* substitution? (2) What proportion is due to the *largest* substitution?

While analytic solutions to these questions do not appear possible, they are easily answered by computer simulation. Some results are shown in Figure 5. The important point is that the first and largest substitution explain a large proportion of the overall increase in fitness. Simulations show that at least 30% of the overall increase in fitness that occurs during a bout of adaptation is due to the first substitution (on average), while at least 50% is due to the largest substitution (on average). The mutational landscape model thus lets us answer questions that cannot be unambiguously answered in Fisher's model.

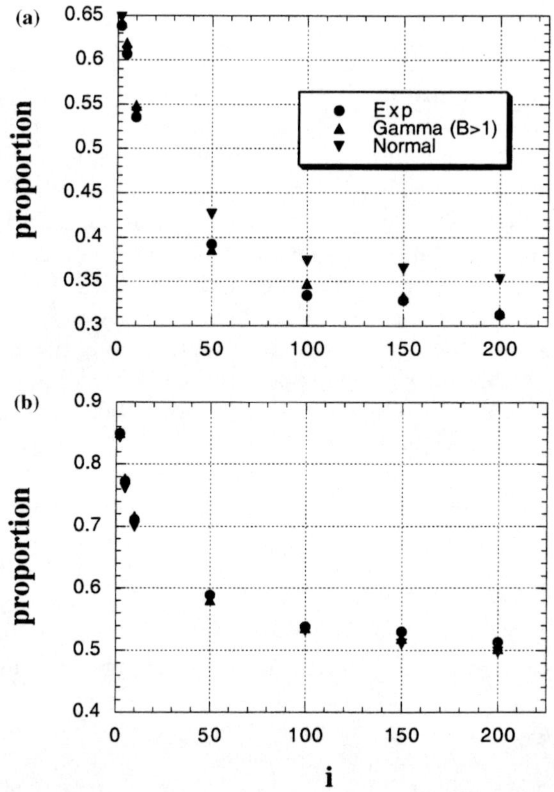

Figure 5. (a) The proportion of the overall gain in fitness due to the *first* substitution in Gillespie's mutational landscape model. (b) The proportion of the overall gain in fitness due to the *largest* substitution in Gillespie's mutational landscape model. The different cases shown refer to exponential, gamma (shape parameter > 1), and normal distributions of allelic fitnesses. These distributions all yield similar results.

Last, we can ask about the factors fixed over an entire adaptive walk to the (locally) best DNA sequence. Computer simulations reveal two surprising results. First, the mean selection coefficients of alleles fixed at subsequent substitutions fall off as an approximate geometric sequence. Second, the overall distribution of fitness jumps over many realizations of adaptive walks to the optimum is approximately exponential (where we ignore factors of very small effect; see Figure 6). These results are both reminiscent of those seen in Fisher's model. It appears then that, despite their fundamental differences, surprisingly similar patterns emerge in the continuous phenotypic and discrete DNA models of adaptation. I consider possible reasons for these shared results below.

What the mutational landscape model doesn't say

Our findings depend on certain assumptions. One is that we study a *single* bout of adaptation, i.e., response to a single environmental change. This is the same assumption made in Fisher's model and is the same assumption that Gillespie (1983, 1984, 1991) made in his earlier work on the mutational landscape model. It is important to note however that, while some of our results depend on this assumption, others do not. In particular, many of our results concern a *single* step in adaptation. These results are more or less unaffected by the assumption that the environment stays unchanged for long stretches of time. But other of our results concern entire adaptive walks and so clearly depend on this assumption. The mutational landscape model does not therefore tell us much if environmental changes occur on a shorter time scale than substitutions. Fortunately this is not a concern in most microbial experimental evolution work. There, one typically exposes a microbe to a single environmental change and studies the burst of substitutions that occur in response.

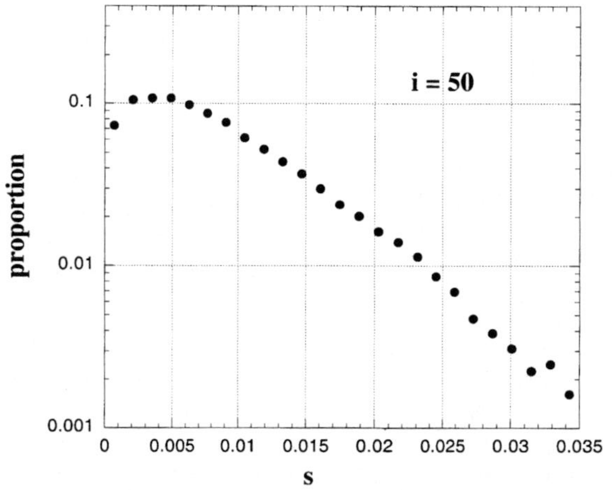

Figure 6. The distribution of selection coefficients among mutations fixed in repeated adaptive walks to the locally best allele in the mutational landscape model (from many realizations of adaptive walks). In the case shown, adaptive walks began at the $i = 50$th best allele and the distribution of allelic fitnesses was gamma with a shape parameter > 1 (yielding a humped distribution).

Another assumption of the mutational landscape model is that the distribution of allelic fitnesses stays the same throughout a bout of adaptation i.e., when the fitnesses of $m = 3L$ mutations are drawn at each step in an adaptive walk, they are drawn from the *same* distribution (Gillespie, 1984, 1991). This assumption differs slightly from that just discussed: even if the environment remains constant throughout an adaptive walk, the distribution of allelic fitnesses may not. Instead, the fitnesses of one-step mutational neighbors might be correlated with that of the present wild-type. Gillespie's model – which considers the simplest case of a 'rugged' landscape (Kauffman, 1993, chapter 2) – does not allow for such correlations. One cannot therefore necessarily extrapolate Gillespie's results or mine to correlated fitness landscapes. Once again, however, note that this limitation does not affect those findings that involve a *single* step in adaptation. (It should also be noted that one result in the genetics of adaptation is known to hold regardless of the ruggedness of the adaptive landscape: Orr (2003b) showed that a minimum of $e - 1$ substitutions (where $e = 2.718...$) are on average required to reach a local optimum when starting from a randomly chosen sequence on any so-called NK adaptive landscape.)

Conclusion: Why do Fisher's and Gillespie's models yield similar results?

I close this brief tour of these adaptation models with two questions. My answers to both are speculative. The first is this: Fisher's model is a model of adaptation because it explicitly considers the fit between a complicated organism and a complicated environment. Gillespie's model does not. How, then, can Gillespie's model be a true model of adaptation? I think the answer is that, although the mutational landscape model does not explicitly consider the fit between organism and environment, it does so implicitly. The point is that, given random mutation in any environment, there will be some distribution of fitness effects among the tiny minority of mutations that is favorable. Extreme value theory tells us what this tail of favorable effects looks like. Extreme value theory, in other words, implicitly captures a point that emerges in a more mechanical way in Fisher's model (all mutations in Fisher's model have a fitness; but because of the constraints of movement in a high dimensional space, only a few have a fitness that exceeds that of the wild-type; these, in other words, occur in the extreme right tail of fitnesses). The critical point is that the mutational landscape model – unlike traditional population genetic ones – does *not* begin with arbitrary selection coefficients. Instead, the distribution of selection coefficients among favorable mutations emerges naturally from the rareness of extreme, highly fit alleles. It is this emergence of selection coefficients that makes the DNA sequence model, like Fisher's, a model of adaptation.

The second question is: Why do Fisher's and Gillespie's models yield some similar results? The models are, after all, fundamentally different, with one considering phenotypic effects in a continuous space and the other fitness effects in a discrete space. Nonetheless in both models effect sizes among fixed favorable mutations fall off as a geometric sequence and the overall distribution of factors fixed during adaptive walks is nearly exponential. The reason for these similarities almost surely has something to do with the above point. But there is another reason: In both models, adaptation is characterized by a kind of repeated re-scaling. In both models, that is, the population confronts essentially the same problem at each substitution, but on a smaller scale. (In Fisher's model, the shrinking scale reflects moving nearer to the optimal phenotype; in Gillespie's model, it reflects moving along the tail of the fitness distribution.) A consequence of this dynamic is that to a good approximation the scale, but not the functional form, of the distribution of factors fixed at each step changes through time. This roughly self-similar behavior appears to give rise to both the geometric sequence and the overall exponential behavior. The biologically significant point is that this self-similarity likely characterizes *any* sensible model of adaptation to a fixed optimum. If so, there is some reason for thinking that the above findings might represent robust properties of adaptation to a fixed optimum.

In summary, it appears that adaptation to a fixed optimum by new mutations may show certain predictable properties. But it is also clear that a large class of biological scenarios – involving moving optima, standing genetic variation, and correlated fitness landscapes – has not been studied. These scenarios are obvious candidates for future work. While it would be pleasing if the same patterns characterized all of these scenarios, this seems unlikely. I suspect, however, that the likely diversity of results is a blessing. If we are to distinguish different forms of adaptation, e.g., that involving new versus. standing variation, we will require that these processes leave different signatures on the genetics of adaptation. The task of evolutionary theory is to determine what these signatures look like.

Acknowledgments

I thank N. Barton, A. Betancourt, J.P. Masly, R. Mauricio, S. Otto, D. Presgraves, D. Schemske, and three anonymous reviewers for helpful comments. This work was supported by NIH Grant 2R01 G51932-06A1.

References

Barton, N., 1998. The geometry of natural selection. Nature 395: 751–752.

Barton, N.H., 2001. The role of hybridization in evolution. Mol. Ecol. 10: 551–568.

Barton, N.H. & P.D. Keightley, 2002. Understanding quantitative genetic variation. Nature Rev. Genet. 3: 11–21.

Bradshaw, H.D., K.G. Otto, B.E. Frewen, J.K. McKay & D.W. Schemske, 1998. Quantitative trait loci affecting differences in floral morphology between two species of Monkeyflower (Mimulus). Genetics 149: 367–382.

Bull, J.J., M.R. Badgett, H.A. Wichman, J.P. Huelsenbeck, D.M. Hillis, A. Gulati, C. Ho & I.J. Molineux, 1997. Exceptional convergent evolution in a virus. Genetics 147: 1497–1507.

Fisher, R.A., 1930. The Genetical Theory of Natural Selection. Oxford University Press, Oxford.

Fisher, R.A., 2000. The Genetical Theory of Natural Selection: *a Complete Variorum Edition*. Oxford University Press, Oxford.
Gerrish, P., 2001. The rhythm of microbial adaptation. Nature 413: 299–302.
Gerrish, P.J. & R.E. Lenski, 1998. The fate of competing beneficial mutations in an asexual population. Genetica 102/103: 127–144.
Gillespie, J., 1984. Molecular evolution over the mutational landscape. Evolution 38: 1116–1129.
Gillespie, J.H., 1991. The Causes of Molecular Evolution. Oxford University Press, Oxford.
Gillespie, J.H., 2002. Why k = 4Nus is silly in Evolutionary Genetics, edited by R.S. Singh. Vol. III.
Gumbel, E.J., 1958. Statistics of Extremes. Columbia University Press, New York.
Hartl, D. & C.H. Taubes, 1998. Compensatory nearly neutral mutations: selection without adaptation. J. Theor. Biol. 182: 303–309.
Kauffman, S.A., 1993. The Origins of Order. Oxford University Press, New York.
Kimura, M., 1983. The Neutral Theory of Molecular Evolution. Cambridge University Press, Cambridge, UK.
Leadbetter, M.R., G. Lindgren & H. Rootzen, 1980. Extremes and Related Properties of Random Sequences and Processes. Springer-Verlag, New York.
Long, A.D., R.F. Lyman, A.H. Morgan, C.H. Langley & T.F. Mackay, 2000. Both naturally occurring insertions of transposable elements and intermediate frequency polymorphisms at the *achaete-scute* complex are associated with variation in bristle number in *Drosophila melanogaster*. Genetics 154: 1255–1269.
Maynard Smith, J., 1962. The limitations of molecular evolution, pp. 252–256 in The Scientist Speculates: an Anthology of Partly-baked Ideas, edited by I.J. Good. Basic Books, Inc., New York.
Maynard Smith, J., 1970. Natural selection and the concept of a protein space. Nature 225: 563–564.
Orr, H.A., 1998a. The population genetics of adaptation: the distribution of factors fixed during adaptive evolution. Evolution 52: 935–949.
Orr, H.A., 1998b. Testing natural selection *versus* genetic drift in phenotypic evolution using quantitative trait locus data. Genetics 149: 2099–2104.
Orr, H.A., 1999. The evolutionary genetics of adaptation: a simulation study. Genet. Res. 74: 207–214.
Orr, H.A., 2000. Adaptation and the cost of complexity. Evolution 54: 13–20.
Orr, H.A., 2001. The 'sizes' of mutations fixed in phenotypic evolution: a response to Clarke and Arthur. Evol. Dev. 3: 121–123.
Orr, H.A., 2002. The population genetics of adaptation: the adaptation of DNA sequences. Evolution 56: 1317–1330.
Orr, H.A., 2003a. The distribution of fitness effects among beneficial mutations. Genetics 163: 1519–1526.
Orr, H.A., 2003b. A minimum on the mean number of steps taken in adaptive walks. J. Theor. Biol. 220: 241–247.
Poon, A. & S.P. Otto, 2000. Compensating for our load of mutations: freezing the meltdown of small populations. Evolution 54: 1467–1479.
Smith, N.G.C. & A. Eyre-Walker, 2002. Adaptive protein evolution in *Drosophila*. Nature 415: 1022–1024.
Welch, J.J. & D. Waxman, 2003. Modularity and the cost of complexity. Evolution 57: 1723–1734.
Wichman, H.A., M.R. Badgett, L.A. Scott, C.M. Boulianne & J.J. Bull, 1999. Different trajectories of parallel evolution during viral adaptation. Science 285: 422–424.
Zeng, Z.-B., J. Liu, L.F. Stam, C.-H. Kao, J.M. Mercer & C.C. Laurie, 2000. Genetic architecture of a morphological shape difference between two Drosophila species. Genetics 154: 299–310.

Genetica (2005) 123: 15–24
© Springer 2005

Testing hypotheses regarding the genetics of adaptation

Patrick C. Phillips
Center for Ecology and Evolutionary Biology, University of Oregon, Eugene, OR 97403-5289 USA (Phone: (541) 346-0916; Fax: (541) 346-2364; E-mail: pphil@uoregon.edu)

Received 20 December 2003 Accepted 15 February 2004

Key words: association mapping, evolution, genomics, microarrays, QTL mapping, quantitative genetics, statistical genetics

Abstract

Many of the hypotheses regarding the genetics of adaptation require that one know specific details about the genetic basis of complex traits, such as the number and effects of the loci involved. Developments in molecular biology have made it possible to create relatively dense maps of markers that can potentially be used to map genes underlying specific traits. However, there are a number of reasons to doubt that such mapping will provide the level of resolution necessary to specifically address many evolutionary questions. Moreover, evolutionary change is built upon the substitution of individual mutations, many of which may now be cosegregating in the same allele. In order for this developing area not to become a mirage that traps the efforts of an entire field, the genetic dissection of adaptive traits should be conducted within a strict hypothesis-testing framework and within systems that promise a reasonable chance of identifying the specific genetic changes of interest. Continuing advances in molecular technology may lead the way here, but some form of genetic testing is likely to be forever required.

Introduction

How should we view historical developments in evolutionary genetics through the particular lens of the genetics of adaptation? Although it is perhaps a bit premature for such pronouncements, one could argue that we are entering a new era of modern evolutionary genetics. The first era, roughly from 1918–1968, was characterized by the theoretical developments in population and quantitative genetics that have laid the foundation for nearly all other work in evolutionary biology (Figure 1, see also Provine, 1971). This period began with the theoretical reconciliation of quantitative and Mendelian genetics by R.A. Fisher (1918) and rapidly expanded into the codification of population genetics theory in the 1920's and 1930's through the work of Fisher, Sewall Wright and J.B.S. Haldane. It runs on through the beginnings of ecological genetics by the likes of E.B. Ford and others and the application of population genetic principles to natural populations led by Theodosius Dobzhansky. It ends with a formalization of earlier models by Gustave Malécot and Motoo Kimura into a framework that set the stage for the utilization of the truly genetic data that was soon to follow (Lewontin, 1974). This period could be classified as theory rich and data poor. Most of the theory that we still utilize today was established before we had any knowledge of the nature of the genetic material, and in this sense these approaches are essentially purely genetic and largely devoid of functional context. Fundamental concepts of genetic entities like loci and alleles have hardly changed in population genetics theory, despite tremendous advances in our knowledge of the physical and molecular properties of genes and genomes.

The second era, from 1968 to 1998, was dominated by an explosion of data, frequently collected in the absence of a compelling theoretical context (Lewontin, 1991). In population genetics, the development of protein electrophoresis

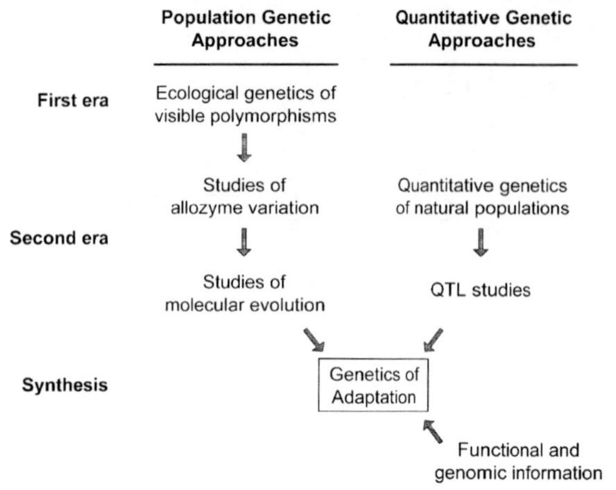

Figure 1. Transitions during the history of population and quantitative genetic approaches to studying the genetics of adaptation. Movement toward a new era of study incorporates these approaches with functional genomic information.

allowed researchers to assess levels of genetic variation in a wide variety of organisms rather than being limited to special cases of known genetic markers (e.g., Drosophila chromosomes) or obviously Mendelizing phenotypes (e.g., snail shell polymorphisms). On the quantitative genetic side of things, the theories originally developed by Fisher and greatly expanded by Wright were finally migrated from agricultural systems into a more formal theory of evolutionary quantitative genetics (e.g., Slatkin, 1970; Lande, 1976; Felsenstein, 1977). Here again, researchers could venture into natural populations to ask questions about levels of genetic variation for ecologically important traits. It seemed that no study of the evolutionary ecology of quantitative traits could be complete without an analysis of underlying genetic variation, because evolutionary change is predicated on its existence. To some extent, both the population and quantitative genetic approaches were victims of their own success. Electrophoretic studies revealed ample levels of genetic variation at most loci, while quantitative genetic studies found significant heritability for most traits. Finding genetic variation for its own sake became a hypothesis-free endeavor. Enough studies of this type have now been performed that one need not actually conduct the studies to know their probable outcome. For the most part, average heterozygosity will vary between 0.05 and 0.2 and heritability will fall somewhere between 0.2 and 0.5. Even if a particular estimate where off by a factor of two or three, would the discussion sections of these particular studies be very different? It is unlikely that they would, which is a testament both to a general lack of precision in these estimates and the lack of a broader hypothesis-testing framework for this work.

Studies of variation per se have developed on one side into much more sophisticated treatments of DNA sequence variation from a molecular evolution viewpoint and on the other side into a formal theory of evolutionary quantitative genetics that treats the entire organism as an integrated whole (Figure 1). Using sequence data, we can address very specific hypotheses regarding historical patterns of selection and rates of evolution of genes of interest, but are frequently far removed from the how, why, what, and where of the adaptive context of that selection. In contrast, in multivariate views of quantitative inheritance, we can measure how selection operates on suites of traits and how trade-offs among traits might structure and constrain the response to selection (Lande, 1988), but are limited to some extent by complexities introduced by the total dimensionality of the system (Charlesworth, 1990) and by the fact that, in order to understand how summary parameters like genetic correlations themselves evolve, we need to have much greater knowledge of the genetic systems underlying these traits (Barton & Turelli, 1989). We are caught between molecular knowledge in the absence of adaptive context and ecological context in the absence of molecular details. One view of the modern challenge to understanding the genetics of adaptation is the need to span this chasm – to be able to move freely from sequence to phenotype to ecological context and, more importantly, to be able to test specific hypotheses at each of these levels.

Are we, then, at the beginning of a self-proclaimed new era? If so, then it is an era that is sure to be dominated by genomic analysis (the 1998 date was chosen because of the publication of the first metazoan genome during this year, The C. elegans Sequencing Consortium, 1998). The hope is to use our new abilities to look at genome-wide patterns of genetic variation and gene function to investigate the genetics of adaptation from multiple perspectives. The fear is that we instead will repeat the mistakes of previous technological transitions and collect information in the absence of definitive hypothesis tests; or worse,

Table 1. Some central questions in the genetics of adaptation

- How many genes underlie specific adaptations?
- What is the distribution of their effects?
- What is the spectrum of new mutations at these genes?
- How do these genes interact with one another?
- Do genes tend to affect traits independently of one another or do genes typically have manifold effects across the whole organism (i.e., pleiotropy)?
- How does natural selection affect the distribution of effects and/or the nature of the interactions?
- Does the response to selection tend to occur more frequently through changes in gene regulation or gene structure/function?
- What is the relationship between loci that generate variation within populations and those responsible for differences among populations?
- How can we combine these insights into an understanding of the evolution of developmental systems, morphology, behavior, etc.?

over-interpret the results that we are capable of collecting right now without appreciating the limitations inherent in our current methods.

Questions and hypotheses

It is not difficult to collect a long list of questions that we would like answered regarding the genetics of adaptation (Table 1). Primary among these are the most basic, like how many genes are involved, what are the distribution of the effects of alleles at these loci, and how does standing variation and mutational input become converted by selection into the adaptive differences that we might observe today? We currently cannot answer these questions for any trait, for any organism, for any natural system. It would therefore seem that we have a long way to go before we can address even the most basic questions in what should be a central area of evolutionary genetics. Many people, of course, are trying to tackle one or another of this broad set of questions, but if we are not careful we will find ourselves in same state as those studying the allozyme variation and heritability a few decades ago: lots of information and precious little context within which to evaluate that information. We can already guess that adaptive changes are sometimes going to be caused by a few loci and sometimes by many more. Some loci are undoubtedly going to have large effects while others will have smaller effects. Sometimes standing variation will be central, other times novel mutations will be essential. Collecting the basic pieces of information underlying the genetics of adaptation is obviously going to be important, but as with earlier revolutions in evolutionary genetics, will our level of resolution be sufficiently adequate to estimate the needed underlying parameters in such a way that estimation alone will be sufficient justification for conducting the work? We can avoid these pitfalls by making sure that the work that we do is conducted within a specific hypothesis-testing framework.

The essential problem with studying adaptation in natural populations is that we have no control over the genetic system. Genomes are vast and important change can potentially be anywhere. How then are we to find the genes of interest? More humbly, how effectively can we address questions related to the genetics of adaptation without actually having our hands on the genetic changes themselves? There are multiple approaches to this problem, each of which provides varying levels of precision (Figure 2). Each major approach can be seen as logical extensions of the two major branches of evolutionary genetics, and it is in their synthesis that we will finally be in a position to address fundamental questions about the genetic basis of adaptive evolution (Figure 1).

Mapping as a paradigm

Number of genes

If we are lucky enough (or choosy enough) to study a character that readily Mendelizes, we can at least to hope to map the gene with some precision. Moreover, we have prima facie evidence that we are dealing with at least one gene of major effect. Although there are important instances of changes of this sort (e.g., Crow, 1957; Peichel, et al., 2001; Nachman, Hoekstra & D'Agostino,

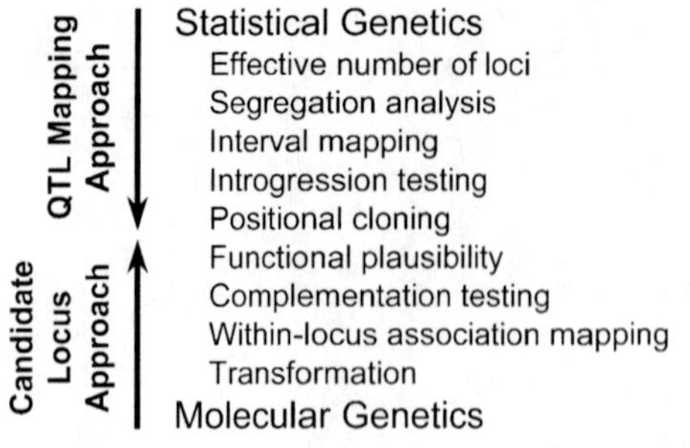

Figure 2. A hierarchical set of methods for determining the genetic basis of adaptive variation. A top-down, statistical genetics approach is built upon QTL mapping, while a bottom-up, molecular genetic approach is built upon identifying specific candidate loci. Confidence in a genetic causation increases as one moves from top to bottom.

2003), we might expect such systems to be outside the norm. Indeed, focus on these single-locus systems has resulted largely from the fact that they are more tractable than systems with more complex genetics. Once we move beyond a single locus, it is extremely difficult to estimate the number of loci affecting a trait simply by observing variation in the trait. In one of the first problems that he addressed, Sewall Wright (in Castle, 1921) derived an estimator for the *minimum effective number of loci* (the number of loci with equal effects) by assuming that two lines being crossed are uniformly divergent for the loci underlying the differences (Figure 2). Although there have been refinements of Wright's original approach (Wright, 1968; Lande, 1981), the method has so many caveats that its overall value beyond demonstrating that a trait is polygenic is questionable (Zeng, Houle & Cockerham, 1990; Zeng, 1992). A slightly more sophisticated approach that combines specific genetic models within the context of a defined pedigree, known as *complex segregation analysis*, is used frequently in human genetics (Figure 2, Khoury, Beaty & Cohen, 1993). Neither of these methods is likely to bring us very close to answering the most basic question of how many loci underlie a given adaptation, much less provide us with any hope of moving us further up the hierarchy of questions (Table 1).

Mapping

One of the more significant developments in evolutionary genetics over the last two decades has been the development of techniques aimed at mapping multiple genes underlying quantitative variation, *quantitative trait locus (QTL) mapping* (Figure 2, Mackay, 2001b). The promise here is that identifying specific regions of the genome responsible for quantitative differences between lines, populations and/or species will allow estimates of some of the fundamental parameters needed to understand the evolution of quantitative characters. While there can be no question that this is the right direction to be heading, we should be very careful not to over interpret the results obtained from such studies. Indeed, it can be argued that mapping per se gets us only slightly further along the road toward answering our fundamental questions than trying to estimate the genic effects directly from variance data.

The real problem is that QTL ('L' = loci) should have been called QTR ('R' = regions). There has been a pull toward creating a central dogma of 'one peak-one gene' in these mapping experiments. If such a one-to-one correspondence where possible, then we would indeed be well on our way to discovering the number of loci underlying specific adaptations. While the attraction of this notion is clear, our current limited experience provides reasons for caution. Mapping is based on linkage disequilibrium between markers that we can measure and QTL of unknown location (Lander & Schork, 1994). Maximizing linkage disequilibrium across the whole genome, as can be accomplished in controlled cross between two extreme populations, greatly enhances the probability that at least one of the markers will be found in association with the QTL of interest (Figure 3). This is a double-edged sword, however, since broad-scale linkage disequilibrium means that a potentially large non-informative chromosomal region surrounding the marker will also be linked to the QTL. This decreases the precision with which the location of the QTL can be identified (Figure 3).

Several decades ago, Coyne (1983) examined the genetic basis of difference in genital morphology between two Drosophila species using a single visible marker per chromosome arm. Each marker did indeed show a significant association with the morphological difference, but rather than conclude that each marker represented a single QTL, Coyne instead reasonably suggested that these differences were likely caused by a potentially large number of

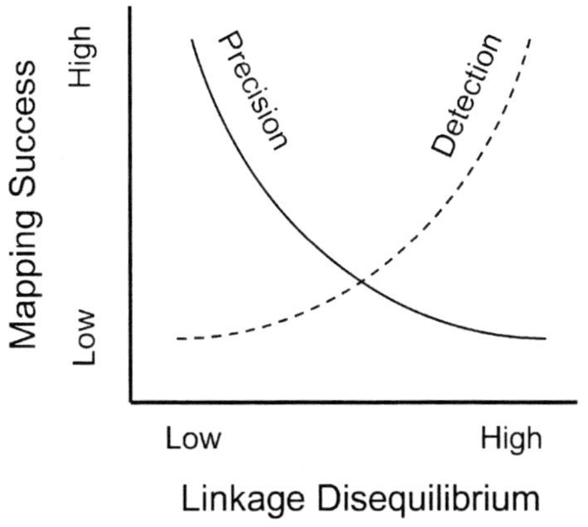

Figure 3. The trade-off between precision and detection as a function of the level of linkage disequilibrium within a population when trying to specific genes. Crosses usually generated in QTL mapping experiments have high levels of linkage disequilibrium and therefore have a large chance of detecting the underlying loci. They may have low precision for identifying where the loci are or even if there are indeed individual loci involved. In contrast, association mapping studies use outbred populations, usually with much lower levels of linkage disequilibrium. These studies require very large samples and very localized dense genetic maps in order to detect the loci involved, but they should in principle allow high precision in identifying the genes, and potentially the nucleotides, involved.

loci since his level of genetic resolution was so crude (a prediction that turned out to be correct, Zeng et al., 2000). We must be careful not to cavalierly equate regions of large effect with genes of large effect when we are in fact frequently barely a few steps beyond Coyne's level of resolution, even with the advent of large numbers of molecular markers. For example, using a high-resolution deletion mapping study of longevity in *Drosophila melanogaster* layered on top of a traditional QTL analysis, Pasyukova, Vieira & Mackay (2000) demonstrated that many of the QTL peaks obtained from a standard cross in fact housed several loci, frequently with opposing effects.

A more fundamental problem for interpreting mapping results is that we may be dealing with a scale of resolution that is simply impenetrable to traditional mapping approaches. Although in many applications of QTL analysis, such as in human health, it may be sufficient to simply identify the locus of interest, in evolutionary studies a 'locus of large effect' and a 'substitution of large effect' should not be equated (Phillips, 1999). The potential confusion is derived from typological definitions of concepts like 'locus' and 'allele' that span the last one hundred years of evolutionary genetics, but which are at odds with modern understanding of genetic change. Two very distinct 'alleles' may segregate in a cross between populations, but the alleles themselves may be the products of many substitution events over the evolutionary history of the divergence of those populations. The concept of 'locus' in theories of the genetics of adaptation may be quite different from traditional definitions of locus – we may frequently need to look for multiple changes within individual genes (e.g., Orr, 2002). The best example of this is Stam and Laurie's (1996) study of functional variation at the ADH locus in *Drosophila melanogaster*. They found that most of the difference in levels of gene expression could indeed be explained by the traditional fast/slow replacement that leads to differences in allozymes, but also that a secondary and very significant effect is generated by an epistatic interaction between two control regions within the gene that is also part of the 'allelic' difference in this case. Even high resolution QTL mapping will not allow us to detect complex changes and interactions occurring within genes. The importance of resolution at this scale is likely to depend on the general ubiquity of complex regulatory systems within genes (Davidson, 2001), but much of the future challenge of the functional genetics of adaptation lies firmly here.

Otto and Jones (2000) provide a method for extrapolating from the estimated number of QTL to the like number of 'true' QTL by assuming certain distributions of effects. Approaches such as this are surely improvements on the Wright-inspired estimators, but these methods will be strongly limited by the resolution of the map, as indicated above.

Distribution of allelic effects

Ignoring the problem of counting genes for a moment, to what extent can we expect to be able to infer the nature of the effects of the genes that we do find, especially with an eye toward estimating the distribution of effects (Orr, 1998)? There are several obstacles here as well. Because QTL are recognized statistically as genomic regions yielding a significant association with phenotypic variation, when sampling errors lead to an overestimate of the size of an effect, that effect is more likely to be classified as a QTL. This leads to an upward bias in estimated effect sizes, which becomes especially magnified in studies with low statistical power (Beavis, 1994, 1998). Lack of resolution can also lead to misestimates of the distribution of effect sizes when more than one gene is located within a QTL region. For example, even when the actual distribution of effects is constant or uniform, it is possible to wrongly infer a negative exponential distribution of effect sizes when a large number of loci are randomly distributed throughout the genome (Bost et al., 2001). When regions of the genome below the mapping resolution threshold accumulate a number of true QTL, the single estimated effect size will be closer to the sum of those QTL than to the effect size of the individual elements (see Noor, Cunningham & Larkin, 2001). Thus, summary distributions of QTL effects are likely to be somewhat suspect in the absence a firm sense of the level of precision in the mapping itself. Finally, complications introduced from genetic interactions among loci (epistasis, Phillips, 1998) are only now starting to be explored because of complications in the analysis and issues of the scale of experiments needed to estimate such a large pool of pair-wise interactions (e.g., Kao & Zeng, 2002).

Mapping to test hypotheses

If mapping serves somewhat poorly for estimating the essential parameters of the genetics of adaptation, then what can it provide us? Apart from serving as an important step on the road toward finding the genes themselves, mapping can be used to test specific hypotheses regarding the genetics of adaptation. The essential point here is that the heart of hypothesis testing involves a comparison of some sort. Comparative mapping in well-articulated circumstances allows one to test the hypothesis of a shared genetic basis of traits across different environments or in different populations. Pleiotropy is the hypothesis of interest when looking at variation within a population, while a parallel response to selection is the focal hypothesis when comparing populations. Unfortunately, much like paternity analysis, it is easier to exonerate a particular region of the genome than to implicate a specific gene. If two QTL regions can be clearly distinguished from one another, then the hypothesis of a shared genetic basis to the traits can be rejected (although it is still possible that similar genes are involved, but with very different effect sizes). If a similar location is identified, however, the precision issue discussed above rears its head again. Is this in fact the same gene being used in each case or simply another locus that happens to be linked to the target QTL region by chance? Fortunately in a comparative context, we can take the precision of our QTL estimates into account and actually test the hypothesis of pleiotropy quantitatively (Cheverud, Routman & Irschick, 1997; Lebreton et al., 1998).

The pleiotropy can be across different traits, such as different morphological features of a flower (Juenger, Purugganan & Mackay, 2000); across time, as in growth and change in body size in mice (Vaughn et al., 1999); and/or across environments such as larval density (Leips & Mackay, 2000) or geographic differences in growing conditions (Weinig et al., 2002). Parallel mapping across multiple divergent populations has rarely been performed, but even between-population crosses can be used to test for similar regions affecting the trait of interest in different genetic backgrounds (e.g., Zeng et al., 2000).

Getting to the genes

Ultimately, tests of pleiotropy and the genetic basis of adaptive differences in general will require finding the genes themselves – indeed the nucleotide changes themselves. At its heart, mapping is a correlational approach. To move closer to causation, it is necessary to verify hypotheses generated by mapping using more conventional genetics. One of the strongest approaches in this regard is *introgression testing* in which a genomic region containing a putative QTL is backcrossed into a common background and retested for its effects (e.g., Laurie et al., 1997). Repeated backcrossing can be used to generate near-isogenic lines, a springboard for the holy grail of QTL mapping, *positional cloning* of the locus (Remington, Ungerer & Purugganan, 2001). The level of effort needed to positionally clone a QTL from mapping data alone is quite daunting. At the very least, significant genomic resources will need to be brought to bear on the problem. While this may be feasible (and justifiable) in many cases in agriculture and human health, the general level of effort needed will remain a significant issue in the evolutionary genetics of non-model species until technology takes another leap forward. For the time being, most studies in evolutionary genetics that use QTL mapping are likely to find themselves marooned upon QTL peaks surrounded by a sea of thousands of possible genes, with little means of identifying or distinguishing among them. This is why it is crucial to choose the level of resolution appropriate (and possible) to the hypothesis being addressed. The field as a whole will gain very little if the majority of studies become stranded halfway between ideals of causal explanation.

The candidate-locus paradigm

The alternative to the top-down approach of QTL mapping is a bottom-up approach based on candidate loci (Figure 2). Here, in-depth knowledge of gene function motivates the selection of a subset of genes that can be used as targets for genetic analysis. A possible first step here is to examine *functional plausibility* by examining differences in DNA sequence among the divergent populations of interest. Since most populations are likely to differ at many nucleotide sites, this is unlikely to be a particularly fruitful exercise, although information of this sort can be used in a broader molecular evolutionary context (e.g., Jovelin, Ajie & Phillips, 2003). The advent of the ability to perform genome-wide functional analyses, such as microarrays, has greatly expanded the set of 'plausible' targets, however (Gibson, 2002). For example, Wayne and McIntyre (2002) have combined gene expression data with QTL mapping results to develop the most likely set of functional targets to pursue as candidate loci in the studies of ovariole number in *D. melanogaster*. At present, these approaches are too new to know whether gene expression differences per se will be useful indicators of underlying genetic divergence. Expression at a given locus can be different due to changes at other loci and the total variance in expression may tend to overwhelm the available signal. Nevertheless, the potential for using this and similar technologies for hypothesis building, especially in non-model organisms, is tremendous (e.g., Oleksiak, Churchill & Crawford, 2002).

The best next step beyond simple plausibility is genetics. An especially powerful approach is to use a *quantitative complementation test* to examine variation in the genetic pathway involving the candidate locus (Doebly, Stec & Gustus, 1995; Long et al., 1996; Lyman & Mackay, 1998). This is an interaction test in which a line with a mutation at a given locus is crossed with natural variants with the aim of assessing allelic variation at that specific locus (Mackay, 2001a). In reality, the response could be due to variation at the locus of interest or a locus that interacts somewhere in the same pathway as the mutation, such that variation is exposed when tested against the mutant background. This approach can be generalized on a genomic scale using deficiency mapping with a very large set of tester lines (Pasyukova, Vieira & Mackay, 2000; Steinmetz et al., 2002).

Finer scale mapping of allelic differences can be addressed using *association mapping* (Figure 2, Mackay & Langley, 1990; Long et al., 1998). Here, QTL mapping is essentially being performed within a locus. In the balance between precision and detection outlined in Figure 3, association mapping is decidedly on the side of precision. The linkage disequilibrium utilized in an association mapping study is that present in the natural population after many generations of recombination. Association mapping looks to detect the resonance signal left behind from the appearance of the unique mutation that is now the target of interest. Because every mutation arises within a unique genomic background, it is in initially in complete linkage disequilibrium with every marker in that genome. Over time recombination will break these associations down until only the closest associations remain. This is how it works in principle. In practice, the pattern of linkage disequilibrium can be non-uniform over a given genomic region. Low detection thresholds suggest that sample sizes will frequently need to be very large for this approach to work in most outbred populations. It is important to note that the studies in which this approach has been more successfully been applied have used crosses to isolate the chromosomal region of interest against a stable genetic background so as to reduce the total level of genetic variation in the system (e.g., Long et al., 1998; Long et al., 2000). Most outbred populations are likely to require samples potentially orders of magnitude larger to overcome the 'needle in the haystack' nature of the entire approach.

Interestingly, a number of the studies that have been able to work down toward the level of individual nucleotides have found their significant associations in control regions and introns rather than in coding regions (Phillips, 1999). This makes identifying the specific changes responsible especially difficult since we currently do not understand the language that describes gene regulation in the same way that we are able to understand how changes in coding regions change gene function (Stern, 2000). This also stands in stark contrast to mapping results for human disease genes, in which a small minority of changes appear to be regulatory in nature (Botstein & Risch, 2003). Resolution of this contrast with additional data will illuminate one of the more interesting long-term questions in evolutionary genetics: evolution via regulatory changes versus structural change (Table 1).

Finally, all of the approaches for finding genes underlying complex adaptations outlined above are essentially circumstantial. Any given study is likely to need to combine a number of different approaches to properly address a causal hypothesis relating to specific gene function. One remaining approach neatly solves this problem through a strong hypothesis test in an experimental context. *Transformation* of one natural allele with another allows for a direct test of allelism while completely controlling for the effects of genetic background. Unfortunately, transformation at this level of precision is difficult even in model systems like Drosophila and *C. elegans*. Yeast is currently the most capable system from genetic manipulation standpoint (Steinmetz et al., 2002). Techniques for transformation in Drosophila have also recently taken a large step forward in the context of testing adaptive gene function (Siegal & Hartl, 1998; Greenberg et al., 2003). Non-model systems being investigated in more meaningful ecological contexts will be hard pressed to meet this standard for the time being.

Conclusions

Running completely through the cycle of causation outlined in Figure 2 is likely to be difficult in most circumstances, and 'proof' that one has actually identified a gene underlying a specific adaptation quantitative trait has only been obtained thus far in a handful of circumstances (Glazier, Nadeau & Aitman, 2002). All of the successful cases have been in either agricultural or model systems. Will finding the actual genes underlying adaptations be feasible in most natural systems? To do so will require generating sufficient genomic resources such that non-model systems essentially serve as their own models. Rapid progress in genomic technology is making this more possible all of the time, but it is important to recognize the cost of this pursuit, both financially and in terms of the large set of potentially more tractable questions that are likely to be abandoned along the way (Lewontin, 1991). Furthermore, ultimate tests of genetic causation rely on actually being able to do genetics – the ability to perform crosses to test specific hypotheses. This will not be feasible in many non-model systems. If we cannot test the hypotheses we are setting out to study, is it worth beginning the endeavor in the first place?

A central question, then, is the extent to which we actually need to identify the specific genes underlying adaptive change to in order to address the big questions in evolutionary genetics. I contend that we do. Indeed I will go further to say that we need to know the specific nucleotide changes responsible. We cannot be distracted by allelism per se but instead need to concentrate on the pattern of substitution of specific variants that have arisen via natural mutations. This will not be easy or even possible in many instances, but the very fact that we are contemplating it suggests that we are indeed entering a new era.

Acknowledgements

I thank Rodney Mauricio for organizing this symposium and for inviting me to participate. This work has been supported by a grant from the National Institutes of Health (GM54185).

References

Barton, N.H. & M. Turelli, 1989. Evolutionary quantitative genetics: How little do we know? Annu. Rev. Genet. 23: 337–370.

Beavis, W.D., 1994. The power and deceit of QTL experiments: lessons from comparative QTL studies, pp. 252–268 in 49th Annual Corn and Sorghum Research Conference, edited by American Seed Trade Association, Washington, DC.

Beavis, W.D., 1998. QTL analyses: power, precision, and accuracy, pp. 145–162 in Molecular Dissection of Complex

Traits, edited by A.H. Paterson. CRC Press, Boca Raton, FL.

Bost, B., D. de Vienne, F. Hospital, L. Moreau & C. Dillmann, 2001. Genetic and nongenetic bases for the L-shaped distribution of quantitative trait loci effects. Genetics 157: 1773–1787.

Botstein, D. & N. Risch, 2003. Discovering genotypes underlying human phenotypes: past successes for mendelian disease, future approaches for complex disease. Nat. Genet. 33S: 228–237.

Castle, W.E., 1921. An improved method of estimating the number of genetic factors concerned in cases of blending inheritance. Science 54: 223.

Charlesworth, B., 1990. Optimization models, quantitative genetics, and mutation. Evolution 44: 520–538.

Cheverud, J.M., E.J. Routman & D.J. Irschick, 1997. Plieotropic effects of individual gene loci on mandibular morphology. Evolution 51: 2006–2016.

Coyne, J.A., 1983. Genetic basis of difference in genital morphology among three sibling species of Drosophila. Evolution 37: 1101–1118.

Crow, J.F., 1957. Genetics of insect resistance to chemicals. Ann. Rev. Entomol. 2: 227–246.

Davidson, E.H., 2001. Genomic Regulatory Systems: Development and Evolution. Academic Press, New York.

Doebly, J., A. Stec & C. Gustus, 1995. *Teosinte branched1* and the origin of mazie: evidence for epistasis and the evolution of dominance. Genetics 141: 333–346.

Felsenstein, J., 1977. Multivariate normal genetic models with a finite number of loci, in proceedings of the Second International Conference on Quantitative Genetics, edited by Iowa State University Press, Ames, IA.

Fisher, R.A., 1918. The correlations between relatives on the supposition of Mendelian inheritance. Trans. Roy. Soc. Edinb. 52: 399–433.

Gibson, G., 2002. Microarrays in ecology and evolution: a preview. Mol. Ecol. 11: 17–24.

Glazier, A.M., J.H. Nadeau & T.J. Aitman, 2002. Finding genes that underlie complex traits. Science 298: 2345–2349.

Greenberg, A.J., J.R. Moran, J.A. Coyne & C.I. Wu, 2003. Ecological adaptation during incipient speciation revealed by precise gene replacement. Science 302: 1754–1757.

Jovelin, R., B.C. Ajie & P.C. Phillips, 2003. Molecular evolution and quantitative variation for chemosensory behaviour in the nematode genus Caenorhabditis. Mol. Ecol. 12: 1325–1337.

Juenger, T., M. Purugganan & T.F. Mackay, 2000. Quantitative trait loci for floral morphology in *Arabidopsis thaliana*. Genetics 156: 1379–1392.

Kao, C.H. & Z.B. Zeng, 2002. Modeling epistasis of quantitative trait loci using Cockerham's model. Genetics 160: 1243–1261.

Khoury, M., T.H. Beaty & B.H. Cohen, 1993. Fundamentals of Genetic Epidemiology. Oxford University Press, New York.

Lande, R., 1976. Natural selection and random genetic drift in phenotypic evolution. Evolution 30: 314–334.

Lande, R., 1981. The minimum number of genes contributing to quantitative variation between and within populations. Genetics 99: 544–553.

Lande, R., 1988. Quantitative genetics and evolutionary theory, pp. 71–84 in Proceeding of the Second International Conference on Quantitative Genetics, edited by Sinauer, Sunderland, Mass.

Lander, E.S. & N.J. Schork, 1994. Genetic dissection of complex traits. Science 265: 2037–2048.

Laurie, C.C., J.R. True, J. Liu & J.M. Mercer, 1997. An introgression analysis of quantitative trait loci that contribute to a morphological difference between *Drosophila simulans* and *D. mauritiana*. Genetics 145: 339–348.

Lebreton, C.M., P.M. Visscher, C.S. Haley, A. Semikhodskii & S.A. Quarrie, 1998. A nonparametric bootstrap method for testing close linkage vs. pleiotropy of coincident quantitative trait loci. Genetics 150: 931–943.

Leips, J. & T.F. Mackay, 2000. Quantitative trait loci for life span in *Drosophila melanogaster*: interactions with genetic background and larval density. Genetics 155: 1773–1788.

Lewontin, R.C., 1974. The Genetic Basis of Evolutionary Change. Columbia University Press, New York.

Lewontin, R.C., 1991. Twenty-five years ago in Genetics: electrophoresis in the development of evolutionary genetics: milestone or millstone? Genetics 128: 657–662.

Long, A.D., S.L. Mullaney, T.F. Mackay & C.H. Langley, 1996. Genetic interactions between naturally occurring alleles at quantitative trait loci and mutant alleles at candidate loci affecting bristle number in *Drosophila melanogaster*. Genetics 144: 1497–1510.

Long, A.D., R.F. Lyman, C.H. Langley & T.F. Mackay, 1998. Two sites in the *Delta* gene region contribute to naturally occurring variation in bristle number in *Drosophila melanogaster*. Genetics 149: 999–1017.

Long, A.D., R.F. Lyman, A.H. Morgan, C.H. Langley & T.F. Mackay, 2000. Both naturally occurring insertions of transposable elements and intermediate frequency polymorphisms at the *achaete-scute* complex are associated with variation in bristle number in *Drosophila melanogaster*. Genetics 154: 1255–1269.

Lyman, R.F. & T.F. Mackay, 1998. Candidate quantitative trait loci and naturally occurring phenotypic variation for bristle number in *Drosophila melanogaster*: the Delta-Hairless gene region. Genetics 149: 983–998.

Mackay, T.F.C. & C.H. Langley, 1990. Molecular and phenotypic variation in the *achaete-scute* region of *Drosophila melanogaster*. Nature 348: 64–66.

Mackay, T.F.C., 2001a. Quantitative trait loci in Drosophila. Nat. Rev. Genet. 2: 11–20.

Mackay, T.F.C., 2001b. The genetic architecture of quantitative traits. Annu. Rev. Genet. 35: 303–339.

Nachman, M.W., H.E. Hoekstra & S.L. D'Agostino, 2003. The genetic basis of adaptive melanism in pocket mice. Proc. Natl. Acad. Sci. USA 100: 5268–5273.

Noor, M.A., A.L. Cunningham & J.C. Larkin, 2001. Consequences of recombination rate variation on quantitative trait locus mapping studies. Simulations based on the *Drosophila melanogaster* genome. Genetics 159: 581–588.

Oleksiak, M.F., G.A. Churchill & D.L. Crawford, 2002. Variation in gene expression within and among natural populations. Nat. Genet. 32: 261–266.

Orr, H.A., 1998. The population genetics of adaptation: the distribution of factors fixed during adaptive evolution. Evolution 52: 935–949.

Orr, H.A., 2002. The population genetics of adaptation: the adaptation of DNA sequences. Evolution 56: 1317–1330.

Otto, S.P. & C.D. Jones, 2000. Detecting the undetected: estimating the total number of loci underlying a quantitative trait. Genetics 156: 2093–2107.

Pasyukova, E.G., C. Vieira & T.F. Mackay, 2000. Deficiency mapping of quantitative trait loci affecting longevity in *Drosophila melanogaster*. Genetics 156: 1129–1146.

Peichel, C.L., K.S. Nereng, K.A. Ohgi, B.L. Cole, P.F. Colosimo, C.A. Buerkle, D. Schluter & D.M. Kingsley, 2001. The genetic architecture of divergence between threespine stickleback species. Nature 414: 901-905.

Phillips, P.C., 1998. The language of gene interaction. Genetics 149: 1167–1171.

Phillips, P.C., 1999. From complex traits to complex alleles. Trends Genet. 15: 6–8.

Provine, W.B., 1971. The Origins of Theoretical Population Genetics. The University of Chicago Press, Chicago.

Remington, D.L., M.C. Ungerer & M.D. Purugganan, 2001. Map-based cloning of quantitative trait loci: progress and prospects. Genet. Res. 78: 213–218.

Siegal, M.L. & D.L. Hartl, 1998. An experimental test for lineage-specific position effects on alcohol dehydrogenase (*Adh*) genes in Drosophila. Proc. Natl. Acad. Sci. USA 95: 15513–15518.

Slatkin, M., 1970. Selection and polygenic characters. Proc. Natl. Acad. Sci. USA 66: 87–93.

Stam, L.F. & C.C. Laurie, 1996. Molecular dissection of a major gene effect on a quantitative trait: the level of alcohol dehydrogenase expression in *Drosophila melanogaster*. Genetics 144: 1559–1564.

Steinmetz, L.M., H. Sinha, D.R. Richards, J.I. Spiegelman, P.J. Oefner, J.H. McCusker & R.W. Davis, 2002. Dissecting the architecture of a quantitative trait locus in yeast. Nature 416: 326–330.

Stern, D.L., 2000. Evolutionary developmental biology and the problem of variation. Evolution 54: 1079–1091.

The C. elegans Sequencing Consortium, 1998. Genome sequence of the nematode *C. elegans*: a platform for investigating biology. Science 282: 2012–2018.

Vaughn, T.T., L.S. Pletscher, A. Peripato, K. King-Ellison, E. Adams, C. Erikson & J.M. Cheverud, 1999. Mapping quantitative trait loci for murine growth: a closer look at genetic architecture. Genet. Res. 74: 313–22.

Wayne, M.L. & L.M. McIntyre, 2002. Combining mapping and arraying: an approach to candidate gene identification. Proc. Natl. Acad. Sci. USA 99: 14903-14906.

Weinig, C., M.C. Ungerer, L.A. Dorn, N.C. Kane, Y. Toyonaga, S.S. Halldorsdottir, T.F. Mackay, M.D. Purugganan & J. Schmitt, 2002. Novel loci control variation in reproductive timing in *Arabidopsis thaliana* in natural environments. Genetics 162: 1875-1884.

Wright, S., 1968. Evolution and the Genetics of Populations. Vol. 1. Genetic and Biometric Foundations. University of Chicago Press, Chicago.

Zeng, Z.B., D. Houle & C.C. Cockerham, 1990. How Informative is Wright's estimator of the number of genes affecting a quantitative character? Genetics 126: 235–247.

Zeng, Z.B., 1992. Correcting the bias of Wright's estimates of the number of genes affecting a quantitative character: a further improved method. Genetics 131: 987–1001.

Zeng, Z.B., J. Liu, L.F. Stam, C.H. Kao, J.M. Mercer & C.C. Laurie, 2000. Genetic architecture of a morphological shape difference between two Drosophila species. Genetics 154: 299–310.

Genetica (2005) 123: 25–37 © Springer 2005

QTL mapping and the genetic basis of adaptation: recent developments

Zhao-Bang Zeng
Departments of Statistics and Genetics & Bioinformatics Research Center, North Carolina State University, Raleigh, NC 27695-7566, USA (E-mail: zeng@statgen.ncsu.edu)

Received 28 July 2003 Accepted 1 May 2004

Key words: genetic architecture, genetic basis of adaptation, genetic correlation, genotype by environment interaction, microarrays, QTL mapping, quantitative trait loci

Abstract

Quantitative trait loci (QTL) mapping has been used in a number of evolutionary studies to study the genetic basis of adaptation by mapping individual QTL that explain the differences between differentiated populations and also estimating their effects and interaction in the mapping population. This analysis can provide clues about the evolutionary history of populations and causes of the population differentiation. QTL mapping analysis methods and associated computer programs provide us tools for such an inference on the genetic basis and architecture of quantitative trait variation in a mapping population. Current methods have the capability to separate and localize multiple QTL and estimate their effects and interaction on a quantitative trait. More recent methods have been targeted to provide a comprehensive inference on the overall genetic architecture of multiple traits in a number of environments. This development is important for evolutionary studies on the genetic basis of multiple trait variation, genotype by environment interaction, host–parasite interaction, and also microarray gene expression QTL analysis.

Abbreviations: CIM – composite interval mapping; EM – expectation and maximization algorithm; IM – interval mapping; MIM – multiple interval mapping; QTL – quantitative trait loci.

Introduction

Quantitative trait loci (QTL) mapping is a genome-wide inference of the relationship between genotype at various genomic locations and phenotype for a set of quantitative traits in terms of the number, genomic positions, effects, interaction and pleiotropy of QTL and also QTL by environment interaction. The primary purpose of QTL mapping is to localize chromosomal regions that significantly affect the variation of quantitative traits in a population. This localization is important for the ultimate identification of responsible genes and also for our understanding of the genetic basis of quantitative trait variation.

Applied to natural populations, most QTL mapping experiments are designed to study the genetic basis of phenotypic differences between different natural populations or between different species (Mackay, 2001; Mauricio, 2001). Starting from two differentiated populations, a cross is usually made between the populations to create a hybrid, and then either backcross the hybrid to the parental population(s) to create backcross population(s) or intercross among hybrids (if possible) to create a F_2 population. Recombinant inbred lines can also be created from the cross and are popular for QTL mapping study. QTL mapping analysis is performed in these segregating populations to locate QTL that are responsible for the difference between the parental populations which could be due to adaptation. QTL mapping analysis in these populations can help us to understand a number of issues that are associated with the genetic basis of adaptation. It can estimate how many QTL that have different alleles between populations and contribute significantly to the population difference. It can estimate where they are located in the genome; what their effects are; how they interact; and how QTL interact with the environment. All these are critically important for the study of the genetic basis of adaptation.

QTL analysis certainly has many limitations. The number of QTL is likely to be downwardly biased estimated due to linkage and limited sample size. There is also likely a bias in the estimation of QTL effect distribution as only QTL with relatively large effects are likely to be detected and some QTL effects may represent the joint effects of multiple closely linked genes. Analysis of epistasis may only detect a part of gene interactions and there could be many other hidden interactions between detected and undetected QTL. Certainly, there is a big gap between QTL that are mapped with a confidence interval in many cM units and genes that are responsible for the variation. Mauricio (2001) discussed some caveats in using these methods for interpreting the genetic basis of adaptation for evolutionary biology studies.

In this article, I review some statistical methods used for QTL mapping analysis, particularly the methods used to map multiple QTL simultaneously for studying QTL epistasis and for estimating the overall genetic architecture of quantitative trait variation. I will use two QTL mapping experiments to illustrate the use of these methods and interpretation of the mapping analysis. One experiment is the study of genetic basis of a morphological shape difference between two *Drosophila* species due to adaptation (Zeng et al., 2000). The other experiment is the study of genetic basis of long-term selection response on wing size of *Drosophila melanogaster* (Weber et al., 1999, 2001). I also describe a method to study details of genetic correlation between multiple traits and to test QTL by environment interaction. In the end, I discuss the connection of this multiple trait QTL analysis with microarray gene expression data and outline an approach in using this method for the construction of genetic effect network between QTL, gene expressions and quantitative trait phenotypes.

Statistical framework

Statistical analysis of QTL mapping works with two data sets. One is the molecular marker data set that provides information of segregation of a genome at various marker positions in a population, and the other is the quantitative trait data set that provides information of segregation and effects of QTL. The connection between the two data sets is QTL. The variation of trait values in a population is partially due to the segregation of QTL alleles, and QTL are linked to some molecular markers. It is this linkage that provides information to localize QTL in a genome.

Let Y denote the trait data and X denote the marker data. In a joint analysis of marker and trait data, we study the joint probability of Y and X

$$\begin{aligned} P(Y,X) &= P(Y|X)P(X) \\ &= \sum_Q P(Y|Q,X)P(X) \\ &= \sum_Q P(Y|Q)P(Q|X)P(X) \end{aligned} \tag{1}$$

This joint probability can be split into two parts. One is $P(X)$ which can be modeled as a function of marker linkage order ω, linkage phases ϕ and recombination frequencies γ between markers. This analysis is the marker linkage analysis and $P(X|\gamma,\phi,\omega)$ is the likelihood of marker data.

The other part is $P(Y|X)$ which represents the QTL analysis, analyzing the conditional probability of trait values Y given marker genotypes X through QTL genotypes Q. $P(Q|X)$ is a function of QTL positions λ in relation to markers, and involves the segregation analysis of QTL given marker genotypes. $P(Y|Q)$ is a link function between QTL genotypes Q and trait phenotypes Y, and can be modeled as a function of QTL effect parameters θ, such as additive, dominance and epistatic effects of QTL and any other parameters that link QTL genotypes to trait phenotypes. Together, λ and θ represent the genetic architecture parameters of QTL. In this form, we generally represent $P(Y|X)$ as

$$P(Y|X,\lambda,\theta) = \sum_Q P(Y|Q,\theta)P(Q|X,\lambda) \tag{2}$$

which is the likelihood of trait data given marker data and is the main focus of this article.

Another statistical approach that has been used for QTL analysis is Bayesian posterior inference. In Bayesian statistics, model parameters are regarded as random variables, and we are concerned with the inference of posterior probability of model parameters. In a joint analysis of trait and markers, the posterior probability is

$$\begin{aligned}
&P(\omega,\phi,\gamma,\lambda,\theta|Y,X)\\
&= P(Y|X,\lambda,\theta)P(X|\omega,\phi,\gamma)P(\omega,\phi,\gamma,\lambda,\theta)/c\\
&= \sum_{Q} P(Y|Q,\theta)P(Q|X,\lambda)P(X|\omega,\phi,\gamma)\\
&\quad \times P(\omega,\phi,\gamma,\lambda,\theta)/c \qquad (3)
\end{aligned}$$

where c is a constant to make the posterior sum to 1 as a probability. This posterior is partitioned into three parts, the prior probability of parameters $P(\omega,\phi,\gamma,\lambda,\theta)$, the likelihood of marker data $P(X|\omega,\phi,\gamma)$, and the likelihood of trait data given marker data $P(Y|X,\lambda,\theta)$.

Multiple interval mapping: map multiple QTL and epistasis

Model and likelihood analysis

In QTL mapping likelihood analysis, we make an inference of genetic architecture of quantitative traits by testing and estimating model parameters θ and λ using likelihood (2). However, this analysis depends on the experimental design. One popular experimental design is to cross two widely separated inbred lines, populations or species, to create a heterozygous F_1 population, and then backcross the F_1 to parental lines to create backcross populations, or alternatively to intercross F_1 to create an F_2 population. Recombinant inbred lines are also popular for QTL mapping. For these standard experimental designs, the number of segregating QTL alleles is restricted to two, and the allelic frequencies of the QTL (as well as markers) and their linkage phases are known, thus greatly simplifying the genetic architecture of the traits.

The first part of the analysis is to calculate the conditional probability of QTL genotypes given observed marker genotypes, $P(Q|X,\lambda)$. For example, for a backcross population, there are two possible genotypes for a QTL, say Q_1Q_1 and Q_1q_1. Given the genotypes of two flanking markers, say X_1X_2/X_1X_2, X_1X_2/X_1x_2, X_1X_2/x_1X_2 and X_1X_2/x_1x_2, we can express the conditional probabilities of QTL genotypes given marker genotypes as a function of relative position of QTL (Q_1) in relation to the flanking markers (X_1 and X_2), λ_1:

$$\begin{aligned}
&P(Q_1Q_1|X_1X_2/X_1X_2); = 1;\\
&P(Q_1q_1|X_1X_2/X_1X_2) = 0\\
&P(Q_1Q_1|X_1X_2/X_1x_2) = 1-\lambda_1;\\
&P(Q_1q_1|X_1X_2/X_1x_2) = \lambda_1\\
&P(Q_1Q_1|X_1X_2/x_1X_2) = \lambda_1;\\
&P(Q_1q_1|X_1X_2/x_1X_2) = 1-\lambda_1\\
&P(Q_1Q_1|X_1X_2/x_1x_2) = 0;\\
&P(Q_1q_1|X_1X_2/x_1x_2) = 1 \qquad (4)
\end{aligned}$$

where $\lambda_1 = r_{X1Q1}/r_{X1X2}$, r_{X1Q1} is the recombination rate between X_1 and Q_1, and r_{X1X2} is the recombination rate between X_1 and X_2 (ignoring the double recombination for simplicity). For multiple loci in multiple different marker intervals, the joint conditional probability of multiple QTL genotypes is simply the product of separate conditional QTL genotype probabilities given marker genotypes under the assumption of no crossing-over interference.

$$P(Q|X,\lambda) = \prod_{r=1}^{m} P(Q_r|X,\lambda_r) \qquad (5)$$

As Q_r has two possible genotypes (Q_rQ_r and Q_rq_r), Q has a total of 2^m possible genotypes (joint configurations of Q_r's).

If two QTL fall into one marker interval, the calculation of the joint probability is more complicated (see Jiang & Zeng, 1995). Jiang and Zeng (1997) provided a general algorithm based on a hidden Markov model to take missing and dominant markers into account for this calculation for many populations derived from a cross between two inbred lines.

The second part of analysis is to fit trait phenotypes to QTL genotypes based on a genetic model, $P(Y|Q,\theta)$, and estimate model parameters θ. In quantitative genetics, the relationship between genotype and phenotype is usually modeled based on a linear model. For m putative QTL in a backcross population with sample size n, we can model a trait value as

$$y_i = \mu + \sum_{r=1}^{m} \alpha_r x_{ir}^* + \sum_{r<s}^{l} \beta_{rs}(x_{ir}^* x_{is}^*) + e_i \qquad (6)$$

for $i = 1,\ldots,n$, where y_i is the trait value of individual i, x_{ir}^* is a genotypic value of putative QTL r (which can be denoted as 1/2 or −1/2 for the two possible QTL genotypes), μ is the mean of model, α_r is the main effect of QTL r, β_{rs} is the epistatic effect between QTL r and s, e_i is a residual effect usually assumed to be normally distributed with mean zero and variance σ^2.

In this model, $\theta = \{\mu, \sigma^2, \mathrm{E}\}$, and

$$\begin{aligned} P(Y|X,\theta,\lambda) &= \prod_{i=1}^{n} P(y_i|x_i,\theta,\lambda) \\ &= \prod_{i=1}^{n}\sum_{j=1}^{2^m} P(Q_j|x_i,\lambda)P(y_i|Q_J,\theta) \\ &= \prod_{i=1}^{n}\left[\sum_{j=1}^{2^m} p_{ij}\frac{1}{\sqrt{2\pi\sigma^2}} \right. \\ &\quad \left. \times \exp\left[-\left(y_i-\mu-D_J E\right)^2 / \left(2\sigma^2\right)\right]\right] \end{aligned} \tag{7}$$

where x_i is the joint marker genotype of individual i, p_{ij} ($j = 1, \ldots, 2^m$) is the conditional probability of the jth joint QTL genotype for individual i given by (5), D_j is the raw vector of x_r^*'s and $(x_{ir}^* x_{is}^*)$'s corresponding to the jth joint QTL genotype, and E is a column vector of α_r's and β_r's. The dimension of D_j and E is the number of QTL effects in the model ($m + l$).

Kao and Zeng (1997) and Zeng, Kao and Basten (1999) described a procedure to obtain maximum likelihood parameter estimates using an expectation/maximization (EM) algorithm. The EM algorithm is an iterative procedure involving an E-step (expectation) and an M-step (maximization) in each iteration. In the $[k + l]$th iteration, the E-step is

$$\pi_{ij}^{[k+1]} = \frac{p_{ij}\frac{1}{\sqrt{2\pi\sigma^{2[k]}}}\exp\left[-\left(y_i-\mu^{[k]}-D_jE^{[k]}\right)^2/\left(2\sigma^{2[k]}\right)\right]}{\sum_{j=1}^{2^m} p_{ij}\frac{1}{\sqrt{2\pi\sigma^{2[k]}}}\exp\left[-\left(y_i-\mu^{[k]}-D_jE^{[k]}\right)^2/\left(2\sigma^{2[k]}\right)\right]} \tag{8}$$

and the M-step is

$$E_r^{[k+1]} = \frac{\sum_i\sum_j \pi_{ij}^{[k+1]} D_{jr}\left[\left(y_i-\mu^{[k]}\right)\right] - \sum_{s=1}^{r-1} D_{js}E_s^{[k+1]} - \sum_{s=r+1}^{m+1} D_{js}E_s^{[k]}}{\sum_i\sum_j \pi_{ij}^{[k+1]} D_{jr}^2} \tag{9}$$

for $r = 1, \ldots, m+1$

$$\mu^{[k+1]} = \frac{1}{n}\sum_i\left(y_i - \sum_j\sum_r \pi_{ij}^{[k+1]} D_{jr}E_r^{[k+1]}\right) \tag{10}$$

$$\begin{aligned} \sigma^{2[k+1]} = \frac{1}{n}\Bigg[& \sum_i\left(y_i-\mu^{[k+1]}\right)^2 - 2\sum_i\left(y_i-\mu^{[k+1]}\right) \\ & \times \sum_j\sum_r \pi_{ij}^{[k+1]} D_{jr}E_r^{[k+1]} \\ & + \sum_r\sum_s\sum_i\sum_j \pi_{ij}^{[k+1]} \\ & \times D_{jr}D_{js}E_r^{[k+1]}E_s^{[k+1]}\Bigg] \end{aligned} \tag{11}$$

where E_r is the rth element of E and D_{jr} is the rth element of D_j. There are many practical issues for the efficient and reliable implementation of this algorithm. Zeng, Kao and Basten (1999) discussed some strategies to alleviate the computational problem involved with 2^m components when m is not small.

Another computational method that has been used for QTL likelihood analysis is imputation (Sen & Churchill, 2001). Instead of directly evaluating the likelihood of the mixture model, the imputation method samples the missing QTL genotypes based on the conditional probabilities (5) and regresses trait values directly to sampled QTL genotypes. However, this has to be evaluated for a number of samples to obtain a reliable estimate of parameters and testing statistic for a given QTL model.

For given positions λ of m putative QTL and $m + l$ QTL effects, the likelihood analysis can proceed as outlined above or through imputation. The task is then to search and select genetic models (number, positions, effects and interaction of QTL) that best fit the data.

QTL model selection

Model selection is a key component of the analysis. It is a basis for interpreting and estimating the genetic architecture of QTL. Several methods have been developed. Kao, Zeng and Teasdale (1999) and Zeng, Kao and Basten (1999) worked out a stepwise search procedure to search for positions and interaction pattern of multiple QTL. This procedure has been implemented in QTL Cartographer (Basten, Weir & Zeng, 1995–2004) and

Windows QTL Cartographer (Wang, Basten & Zeng, 1999–2004).

Carlborg, Andersson and Kinghorn (2000) and Nakamichi, Ukai and Kishino (2001) used genetic algorithms for QTL model search. Satagopan et al. (1996) and Sillanpaa and Arjas (1998) worked on Bayesian methods using Markov chain Monte Carlo to search for number and positions of multiple QTL.

The stepwise search procedure outlined in Kao, Zeng and Teasdale (1999) and Zeng, Kao and Basten (1999) and implemented in QTL Cartographer has several interactive steps:

1. Initial model selection: In order to save computation time, some approximate and efficient statistical methods can be used to select an initial model for subsequent analysis. One method is to use composite interval mapping (CIM) (Zeng, 1994) for initial model selection. Another method is to use a forward or backward stepwise regression or a combined forward–backward stepwise regression on markers to select a subset of significant markers. For this analysis, it is found that using a stopping rule based on an F-to-drop or F-to-enter statistic with $\alpha = 0.01$ is generally satisfactory. All these procedures are implemented in Windows QTL Cartographer.
2. Optimize QTL positions: With an initial model or subsequent update of a QTL model, it is always desirable to update QTL position estimates using multiple interval mapping (MIM). Generally, it is reasonably sufficient to search and update position for each QTL in turn, conditioned on the current estimates for other QTL positions, based on likelihood. This process can be repeated.
3. Search new QTL: Scan the genome (except of the vicinity areas of current QTL positions) for the best position of a new QTL conditional on other QTL effects and interactions. Decision whether to add this QTL into the model depends on model selection criterion.
4. Select QTL epistasis: When a QTL model (number and positions) is changed, it may be necessary or desirable to update significant interaction components among QTL. This can be achieved by a backward stepwise search if possible, from possible interaction components, or a combined forward and then backward search.

Sometimes it may be worthwhile to attempt to search for significant epistatic effects between selected and unselected QTL positions. This may be performed in a stepwise manner by searching for the largest epistatic effect(s) between a current QTL position and an unselected genomic position at 1 or 2 cM intervals, and testing for significance. Of course, numerical calculation can be very intensive for this analysis.

The stepwise search may fail to uncover QTL in close repulsion linkage or to identify complex epistasis that involve multiple components. Sometimes, it may be necessary to employ chunkwise selection (Kao, Zeng & Teasdale, 1999) to improve model fitting. Although this procedure is difficult to implement automatically, it can be performed interactivelly using Windows QTL Cartographer.

The issue of model selection criterion is a complex one (Zeng, Kao & Basten, 1999; Broman & Speed, 2002). A number of criteria have been used to guide QTL model selection, such as Akaike information criterion, Bayes information criterion, residual bootstrap/permutation test. Individual QTL effects can also be tested based on a likelihood ratio test conditional on other QTL effects. However, it is still not clear how to take into account some biological and experimental information, such as heritability, marker coverage and sample size, in setting up more appropriate model selection criteria for QTL mapping analysis. More research is needed.

Given the identification of QTL, MIM provides a comprehensive way to estimate genetic architecture parameters for the difference between parental populations and also for the segregating population. It provides a cohesive estimate of additive, dominance and epistatic effects of QTL and the partition of genetic variance explained by QTL: how much directly due to which QTL (additive and dominance effects), how much due to epistasis, and how much through linkage or linkage disequilibrium (Zeng, Kao & Basten, 1999). This estimation is a strong point for the MIM analysis method. The method can also provide an efficient estimation or prediction of genotypic values for individuals based on marker data, which can be used for marker-assisted selection.

QTL mapping examples

Genetic architecture of a morphological shape difference between two Drosophila *species*

As examples, two large scale QTL mapping experiments in *Drosophila* are described here. In a study of genetic architecture of a morphological shape difference between two *Drosophila* species (Zeng et al., 2000), *Drosophila simulans* and *D. mauritiana* were crossed to make F_1 hybrids. Because F_1 males are sterile, females of the F_1 population were backcrossed to each of the parental lines to produce two backcross populations, each about 500 individuals. The trait is the morphology of the posterior lobe of the male genital arch analyzed as the first principal component in an elliptical Fourier analysis (Liu et al., 1996). Both the parental difference (35 environmental standard deviations) and the heritability (>0.9) of the trait in backcross populations are very large, providing a very favorable situation for QTL mapping. QTL analysis using MIM gives evidence of 19 QTL (based on the joint analysis in two backcrosses) distributed on the three major chromosomes, X, II and III (Figure 1). The additive effect estimates range from 1.0 to 11.4% of the parental difference. The greatest additive effect estimate is about four environmental standard deviations, but could represent multiple, closely linked QTL. Dominance effect estimates vary among loci from essentially no dominance to complete dominance, and *mauritiana* alleles tend to be dominant over *simulans* alleles. Epistasis appears to be relatively unimportant as a source of variation. All but one of the additive effect estimates have the same sign, which means that one species has nearly all the plus alleles and the other nearly all the minus alleles. This result is unexpected under most evolutionary scenarios, and suggests a history of strong directional selection acting on the posterior lobe.

Genetic basis of divergent selection response on wing size in D. melanogaster

The second experiment is about the genetic basis of divergent selection response on wing shape in *D. melanogaster*. Starting from a natural population, two selection lines were maintained with one selecting for high value and one for low value of the trait measurement. After 15 generations of intensive divergent selection, the wing shape, measured by an index incorporating two dimensions, differs in the high and low lines by 20 standard deviations (Weber, 1990). From the cross of the high and low lines, 519 third chromosome recombinant isogenic lines were created (Weber et al., 1999) in which marker and QTL alleles are segregating in the third chromosome and not in the other chromosomes. Using 65 *in situ*-labeled transposable elements as markers, 11 QTL were estimated by using MIM (Figure 2A) with additive

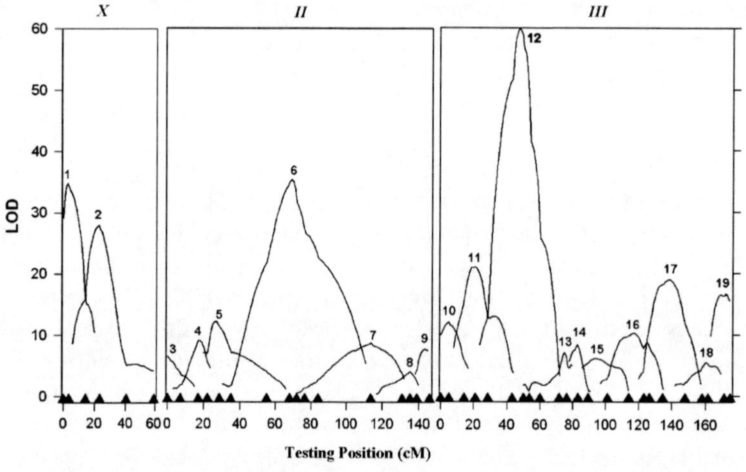

Figure 1. Genetic mapping of QTL on a morphological shape difference between *Drosophila simulans* and *D. mauritiana*. LOD score curves of a MIM analysis for each of 19 QTL are shown on a linkage map of the three major chromosomes. Marker positions are given by triangles.

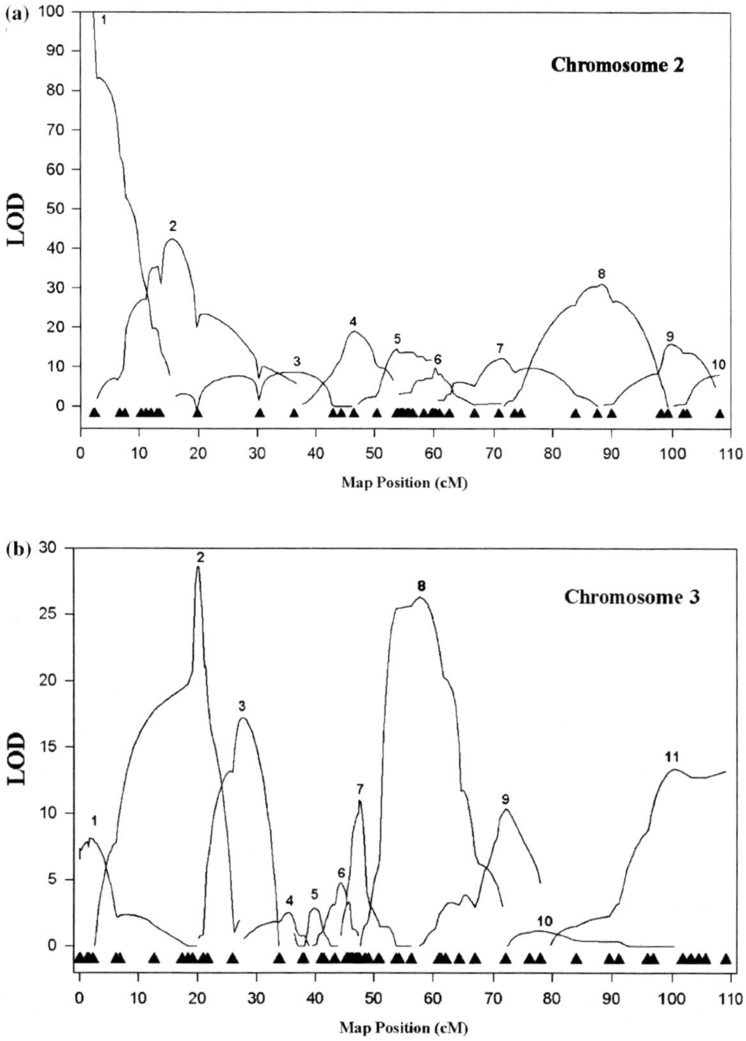

Figure 2. Genetic mapping of QTL on wing shape from a long-term divergent selection in (A) chromosome II and (B) chromosome III of *Drosophila melanogaster*. LOD score curves of a MIM analysis are shown on a linkage map of chromosomes II and III. Marker positions are given by triangles.

effect estimates ranging from 2.3 to 18.9% and added up to 99% of the parental line difference due to the third chromosome. All but one of the additive effect estimates have the same sign. Together, the 11 additive effects explain 0.947 of the total phenotypic variance with 0.274 due to the variance of additive effects and 0.673 due to the covariances between additive effects. There are nine QTL pairs that show significant additive by additive interaction effects. However, epistatic effect estimates are about equally positive and negative, and the nine epistatic effects explain only 0.012 of the total variance (0.072 due to the variance of epistatic effects and −0.060 due to the covariance between epistatic effects). The covariances between additive and epistatic effects, expected to be zero asymptotically, are negative and very small (−0.004) due to sampling. Thus the model explains 0.955 of the total phenotypic variance in the third chromosome recombinant isogenic lines.

To study QTL on the second chromosome, 701 second chromosome recombinant isogenic lines were created from the same high and low selection lines (Weber et al., 2001). Based on 47 markers and a MIM analysis with the residual permutation test as the model selection criterion, 10 QTL were detected (Figure 2B). The estimated additive effects are all in one direction, ranging in magnitude from 5 to 21% of the phenotypic difference between the two parental genotypes on this chromosome, and sum to 99.1% of the difference. There are 14 QTL pairs that show significant additive by additive interaction effects. Again, we observed the same pattern that epistatic effect estimates are about equally positive and negative. The additive effects together explain 0.951 of the total phenotypic variance, and the epistatic effects together explain only 0.003 of the total variance. The covariances between the additive effects and epistatic effects are almost zero as expected. The model explains 0.954 of the total variance in this recombinant population. The genetic architectures on the second and third chromosomes seem to be quite comparable in terms of number and distribution of QTL and QTL interaction pattern. It is interesting to observe that there are very significant epistatic effects between QTL from various statistical tests, and yet the total variance explained by epistasis is small. The sum of genetic variances due to individual epistatic effects is actually very substantial. But there are significant amount of negative covariances between different epistatic effects due to almost equal amount of plus and minus epistatic effects and linkage disequilibrium that cancel out much of the epistatic variances. This epistatic variation hidden by linkage disequilibrium could be released into the population as linkage disequilibrium deceases with further recombination. This epistatic pattern is consistent in both data sets for the second and third chromosomes. It is however not clear how common this epistatic pattern is for other traits and organisms.

Multiple trait QTL analysis: studying the genetic basis of trait correlations

Data structure, genetic models and likelihood analysis

Most QTL mapping experiments have observations on multiple traits, either for the purpose to study different attributes of a general biological character such as different measurements for a shape, different fitness components or a phenotype at different developmental stages, or for the purpose of studying genotype by environment interaction by regarding trait phenotypes in different environments as different trait states. Certainly, it would be important to take the information of multiple traits or multiple trait states in different environments into account in QTL mapping analysis. Such a multiple trait QTL analysis could improve the statistical power to detect QTL and improve the resolution to estimate QTL positions and effects. Probably more importantly, it provides a basis and formal procedures to test a number of biologically interesting hypotheses concerning the nature of genetic correlations between different traits, such as pleiotropic effects of QTL and QTL by environment interactions, and provide a framework for a comprehensive estimation about the genetic architecture of quantitative traits including the structure of genetic correlations between traits. In general, data on multiple trait QTL analysis can be classified in two categories. For the first category, multiple traits are measured on the same individuals. Trait (Y) and marker (X) data matrices may look like the following (for t traits, f markers, n individuals)

$$Y = \begin{bmatrix} y_{11} & y_{12} & \cdots & y_{1n} \\ y_{21} & y_{22} & \cdots & y_{2n} \\ \vdots & \vdots & \ddots & \vdots \\ y_{t1} & y_{t2} & \cdots & y_{tn} \end{bmatrix}$$

$$\text{and} \quad X = \begin{bmatrix} x_{11} & x_{12} & \cdots & x_{1n} \\ x_{21} & x_{22} & \cdots & x_{2n} \\ \vdots & \vdots & \ddots & \vdots \\ x_{f1} & x_{f2} & \cdots & x_{fn} \end{bmatrix}$$

For the second category, multiple traits or trait states are measured on different individuals. Trait and marker data matrices may look like the following (with one set of traits measured in population one with n_1 individuals and f_1 markers, and another set measured in another population with n_2 individuals and f_2 markers)

$$Y_1 = \begin{bmatrix} y_{11} & y_{12} & \cdots & y_{1n_1} \\ y_{21} & y_{22} & \cdots & y_{2n_1} \\ \vdots & \vdots & \ddots & \vdots \\ y_{t_11} & y_{t_12} & \cdots & y_{t_1n_1} \end{bmatrix}$$

$$\text{and} \quad X_1 = \begin{bmatrix} x_{11} & x_{12} & \cdots & x_{1n_1} \\ x_{21} & y_{22} & \cdots & x_{2n_1} \\ \vdots & \vdots & \ddots & \vdots \\ x_{f_1 1} & x_{f_1 2} & \cdots & x_{f_1 n_1} \end{bmatrix}$$

$$Y_2 = \begin{bmatrix} y_{11} & y_{12} & \cdots & y_{1n_2} \\ y_{21} & y_{22} & \cdots & y_{2n_2} \\ \vdots & \vdots & \ddots & \vdots \\ y_{t_2 1} & y_{t_2 2} & \cdots & y_{t_2 n_2} \end{bmatrix}$$

$$\text{and} \quad X_2 = \begin{bmatrix} x_{11} & x_{12} & \cdots & x_{1n_2} \\ x_{21} & x_{22} & \cdots & x_{2n_2} \\ \vdots & \vdots & \ddots & \vdots \\ x_{f_2 1} & x_{f_2 2} & \cdots & x_{f_2 n_2} \end{bmatrix}$$

This can represent several situations, for example, the same traits measured in two backcrosses (B_1 and B_2) on different individuals in which case a test on QTL by backcross interaction is a test about dominance of QTL, the same trait measured in two sexes in which case a test on QTL by sex interaction can be performed, or different groups of individuals are planted in two or multiple geographic locations.

Jiang and Zeng (1995) have studied multiple trait QTL mapping analysis methods formulated under the framework of CIM. This method has been implemented in QTL Cartographer (Basten, Weir & Zeng, 1995–2004) and also Windows QTL Cartographer (Wang, Basten & Zeng, 1999–2004). Recently, we have been working on extending MIM to multiple traits to provide a comprehensive estimation of genetic correlation between different traits and its partition to different QTL due to pleiotropy or linkage. Here, I outline this MIM on multiple traits, although the details of this method will be published elsewhere.

For m putative QTL of T traits in S environments/populations, the MIM model (for a backcross population) is defined by

$$y_{sti} = \mu_{st} + \sum_{r=1}^{m} \alpha_{str} x^*_{sir} + e_{sti} \tag{12}$$

where y_{sti} is the phenotypic value of trait t for individual i in environment/population s; i indexes individuals of the sample ($i = 1, 2, \ldots, n_s$); t indexes traits ($t = 1, 2, \ldots, T$); s indexes environments/populations ($s = 1, 2, \ldots, S$); μ_{st} is the mean of the model; α_{str} is the effect of putative QTL r on trait t in population s; x^*_{sir} is a coded variable denoting the genotype of putative QTL r (defined by 1/2 or −1/2 for the two genotypes) for individual i in population s, which is unobserved but can be inferred from maker data in sense of probability; e_{sti} is a residual effect of the model assumed to be multivariate normal distributed with mean vector 0 and variance matrix V_s.

The likelihood function of the data given the model is a mixture of normal distributions

$$L = \prod_{s=1}^{S} \prod_{i=1}^{n_s} \left[\sum_{j=1}^{2^m} p_{sij} \phi\left(y_{si} | \mu_s + A_s D_{sj}, V_s\right) \right] \tag{13}$$

where p_{sij} is the probability of each multilocus genotype conditional on marker data; A_s is a matrix of QTL parameters (α's) for population s; D_{sj} is a vector specifying the configuration of x^*'s associated with each α for the jth QTL genotype; ϕ $(y|\mu, V)$ denotes a multivariate normal density function for y with mean vector μ and variance matrix V.

This likelihood can also be evaluated through an EM algorithm. In the $[k + l]$th iteration, the E-step is to update the conditional probabilities of multiple locus QTL genotypes given marker genotypes and phenotypic trait values

$$\pi_{sij}^{[k+1]} = \frac{p_{sij} \phi\left(y_{si} | \mu_s^{[k]} + A_s^{[k]} D_{sj}, V_s^{[k]}\right)}{\sum_{j=1}^{2^m} p_{sij} \phi\left(y_{si} | \mu_s^{[k]} + A_s^{[k]} D_{sj}, V_s^{[k]}\right)} \tag{14}$$

and the M-step is to update estimates of model parameters

$$\alpha_{str}^{[k+1]} = \frac{\sum_s \sum_i \sum_j \pi_{sij}^{[k+1]} D_{sjr} \left[\left(y_{sti} - \mu_{st}^{[k]} \right) - \sum_{u=1}^{r-1} D_{sju} \alpha_{stu}^{[k+1]} - \sum_{u=r+1}^{m} D_{sju} \alpha_{stu}^{[k]} \right]}{\sum_s \sum_i \sum_j \pi_{sij}^{[k+1]} D_{sjr}^2} \tag{15}$$

$$\mu_{st}^{[k+1]} = \frac{1}{n_s} \sum_i \left(y_{sti} - \sum_j \sum_r \pi_{sij}^{[k+1]} D_{sjr} \alpha_{str}^{[k+1]} \right) \tag{16}$$

$$v_{stt'}^{[k+1]} = \frac{1}{n_s}\left[\sum_i \sum_j \pi_{sij}^{[k+1]} \times \left(y_{sti} - \mu_{st}^{[k+1]} - \sum_r D_{sjr}\alpha_{str}^{[k+1]}\right) \times \left(y_{st'i} - \mu_{st'}^{[k+1]} - \sum_r D_{sjr}\alpha_{st'r}^{[k+1]}\right)\right] \quad (17)$$

where α_{str} is the effect of QTL r on trait t in environment s and D_{sjr} is the rth element of D_{sj}. This algorithm is very stable even with a large number of parameters as in this case. However, one problem with the algorithm is its slow convergence, particularly in this case with a large number of parameters. We have studied other alternative and more efficient algorithms and found that an algorithm based on the generalized EM combined with Newton–Ralphson algorithm provides a good balance of stability and efficiency for this application (Qin & Zeng, unpublished data).

Model selection

It is very tricky to perform model selection on multiple traits. Model selection can proceed as in MIM in a similar way as outlined in MIM. In this case, when a QTL is selected, its effects are fitted and estimated for all traits in all environments or populations, regardless whether the QTL effect is significant for a particular trait in a particular environment. Steps are as follows:

1. Initial model: Use multivariate backward stepwise regression on markers to select an initial model.
2. Optimize the estimates of QTL positions based on the currently selected model.
3. Scan the genome to determine the best position for adding a new QTL.
4. Repeat (2) and (3) for a few times to select a few competing models.
5. If epistasis is considered, select significant epistatic terms.
6. Select the final model based on some information criterion.

Partition of genetic correlation: pleiotropy versus linkage

Given a selected genetic model, we can estimate the genetic variance and covariance explained by QTL for different traits and partition the genetic correlation between traits into components due to pleiotropic effects of QTL and those due to linkage disequilibrium. The genetic variance explained by QTL for trait t in a particular environment can be estimated as and partitioned into the following components.

$$\begin{aligned}\hat{\sigma}_{g_t}^2 &= \sum_r \sum_{r'}\left[\frac{1}{n}\sum_{i=1}^n \sum_j \hat{\pi}_{ij}(D_{jr} - \bar{D}_r) \times (D_{jr'} - \bar{D}_{r'})\hat{\alpha}_{tr}\hat{\alpha}_{tr'}\right] \\ &= \sum_r \left[\frac{1}{n}\sum_{i=1}^n \sum_j \hat{\pi}_{ij}(D_{jr} - \bar{D}_r)^2 \hat{\alpha}_{tr}^2\right] \\ &\quad + \sum_{r \neq r'}\left[\frac{1}{n}\sum_{i=1}^n \sum_j \hat{\pi}_{ij}(D_{jr} - \bar{D}_r) \times (D_{jr'} - \bar{D}_{r'})\hat{\alpha}_{tr}\hat{\alpha}_{tr'}\right] \\ &= \sum_r \hat{\sigma}_{\alpha_{tr}}^2 + \sum_{r \neq r'} \hat{\sigma}_{\alpha_{tr}\alpha_{tr'}} \end{aligned} \quad (18)$$

Similarly, the genetic covariance between trait t and t' in a particular environment can be estimated as and partitioned into the corresponding components.

$$\begin{aligned}\hat{\sigma}_{g_{tt'}} &= \sum_r \sum_{r'}\left[\frac{1}{n}\sum_{i=1}^n \sum_j \hat{\pi}_{ij}(D_{jr} - \bar{D}_r) \times (D_{jr'} - \bar{D}_{r'})\hat{\alpha}_{tr}\hat{\alpha}_{t'r'}\right] \\ &= \sum_r \left[\frac{1}{n}\sum_{i=1}^n \sum_j \hat{\pi}_{ij}(D_{jr} - \bar{D}_r)^2 \hat{\alpha}_{tr}\hat{\alpha}_{t'r}\right] \\ &\quad + \sum_{r \neq r'}\left[\frac{1}{n}\sum_{i=1}^n \sum_j \hat{\pi}_{ij}(D_{jr} - \bar{D}_r) \times (D_{jr'} - \bar{D}_{r'})\hat{\alpha}_{tr}\hat{\alpha}_{t'r'}\right] \\ &= \sum_r \hat{\sigma}_{\alpha_{tr}\alpha_{t'r}} + \sum_{r \neq r'} \hat{\sigma}_{\alpha_{tr}\alpha_{t'r'}} \end{aligned} \quad (19)$$

Thus the genetic correlation between the traits t and t' can be estimated as

$$\hat{\gamma}_{g_{tt'}} = \frac{\hat{\sigma}_{g_{tt'}}}{\sqrt{\hat{\sigma}_{g_t}^2 \hat{\sigma}_{g_{t'}}^2}} \quad (20)$$

Then the part in $\hat{\gamma}_{g_{tt'}}$ that is due to the pleiotropic effect of QTL r (α_{tr} and $\alpha_{t'r}$) can be estimated as

$\hat{\sigma}_{\alpha_{tr}\alpha_{t'r}} / \sqrt{\sigma^2_{g_t}\sigma^2_{g_{t'}}}$. Similarly, the part in $\hat{\gamma}_{g_{tt'}}$ that is due to linkage disequilibrium between QTL r and r' can be estimated as $\left(\hat{\sigma}_{\alpha_{tr}\alpha_{t'r'}} + \hat{\sigma}_{\alpha_{tr'}\alpha_{t'r}}\right) / \sqrt{\sigma^2_{g_t}\sigma^2_{g_{t'}}}$. This provides a comprehensive estimation of genetic correlation between traits and its partition to individual QTL for us to assess relative importance of pleiotropy and linkage on the correlation.

Testing QTL by environment interactions

Statistical tests on hypotheses of QTL by environment interactions can help us to understand and interpret the genetic architecture of quantitative traits in those environments/populations. There are several ways to test QTL by environment interaction. As elaborated by Falconer (1952), the genetic correlation between trait measurements in different environments (or different trait states) is a measurement of genotype by environment interaction, with a perfect correlation indicating no interaction. When all QTL have the same effects in different environments, the genetic correlation is perfect. Thus a genome-wide test of QTL by environment interaction can be performed through the following likelihood rate test.

- Genome-wide test of QTL by environment interaction between trait states t and t' in population s:

$$H_0 : A_{st} = A_{st'} (\text{no interaction}) \quad \textit{versus } H1 : A_{st} \neq A_{st'} \ (\text{interaction})$$

with

$$LR = -2 \ln \frac{L_0\left(\hat{\mu}_s, \hat{A}_{st} = \hat{A}_{st'}, \hat{V}_s\right)}{L_1\left(\hat{\mu}_s, \hat{A}_{st}, \hat{A}_{st'}, \hat{V}_s\right)}$$

- Genome-wide test of QTL by environment interaction for trait t between environments s and s':

$$H_0 : A_{st} = A_{s't} (\text{no interaction}) \quad \textit{versus } H1 : A_{st} \neq A_{s't} \ (\text{interaction})$$

with

$$LR = -2 \ln \frac{L_0\left(\hat{\mu}, \hat{A}_{st} = \hat{A}_{s't}, \hat{V}\right)}{L_1\left(\hat{\mu}, \hat{A}_{st}, \hat{A}_{s't}, \hat{V}\right)}$$

The significance value for the test can be assessed through a residual permutation test similar to that explained in Zeng, Kao and Basten (1999). For this test, genotypic values of the trait in different environments are estimated through the constrained likelihood under the null hypothesis and subtracted from the observed phenotypic values. The residues are permuted. Then the likelihood rate test is performed in a number of permuted samples to empirically estimate the distribution of the test statistic at the null hypothesis. This likelihood ratio test on QTL by environment interaction can also be performed on individual QTL or a subset of QTL with the tested QTL effects constrained to be the same for different environments at the null hypothesis and unconstrained at the alternative hypothesis.

Implications for microarray gene expression QTL analysis

Mapping gene expression QTL (eQTL) has recently become an interesting research topic, largely due to the feasibility in performing a relatively large scale microarray typing on multiple genotypes. Several studies have already been published combining gene expression microarray data with molecular marker data to map eQTL (Brem et al., 2002; Eaves et al., 2002; Schadt et al., 2003). In these studies, expression profiles of a number of genes are typed from selected tissues in each individual or line together with molecular marker genotypes and quantitative trait phenotypes. In the eQTL mapping analysis, gene expression profiles are regarded as phenotypes and QTL that affect the gene expressions are mapped.

Largely due to still relatively small sample size (30–100), these studies take a relative simple approach in eQTL mapping analysis, basically performing simple interval mapping (IM) on each gene expression one by one with permutation to assess the genome-wide significance. These studies are the first to show the feasibility of using eQTL mapping to identify genes or genome regions that regulate gene expressions. Some identified eQTL are in the same genomic region that the expressed genes are located with known regulatory genes nearby (Brem et al., 2002; Schadt et al., 2003). There are also many eQTL that are located in other genomic regions with no obvious candidate regulatory genes.

There are a number of ways to perform QTL mapping analysis with this kind of data. The simplest way is to associate each trait or gene expression to each marker by a regression analysis. The method of Lander and Botstein (1989)'s IM uses the same simple regression model, but creates a genome scan to search for QTL. When the sample size is small which has been the case in the few published eQTL mapping studies, few QTL can be detected to be significant for each gene expression and this IM approach may be adequate for data analysis. However, when the sample size is reasonably large, as in many typical QTL mapping experiments, statistical methods using multiple marker information such as CIM (Zeng, 1994) and MIM (Kao, Zeng & Teasdale, 1999) can help to improve statistical power to detect eQTL and also to resolve multiple eQTL including multiple linked eQTL. Compared to MIM, CIM is much easier and simpler to perform with a statistical power, though less than, but not far from MIM for QTL detection. Computationally, MIM is more intensive, requiring a model search in the multiple dimensional parameter space. But MIM has much nicer properties than CIM in the joint estimation of multiple QTL effects, given a genetic model, and also allows the evaluation of QTL epistasis.

However, it is the multiple trait MIM that has the ability to explore the genetic basis and network of correlation between multiple gene expressions and traits. By taking pair-wise or multiple gene expressions together for eQTL analysis, we can test and infer whether the expressions of different genes are co-regulated, i.e., whether eQTL have the similar effects on the expressions of different genes or eQTL affect different gene expressions differently. The overall level of this co-regulation can be measured through a single quantity, the genetic correlation between a pair of gene expressions. We may use this measure to classify the level of gene co-regulation, such as 1–0.75 for high synergistic co-regulation; 0.75–0.50 for medium synergistic co-regulation; 0.50–0.25 for low synergistic co-regulation; 0.25 to −0.25 for little or no co-regulation; −0.25 to −0.5 for low antagonistic co-regulation; −0.50 to −0.75 for medium antagonistic co-regulation; and −0.75 to −1.0 for high antagonistic co-regulation. Also not only can this analysis determine the overall level of gene co-regulation, it can also further partition the genetic correlation into individual eQTL, and to estimate how much the genetic correlation is due to pleiotropic effects of eQTL (true co-regulation) and how much due to linkage of different eQTL (just genetic association due to linkage disequilibrium). This joint detailed analysis can provide a much more defined dissection on the genetic basis of association between different gene expressions, between different traits, and between gene expressions and traits. It is in this respect that the joint inference of the genetic effect network has a much more defined meaning, rather than just a phenotypic correlation between gene expressions or between gene expressions and traits.

Conclusion

QTL mapping has been used by a number of evolutionary studies to study the genetic basis of adaptation by mapping individual QTL that explain the differences between differentiated populations and also estimating their effects and interaction in the mapping population. This analysis can provide many information and clues about the evolutionary history of populations and causes of the population differentiation. QTL mapping analysis methods and associated computer programs provide us tools for such an inference on the genetic basis and architecture of quantitative trait aviation in a mapping population. Current methods have the capability to separate and localize multiple QTL and estimate their effects and interaction on a quantitative trait. More recent methods have been targeted to provide a comprehensive inference on the overall genetic architecture of multiple traits in a number of environments. This development is important for evolutionary studies on the genetic basis of multiple trait variation, genotype by environment interaction, host–parasite interaction, and also microarray gene eQTL analysis.

Acknowledgements

This work was partially supported by a grant GM45344 from the National Institute of Health and a grant from USDA Plant Genome Program.

References

Basten, C., B.S. Weir & Z.-B. Zeng, 1995–2004. QTL Cartographer. Department of Statistics, North Carolina State University, Raleigh, NC (http://statgen.ncsu.edu/qtlcart/cartographer.html).

Brem, R.B., G. Yvert, R. Clinton & L. Kruglyak, 2002. Genetic dissection of transcriptional regulation in budding yeast. Science 296: 752–755.

Broman, K.W. & T.P. Speed, 2002. A model selection approach for the identification of quantitative trait loci in experimental crosses. J. R. Stat. Soc. B 64: 641–656.

Carlborg, O., L. Andersson & B. Kinghorn, 2000. The use of a genetic algorithm for simultaneous mapping of multiple interacting quantitative trait loci. Genetics 155: 2003–2010.

Eaves, I.A., L.S. Wicker, G. Ghandour, P.A. Lyons, L.B. Peterson, J.A. Todd & R.J. Glynne, 2002. Combining mouse congenic strains and microarray gene expression analyses to study a complex trait: the NOD model of Type I diabetes. Genome Res. 12: 232–243.

Falconer, D.S., 1952. The problem of environment and selection. Am. Nat. 86: 293–298.

Jiang, C. & Z.-B. Zeng, 1995. Multiple trait analysis of genetic mapping for quantitative trait loci. Genetics 140: 1111–1127.

Jiang, C. & Z.-B. Zeng, 1997. Mapping quantitative trait loci with dominant and missing markers in various crosses from two inbred lines. Genetica 101: 47–58.

Kao, C.-H. & Z.-B. Zeng, 1997. General formulae for obtaining the MLEs and the asymptotic variance-covariance matrix in mapping quantitative trait loci when using the EM algorithm. Biometrics 53: 653–665.

Kao, C.-H., Z.-B. Zeng & R. Teasdale, 1999. Multiple interval mapping for quantitative trait loci. Genetics 152: 1023–1216.

Lander, E.S. & D. Botstein, 1989. Mapping Mendelian factors underlying quantitative traits using RFLP linkage maps. Genetics 121: 185–199.

Liu, J., J.M. Mercer, L.F. Stam, G.C. Gibson, Z.-B. Zeng & C.C. Laurie, 1996. Genetic analysis of a morphological shape difference in the male genitalia of *Drosophila simulans* and *D. mauritiana*. Genetics 142: 1129–1145.

Nakamichi, R., Y. Ukai & H. Kishino, 2001. Detection of closely linked multiple quantitative trait loci using a genetic algorithm. Genetics 158: 463–475.

Mackay, T.F.C., 2001. Quantitative trait loci in *Drosophila*. Nat. Rev. Genet. 2: 11–20.

Mauricio, R., 2001. Mapping quantitative trait loci in plants: uses and caveats for evolutionary biology. Nat. Rev. Genet. 2: 370–381.

Satagopan, J.M., B.S. Yandell, M.A. Newton & T.C. Osborn, 1996. A Bayesian approach to detect quantitative trait loci using Markov chain Monte Carlo. Genetics 144: 805–816.

Schadt, E.E., S.A. Monks, T.A. Drake, A.J. Lusis, N. Che, V. Colinayo, T.G. Ruff, S.B. Milligan, J.R. Lamb, G. Cavet, P.S. Linsley, M. Mao, R.B. Stoughton & S.H. Friend. 2003. Genetics of gene expression surveyed in maize, mouse and man. Nature 422: 297–302.

Sen, S. & G.A. Churchill, 2001. A statistical framework for quantitative trait mapping. Genetics 159: 371–387.

Sillanpaa, M.J. & E. Arjas, 1998. Bayesian mapping of multiple quantitative trait loci from incomplete inbred line cross data. Genetics 148: 1373–1388.

Wang, S., C. Basten & Z-.B Zeng, 1999–2004. WINDOWS QTL Cartographer. Department of Statistics, North Carolina State University, Raleigh, NC (http://statgen.ncsu.edu/qtlcart/cartographer.html).

Weber, K.E., 1990. Selection on wing allometry in *Drosophila melanogaster*. Genetics 126: 975–989.

Weber, K., R. Eisman, S. Higgins, L. Morey, A. Patty, M. Tausek & Z.-B. Zeng, 2001. An analysis of polygenes affecting wing shape on chromosome 2 in *Drosophila melanogaster*. Genetics 159: 1045–1057.

Weber, K., R. Eisman, L. Morey, A. Patty, J. Sparks, M. Tausek & Z.-B. Zeng, 1999. An analysis of polygenes affecting wing shape on chromosome 3 in *Drosophila melanogaster*. Genetics 153: 773–786.

Zeng, Z.-B., 1994. Precision mapping of quantitative trait loci. Genetics 136: 1457–1468.

Zeng, Z.-B., C.-H. Kao & C.J. Basten, 1999. Estimating the genetic architecture of quantitative traits. Genet. Res. 74: 279–289.

Zeng, Z.-B., J. Liu, L.F. Stam, C.-H. Kao, J.M. Mercer & C.C. Laurie, 2000. Genetic architecture of a morphological shape difference between two *Drosophila* species. Genetics 154: 299–310.

Genetica (2005) 123: 39–47 © Springer 2005

Sex differences in recombination and mapping adaptations

Patrick D. Lorch
Zoology Department, University of Toronto, Ramsay Wright Labs, 25 Harbord St. Toronto, Ontario M5S 3G5, Canada (Phone: (919)843-2320; Fax: (919)962-1625; E-mail: plorch@email.unc.edu); Current address: Biology Department, University of North Carolina, CB 3280, Coker Hall, Chapel Hill, NC 27599-3280, USA

Received 15 April 2002 Accepted 15 September 2003

Key words: condition dependence, good genes, QTL mapping, sexual selection, sexual conflict

Abstract

Since the raw material of marker based mapping is recombination, understanding how and why recombination rates evolve, and how we can use variation in these rates will ultimately help to improve map resolution. For example, using this variation could help in discriminating between linkage and pleiotropy when QTL for several traits co-locate. It might also be used to improve QTL mapping algorithms. The goals of this chapter are: (1) to highlight differences in recombination rates between the sexes, (2) describe why we might expect these differences, and (3) explore how sex difference in recombination can be used to improve resolution in QTL mapping.

Sex differences in recombination

Sex differences in recombination rates generally are seen as differences in linkage maps (Figure 1). Since the physical size of chromosomes in each sex is assumed to be equal, sex differences in recombination result from different amounts of recombination during meiosis. These sex differences become apparent whenever mapping studies are conducted in such a way that recombination rates can be estimated separately for each sex. Taking a backcross design as an example (see Korol, Preygel & Preygel, 1994), the F1 generation produced by crossing two different inbred lines can be used as both sires and dams (pollen parent and seed parent) in the backcross to original inbred parentals. Sex difference in recombination can then be seen in the linkage maps produced from the two sets of backcross offspring. This is because inbred backcross parents should be homozygous at almost all loci, so any recombination occurs in the F1 parent. If half of your backcrosses use F1 dams and the other F1 sires, you can estimate linkage maps separately for each sex.

A survey of published literature shows that sex differences in recombination rates are widespread (for reviews see Callan & Perry, 1977; Trivers, 1988; Burt, Bell & Harvey, 1991; Singer et al., 2002). Table 1 and Figure 2 summarize all the data to date (The Appendix shows data collected since Burt, Bell & Harvey (1991) in a format similar to their appendix.). Where sex differences in recombination have been estimated, we can distinguish between species where both sexes experience some recombination (chiasmate species) and species where one sex has no recombination (achiasmate species). In chiasmate species 45 cases show more female than male recombination, 21 cases show more male than female recombination and 9 cases show no sex difference (Cano & Santos, 1990; Burt, Bell & Harvey, 1991; van Oorschot et al., 1992; Korol Preygel & Preygel, 1994; Lagercrantz & Lydiate, 1995; Kearsey et al., 1996). In achiasmate species 5 cases show female recombination, 8 cases show male recombination, and whenever there are heterogametic sex chromosomes, the heterogametic sex has no recombination (Burt, Bell & Harvey, 1991).

Whatever the causes of these sex differences, they provide a useful example of variation in recombination rates for two reasons. First, the

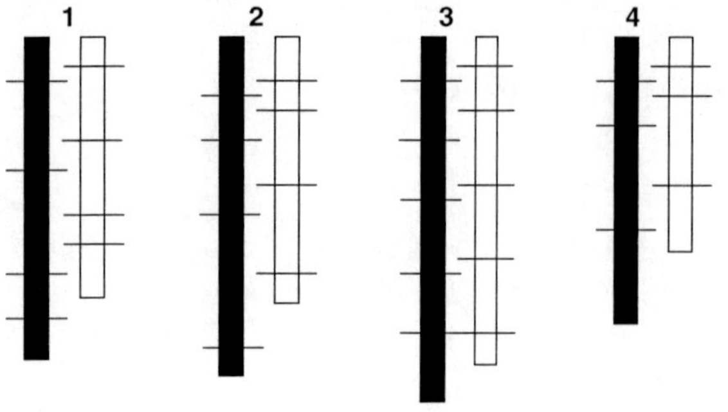

Figure 1. Typical pattern of sex-specific maps for four linkage groups in a hypothetical species. Male and female chromosomes should be of equal length, but maps often show sex differences. Bars show genetic marker loci. Distance between markers indicates larger numbers of recombination events between markers. Typically, female maps (black) are larger than male maps (white) due to more and/or less-localized recombination events.

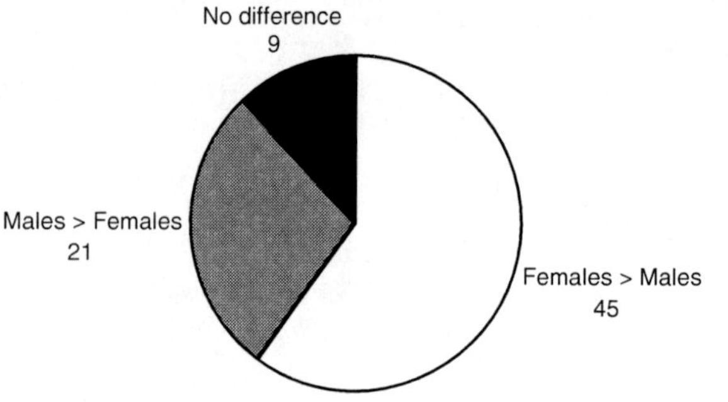

Figure 2. Summary of species where sex differences in recombination have been estimated. For chiasmate species, based on data in Burt, Bell and Harvey (1991) and the Appendix.

Table 1. Breakdown of sex differences in recombination for 75 species by taxon. Lists chiasmate species, based on data in Burt, Bell and Harvey (1991) and the Appendix

Taxon	F > M	M > F	F = M	Comments
Animals				
Platyhelminthes	2	1	0	
Insecta	2	9	3	All Orthoptera
Amphibia	4	2	0	
Mammalia	7	4	1	
Pisces	2	0	0	
Aves	2	0	0	
Plants				
Monocotyledonae	20	3	4	
Dicotyledonae	2	1	1	
Orchidaceae	4	1	0	
Total	45	21	9	

evolution of modifiers of recombination has been studied extensively in the context of the evolution of sex. This means that we have basic theory for understanding how recombination rates can be modified, albeit few specifics about how sex difference can arise. Second, by modifying breeding designs we may be able to exploit sex differences in recombination to improve map resolution and QTL discrimination (Singer et al., 2002). This is not to say that other forms of variation in recombination rates cannot also be used to improve maps, only that since QTL mapping involves crosses and algorithmic estimation of QTL location relative to a marker-based map, sex difference may provide a particularly useful form of variation in recombination rates. To make this second point clear we need to consider what we know about how recombination rates evolve.

How recombination rates can evolve

The evolution of recombination is difficult to study because recombination affects the way genes on the same chromosome interact. As evolution proceeds, recombination does three things, the first two of which directly conflict. It can bring together alleles on one chromosome with positive effects on fitness, allowing one parent to pass along sets of alleles that survived natural and sexual selection in the parents. Recombination can then break up these beneficial associations in the very next generation. It can also bring together deleterious alleles, allowing them to be more efficiently eliminated by selection. The complicated balance between these three processes will determine whether selection acts to increase or decrease recombination rates for a given region of a chromosome (Barton, 1995). Selection can act to increase recombination between some genes under some circumstances and to decrease recombination between another (possibly overlapping) set of genes under other circumstances.

Since the evolution of recombination rates depends on gene interactions, the nature of interactions must be taken into account. In other words, it is important to know whether epistatic effects of groups of alleles on fitness are positive or negative, increasing fitness more or less than the independent affect of alleles at each locus. If we consider a pair of alleles that interact to affect fitness, strong epistatic interactions and strong selection will generally select for decreased recombination (Barton, 1995; Otto & Michalakis, 1998; Phillips, Otto & Whitlock, 2000). This is because recombination increases the likelihood of bringing together strongly deleterious mutations. Under this scenario, selection for increased recombination can only occur when epistasis and selection are weak relative to rates of recombination. When this is the case, Figure 3 (Barton, 1995; Phillips, Otto & Whitlock, 2000) shows when selection will increase or decrease recombination.

This picture predicts when an allele that increases recombination rate between a focal set of alleles will increase or decrease in frequency. For example, recombination rate is more likely to increase between members of a set of alleles if they exhibit negative epistasis and relatively strong negative fitness effects (gray region on left in Figure 3). This is because with these parameters, selection reduces genetic variance for fitness while less effectively removing individuals with multiple deleterious mutations. Recombination creates offspring with fewer than average deleterious mutations, favoring the evolution of increased recombination. Similarly, when interacting genes increase fitness, but show negative epistasis (gray region on right in Figure 3), selection favors recombination which breaks up groups of alleles interacting with negative epistasis. Recombination rate is likely to decrease when alleles interact with relatively strong positive epistasis (upper part of Figure 3). Differences in the way sets of loci along each chromosome interact can lead to the recombination hot-spots and dead-spots seen empirically. This picture was developed to understand the evolution of sex generally, and it treats the effects of sets of alleles in males and females as the same. However, we can use this picture as the basis for understanding how sex differences in recombination can evolve. First we must consider how selection on recombination may differ in males and females.

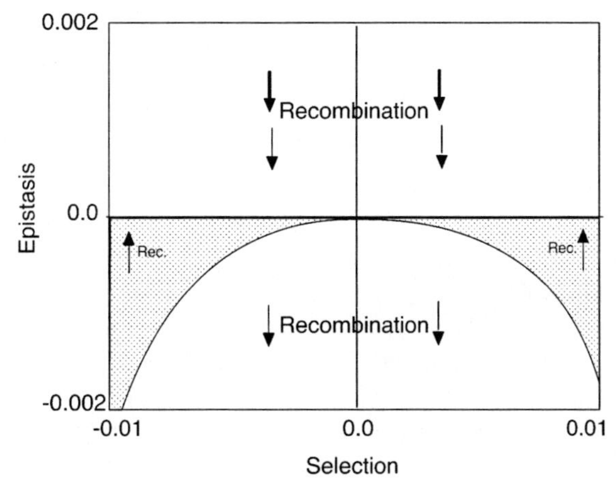

Figure 3. Evolution of recombination rate without considering sex differences, for weak selection and weak epistasis. Whether selection will act to increase or decrease recombination between members of a set of loci depends on the combined effect of the alleles at each locus on fitness and on the nature of epistatic interactions between these alleles. Epistasis is positive when the effect of focal alleles together is greater than the product of the independent effects of those alleles and negative when the combined effect of the alleles is less than the product of each separate effect. (Modified from Phillips et al., 1999.)

How sex differences in recombination can evolve

Korol, Preygel and Preygel (1994) list three hypotheses to explain the evolution of sex differences in recombination rates. The first two fail to explain large fractions of the pattern of sex differences seen in nature. First, higher metabolic activity in females and the resulting increased rate of oxidative damage during oogenesis may require higher rates of recombinational repair in females (Bernstein, Hopf & Michod, 1988, p. 151). This hypothesis does not explain cases where there is higher recombination in males (21 of 75 species), particularly in Orthoptera (9 of 14), and Lepidoptera and Trichoptera (all 7; Cano and Santos, 1990; Burt, Bell & Harvey, 1991). Second, selection for linkage of genes involved in sex determination and differentiation can lead to sex difference in recombination (Haldane, 1992; Nei, 1969). Once there is more than one gene involved in sex determination, there will be strong selection to link these genes together. A modifier of recombination which reduced recombination throughout the genome should increase in frequency. This hypothesis predicts that the sex with lower (or no) recombination will be the heterogametic sex. In achiasmate species this prediction is always supported (in 13 species; Burt, Bell & Harvey, 1991). However, in chiasmate species, the prediction often does not hold (14 of 25 species; Cano & Santos, 1990; Burt, Bell & Harvey, 1991; van Oorschot, 1992). Though these hypotheses may play a role in the evolution of sex differences in recombination, they are not sufficient to explain all of the known variation.

The third hypothesis is that sexual selection can cause sex difference in recombination rate. Sexual selection can result in only a subset of the gametes of one sex (typically males) contributing to offspring, either due to mate selection (Bull, 1983; Trivers, 1988) or due to gamete selection (Korol, Preygel & Preygel, 1994). Trivers pointed out that typically, sexual selection may lead to selection for decreased recombination in male meiosis so that successful males will tend to pass along sets of successful alleles to offspring. This last hypothesis has the potential to explain more of the known variation in recombination rates with respect to sex than the other two hypotheses. All the cases where male recombination exceeds females would seem to go against this hypothesis, but as Trivers (1988) pointed out, these cases appear to be associated with large male parental investment and/or excessive male mating effort. Both of these forms of male investment can reduce the intensity of sexual selection on males and even lead to sexual selection being stronger on females (Jones et al., 2000; Jones, Walker & Avise, 2001). So the sexual selection hypothesis may also explain the cases of higher male than female recombination rates. No quantitative analyses comparing the intensity of sexual selection and the direction and magnitude of sex differences in recombination rates have been done. Such analyses in several species would constitute a strong test of the sexual selection hypothesis.

Only one test of the sexual selection hypothesis has been attempted (Burt, Bell & Harvey, 1991). This was a weak test for several reasons. Unfortunately, for the species where a sex difference in recombination has been measured, the relative intensity of sexual selection is generally not known. This lack of information hindered Burt et al. from testing anything but a very weak prediction based on Trivers' hypothesis, that sex differences in recombination should be ordered:

$$\text{dioecious animals} > \text{hermaphroditic plants} > \text{hermaphroditic animals.}$$

They found no support for this prediction. In the 54 species studied, average sexual dimorphism in recombination rates did not differ between the three ecological groups. However, the variation in the intensity of sexual selection within dioecious animals probably far exceeds the variation between the above groups. A far stronger test would be to compare recombination rates between populations or species with known difference in sexual selection intensity. There is also evidence that the sex with lower recombination rates often limits recombination to the tips of chromosomes, reducing the effect of recombination (e.g., *Triturus helveticus* males have fewer and more terminal crossovers than females while in *T. cristatus*, the reverse is seen; Watson & Callan, 1963). This sort of data is not often reported and was not used by Burt, et al. Finally, taxon sampling is clearly a problem in Burt et al. – all 4 hermaphroditic animals were flatworms. This bias persists even when we add recent data on sex differences to the data collected by Burt et al. (e.g., 25 of 36 plants are from Liliaceae, 15 of 24 insects are from Orthoptera). More data are needed from a wide range of taxa so that conclusions are not biased by peculiarities of the biology of a few taxa (Coddington, 1992). Clearly, a more powerful test of the sexual selection hypothesis is needed.

Sexual selection and condition dependence

Trivers (1988) was fairly vague about exactly how sexual selection could lead to reduced recombination. He said only that '... autosomal genes enjoying reproductive success on the male side are a more restricted sample of the original set of genes with which the generation began than are the genes in breeding females.' And, 'Insofar as the actual *combinations* in which a male's genes appear are important to their success, then he will be selected to reduce rates of recombination (compared to females) in order to preserve these beneficial combinations.' (Trivers' italics; Trivers, 1988) We can set Trivers' idea in the context of recent theory on both evolution of recombination rates and sexual selection to build a more precise model of how sex difference in recombination might evolve.

When sexual selection is acting more strongly on males than females, we would not expect selection on recombination rate between members of a given set of alleles to be the same in each sex. There are two ways to visualize this. First, with sexual selection, separate plots for each sex of the relationship between recombination rate, epistasis and selection (Figure 3) might show that the gray region is larger for females than for males. For example, if females gain more than males from recombination, the gray region of the female plot would be expected to be larger. As Barton (1995) points out, more theory is needed to understand just how sexual selection will affect sex specific pressures on recombination rate. Second, the selective and epistatic effect of a set of alleles may not be the same for each sex when sexual selection is acting (Chippendale, Gibson & Rice, 2001). So, where a given set of alleles falls on these sex-specific plots may not be the same for each sex.

The second point above is true because of the nature of sexual selection, particularly when sexually selected traits (display traits) become dependent on condition (resources available for allocation to fitness enhancing traits; Rowe & Houle, 1996). Under strong sexual selection, exaggeration of display traits will stop if only a small number of genes are involved in display trait expression. This makes examples of extreme exaggeration of display traits difficult to understand (the 'lek paradox'; Borgia, 1979). Rowe & Houle (1996) showed that continued exaggeration of display is possible if genetic variance in condition is 'captured' into display expression by evolving changes in life history allocation patterns. Display then becomes 'condition dependent' or 'indicates condition'. Once this happens, selection on genes related to condition is more intense in one sex than the other (sexual selection combines with existing natural selection). Selection coefficients in males will be greater than in females when sexual selection is stronger on males than females (the typical situation). This can cause divergence (along the x-axis) of the points representing the effect of a set of alleles on recombination (Figure 3).

Epistasis is also likely to typically be stronger in males than in females. If a trait that underlies condition has some optimal value so that fitness falls off as trait value deviates from the optimum (e.g., a Gaussian function), the genes affecting that trait interact epistatically. A mutation that decreases trait value will increase fitness for some individuals and decrease fitness in others, depending on where they are in relation to the mean. By definition this is epistasis – the effects on fitness of a mutation at a given locus depend on what alleles are present at other loci affecting the trait. If selection on this trait is more intense in males, fitness will fall off more quickly with deviations from the optimum. This means that for a given set of genes, epistasis (whether positive or negative) will typically be stronger for males than for females, causing divergence (along the y-axis) of the points representing the effect of a set of alleles on recombination (Figure 3). When sexual selection is stronger on females than males, we might expect the opposite pattern of divergence along both axes as described above.

Since many (if not most) genes are likely to contribute in some way to condition, condition dependent sexual selection has the potential to affect recombination rates throughout the genome.

Consequences for mapping adaptations

Whatever the reason for the pattern, the fact is that in many organisms, sex differences in recombination rates exist. Can we use them to improve QTL mapping? In general the answer must be yes (e.g., Singer et al., 2002). Whether the improvements will be better or cheaper than simply increasing marker density remains to be seen. However, we may also be able to improve QTL placement algorithms by taking into account sex difference in recombination. This may allow gains in precision that increasing marker density cannot provide.

Even if differences between the sexes in recombination rate are not consistent across the genome (Lagercrantz & Lydiate, 1995), setting up crosses both ways with respect to sex (as in the backcross example described earlier) can bring gains in resolution. For given regions of the genome, the cross with the highest recombination rate (largest map distances) can be used for estimating QTL location. This should bring an improvement in map resolution (see example in Box 1 for a demonstration).

Box 1. An example using QTL Cartographer.

No current QTL programs allow for separate estimation of male and female recombination rates (though some are being developed; Korol, personal communication). However, to see some of the effects that a consistent sex difference in recombination could have, you can use QTL Cartographer (http://statgen.ncsu.edu/qtlcart/cartographer.html) to generate two linkage maps, identical except for inter-marker distances (Figure 4; using *Rmap* with the same random number seed and different average inter-marker distances). You can then randomly place QTL onto the chromosomes (using *Rqtl* with the same random number seed). One is then able to generate simulated QTL data for a hypothetical cross using both maps. When you use these data to estimate QTL location, you will see that the data set based on the larger map (higher recombination rate) can give better QTL placement and resolution (Figure 5).

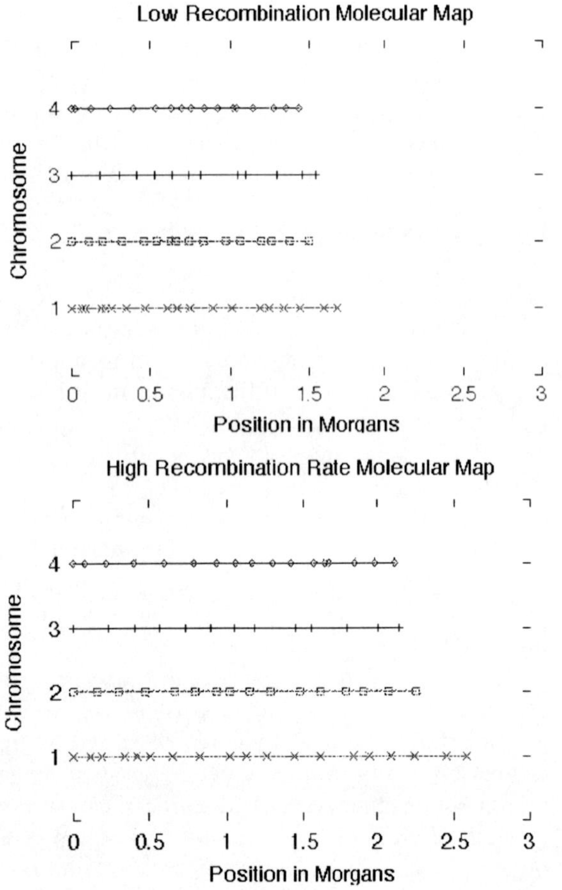

Figure 4. Maps generated using *Rmap* with 14 ± 2.5 markers (± SD), the mean inter-marker distance set to 10 ± 4 cm for low and 15 ± 4 cm for high recombination maps, and other settings left as the default values.

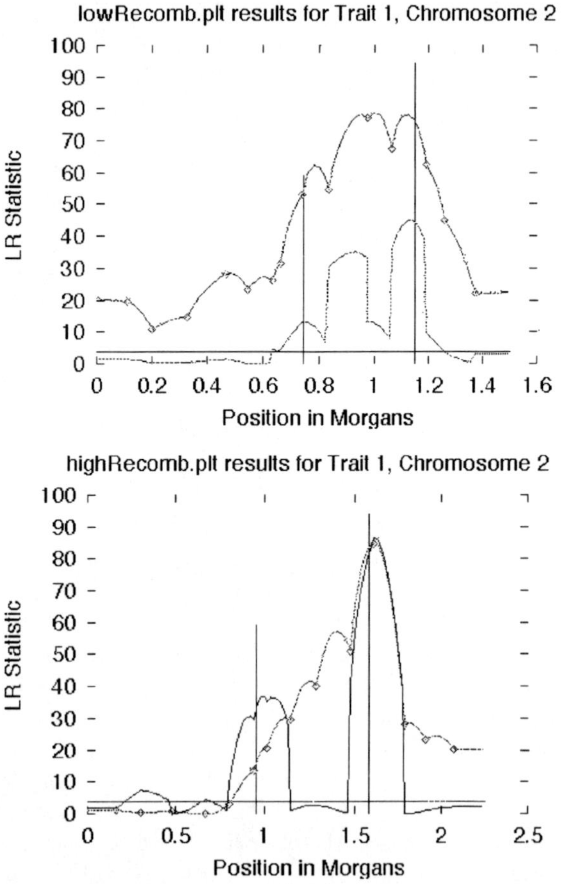

Figure 5. Likelihood ratio statistics showing QTL estimation based on low and high resolution maps. The curve with symbols is interval mapping, not controlling for residual genetic background. The other curve is using composite interval mapping and QTL Cartographer's model 6 to control for genetic background. Vertical bars represent actual QTL locations (using *Rqtl*) and the horizontal line is the default significance threshold (no resampling). Composite interval mapping correctly locates 2 QTL using the higher recombination rate map (lower plot), whereas the other plot shows both QTL under one peak. (Step size = 2 cm background parameters = 5, window size = 10).

Improving QTL estimation algorithms

Improvement in QTL discrimination may come by including sex differences in recombination into the likelihood function used to estimate QTL location relative to markers on a linkage map. The simplest way to do this, taking the backcross design as an example again, would be to use the larger estimates of recombination fractions from the crosses using each sex as the F1 parent. In other words, each type of cross will yield a different estimate of the distance between a given pair of markers (recombination fraction, or the relative frequency of recombination between the markers). Using the larger recombination fraction estimate for each interval will improve QTL map resolution. So for composite interval mapping (CIM), the linear regression equation used to estimate QTL positions (Liu 1998, p. 444) is

$$y_j = b_0 + b_i X_{ij} + \sum_{k \neq i, i+1} b_k X_{kj} + \varepsilon_j, \qquad (1)$$

where y_j is the quantitative trait value for individual j, b_0 is the intercept of the model, b_i is the effect of a potential QTL between markers i and $i+1$, b_k is the effect of a potential QTL relative to markers other than i and $i+1$, X_{ij} and X_{kj} are dummy variables taking 0 or 1 depending on the marker genotype of individual j, and for X_{ij} on the recombination fraction of each genotype (see equation 14.59 in Liu, 1998). ε_j is the residual from the model. Equation (1) is the basis for a likelihood function that is used to derive maximum likelihood estimates of QTL positions (equation 14.60 in Liu, 1998). These position estimates depend on r_1/r, where r_1 is the recombination between a putative QTL and marker 1, and r is the recombination fraction between marker 1 and 2. The r values are themselves estimated as part of the iterative maximum likelihood procedure. By estimating these recombination fractions separately for each kind of cross (e.g., for the backcross, using each sex as the F1 parent) and then using the larger values, we should improve our power to detect QTL and to discriminate between two QTL that are closely linked.

More work needs to be done to determine whether sex differences in recombination can be used to improve other aspects of QTL algorithms. For example it may be that sex differences in recombination will affect which method of controlling the residual genetic background (Zeng, 1994; Basten, Weir & Zeng, 2002) works best.

Appendix

Table A1. Sex differences in recombination for diploid chiasmate species (both sexes have recombination)

Taxon	Sexual system	*n*	Xta		Map ratio	Diff.	Comments	Reference
			Male	Female				
Insecta: Orthoptera: Acrididae (Gomphocerinae)								
Euchorthippus chopardi	d-XO/XX	8	11.62	10.48	–	m	1	(Cano & Santos, 1990)
Euchorthippus pulvinatus	d-XO/XX	8	11.81	11.06	–	m	1	(Cano & Santos, 1990)
Chorthippus vagans	d-XO/XX	8	11.25	10.56	–	m	1	(Cano & Santos, 1990)
Chorthippus parallelus	d-XO/XX	8	13.38	11.81	–	m	1	(Cano & Santos, 1990)
Chorthippus jucundus	d-XO/XX	8	12.26	12.65	–	N	1	(Cano & Santos, 1990)
Omocestus panteli	d-XO/XX	8	11.8	11.26	–	m	1	(Cano & Santos, 1990)
Chordata: Mammalia								
Homo sapiens (Primates)	d-XY/XX	22	–	–	1:1.5	f		(Dib et al., 1996)
Mus musculus (Rodentia)	d-XY/XX	19	–	–	1:1.4	f		(Dietrich et al., 1996)
Canis familiaris (Carnivora)	d-XY/XX	36	–	–	1:1.4	f		(Neff et al., 1990)
Sus domesticus (Artiodactyla)	d-XY/XX	?	–	–	1:1.4	f		(Marklund et al., 1996)
Ovis aries (Artiodactyla)	d-XY/XX	26	–	–	1.26:1	m	2	(Crawford et al., 1995)
Bos Taurus (Artiodactyla)	d-XY/XX	29	–	-	–	N		(Kappes et al., 1997)
Monodelphis domestica (Marsupialia)	d-XY/XX	7	–	–	1.6:1	m	1	(Hayman, Moore & Evans, 1988)
Trichosurus vulpecula (Marsupialia)	d-XY/XX	9	18.14	12.16	1.44:1	m		(Hayman & Rodger, 1990)

Table A1. (*Continued.*)

Taxon	Sexual system	*n*	Xta		Map ratio	Diff.	Comments	Reference
			Male	Female				
Chordata: Pisces								
Oncorhynchus mykiss (Salmoniformes)	d-XY/XX	29	–	–	3.25:1	f	2	(Sakamoto et al., 2000)
Oryzias latipes (Cypriniformes)	d-XY/XX	–	–	–	–	f	3	(Matsuda et al., 1999)
Chordata: Aves								
Gallus domesticus (Galliformes)	d-ZW/ZZ	38	–	–	–	N	4	(Groenen et al., 2000)
Columba livia (Columbiformes)	d-ZW/ZZ	38	–	–	–	N	4,5	(Pigozzi Solari, 1999)
Angiospermae: Monocotyledonae								
Zea mays (Poaceae)	h	8	–	–	1:1.70	m		(Robertson, 1984)
Angiospermae: Dicotyledonae								
Lycopersicon esculentum/ pennellii (Solanaceae)	h	12	–	–	1.19:1	f		(de Vicente & Tanksley, 1991)
Brassica nigra (Brassicaceae)	h	8	–	–	–	N		(Lagercrantz & Lydiate, 1995)
Brassica oleracea (Brassicaceae)	h	9	–	–	1.66:1	f		(Kearsey et al., 1996)

Sexual system (h = hermaphrodite; d = dioecious showing sex chromosome system); *n* is the haploid number of autosomes; Xta frequency is the total number of chiasmata formed; Map ratio indicates ratio of total male to female map distance; Diff. indicates authors claim of sex difference (m = males greater than females; f = females greater; N = not different; parentheses indicate no statistical test).
1 = Sex chromosome in females cannot be distinguished from autosomes so the former are assumed to have the mean chiasma frequency.
2 = Map ratio may include sex chromosomes.
3 = Based on one chromosome and/or a small number of markers only.
4 = Actual map ratio not reported.
5 = Synaptonemal complexes and recombination nodule number used rather than chiasma frequency.

References

Barton, N.H., 1995. A general model for the evolution of recombination. Genet. Res., 65: 123–144.

Basten, C.J., B.S. Weir & Z.-B. Zeng, 2002. QTL Cartographer. Version 1.16 edition.

Bernstein, H., F. Hopf & R.E. Michod, 1988. Is meiotic recombination an adaptation for repairing DNA, producing genetic variation, or both?, Chapt. 9, pp. 139–160, In: The Evolution of Sex, edited by R.E. Michod & B.R. Levin Sinauer Associates Inc., Sunderland, MA.

Borgia, G., 1979. Sexual selection and the evolution of mating systems, pp. 19–80. In: Sexual Selection and Reproductive Competition in Insects edited by M. Blum & A. Blum. Academic Press, Inc., New York.

Bull, J.J., 1983. Evolution of Sex Determining Mechanisms. Menlo Park, Benjamin/Cummings California

Burt, A., G. Bell, & P.H. Harvey, 1991. Sex differences in recombination. J. Evol. Biol. 4: 259–277.

Callan, H.G. & P.E. Perry, 1977. Recombination in male and female meiocytes contrasted. Philos. T. Roy. Soc. Lond., B 277: 387–411.

Cano, M.I. & J.L. Santos, 1990. Chiasma frequencies and distributions in gomphocerine grasshoppers: a comparative study between sexes. Heredity 64: 17–23.

Chippendale, A.K., J.R. Gibson, & W. Rice, 2001. Negative genetic correlation for adult fitness between sexes reveals ontogenetic conflict in Drosophila. Proc. Natl. Acad. Sci., USA 98: 1671–1675.

Coddington, J.A., 1992. Avoiding phylogenetic bias. Trends Ecol. Evol. 7: 68–69. (review of Harvey and Pagel 1991 the comparative method in evolutionary biology.)

Crawford, A.M., K.G. Dodds, A.J. Ede, C.A. Pierson, G.W. Montgomery, H.G. Garmomway, A.E. Beattie, K. Davies, J.F. Maddox, S.W. Kappes, R.T. Stone, T.C. Nguyen, J.M. Penty, E.A. Lord, J.E. Broom, J. Buitkamp, W. Schwaiger, J.T. Epplen, M.E. Matthew, P. & Matthews, D.J. Hulme, K.J. Beh, R.A. McGraw & C.W. Beattie, 1995. An autosomal genetic linkage map of the sheep genome. Genetics 140: 703–724.

de Vicente, M.C. & S.D. Tanksley, 1991. Genome-wide reduction in recombination of backcross progeny derived from male versus female gametes in an interspecific cross of tomato. Theor. Appl. Genet. 83: 173–178.

Dib, C., S. Faure, C. Fizames, D. Samson, N. Drouot, A. Vignal, P. Millasseau, S. Marc, J. Hazan, E. Seboun, M. Lathrop, G. Gyapay, J. Morissette & J. Weissenbach, 1996. A comprehensive genetic map of the human genome based on 5,264 microsatellites. Nature 380(6570): 152–154.

Dietrich, W.F., J. Miller, R. Steen, M.A. Merchant, D. Damron-Boles, Z. Husain, R. Dredge, M.J. Daly, K.A. Inballs, T.J. O'Conner, C.A. Evans, M.M. DeAngelis, D.M. Levinson, L. Kruglyak, N. Goodman, N.G. Copeland, N.A. Jenkins, T.L. Hawkins, L. Stein, D.C. Page & E.S. Lander, 1996. A comprehensive geneic map of the mouse genome. Nature 380(6570): 149–152.

Groenen, M.A., H.H. Cheng, N. Bumstead, B.F. Benkel, W.E. Briles, T. Burke, D.W. Burt, L.B. Crittenden, J. Dodgson, J. Hillel, S. Lamont, A. Ponce de Leon, M. Soller, H. Takahashi & A. Vignal, 2000. A consensus linkage map of the chicken genome. Genome Res. 10: 137–147.

Haldane, J.B.S., 1922. Sex ratio and unisexual sterility in hybrid animals. J. Genet. 12: 101–109.

Hayman, D.L., H.D.M. Moore & E.P. Evans, 1988. Further evidence of novel sex differences in chiasma distribution in marsupials. Heredity 61: 455–458.

Hayman, D.L. & J.C. Rodger, 1990. Meiosis in male *Trichsurus vulpecula* (Marsupialia). Heredity 64: 251–254.

Jones, A.G., G. Rosenqvist, A. Berglund, S.J. Arnold & J.C. Avise, 2000. The Bateman gradient and the cause of sexual selection in a sex-role-reversed pipefish. Proc. R. Soc. Lond. B, Biol. Sci. 267: 677–680.

Jones, A.G., D. Walker & J.C. Avise, 2001. Genetic evidence for extreme poyandry and extraordinary sex-role reversal in a pipefish. Proc. R. Soc. Lond., B 268: 2531–2535.

Kappes, S.M., J.W. Keele, R.T. Stone, R.A. McGraw, T.S. Sonstegard, T.P.L. Smith, N.L. Lopez-Corrales & C.W. Beattie, 1997. A second-generation linkage map of the bovine genome. Genome Res. 7: 235–249.

Kearsey, M.J., L.D. Ramsay, D.E. Jennings, D.J. Lydiate, E.J.R. Bouhuon & D.F. Marshall, 1996. Higher recombination frequencies in females compared to male meioses in Brassica oleracea. Theor. Appl. Genet. 92: 363–367.

Korol, A.B., I.A. Preygel & S.I. Preygel, 1994. Recombination Variability and Evolution: Algorithms of Estimation and Population-Genetic Models. Chapman and Hall, London.

Lagercrantz, U. & D.J. Lydiate, 1995. RFLP mapping in Brassica nigra indicates differeing recombination rates in male and female meiosis. Genome 38: 255–264.

Liu, B.H., 1998. Statistical Genomics: Linkage, Mapping, and QTL Analysis. CRC Press, Boca Raton.

Marklund, L., M. Johansson Moller, B. Høyheim, W. Davies, M. Fredholm, R.K. Juneja, P. Mariani, W. Coppieters, H. Ellegren & L. Andersson, 1996. A comprehensive linkage map of the pig based on a wild pig–Large White intercross. Anim. Genet. 27: 255–269.

Matsuda, M., S. Sotoyama, S. Hamaguchi & M. Sakaizumi, 1999. Male-specific restriction of recombination frequency in the sex chromosomes of the medaka, Oryzias latipes. Genet. Res. 73: 225–231.

Neff, M.W., K.W. Broman, C.S. Mellersh, K. Ray, G.M. Acland, G.D. Aguirre, J.S. Ziegle, E.A. Ostrander & J. Rine, 1990. A second-generation genetic linkage map of the domestic dog, Canis familiaris. Genetics 151: 803–820.

Nei, M., 1969. Linkage modification and sex difference in recombination. Genetics 63: 681–699.

Otto, S.P. & Y. Michalakis, 1998. The evolution of recombination in changing environments. Trends Ecol. Evol. 13: 145–151.

Phillips, P.C., S.P. Otto & M.C. Whitlock, 2000. Beyond the average: the evolutionary importance of gene interactions and variability of epistatic effects, Chapter 2, pp. 20–38. In: Epistasis and the Evolutionary Process, edited by J.B. Wolf, E.D. Brodie & M.J. Wade Oxford University Press, New York, NY.

Pigozzi, M. & A. Solari, 1999. Equal frequencies of recombination nodules in both sexes of the pigeon suggest a basic difference with eutherian mammals. Genome 42: 315–321.

Robertson, D.S., 1984. Different frequency in the recovery of crossover products from male and female gametes of plants hypoploid for B-A translocations in Maize. Genetics 107: 117–130.

Rowe, L. & D. Houle, 1996. The lek paradox and the capture of genetic variance by condition dependent traits. Proc. R. Soc. Lond. B, Biol. Sci. 263: 1415–1421.

Sakamoto, T., R.G. Danzmann, K. Gharbi, P. Howard, A. Ozaki, S. Kean Khoo, R.A. Woram, N. Okamoto, M.M. Ferguson, L.-E. Holm, R. Guyomard & B. Hoyheim, 2000. A microsatellite linkage map of rainbow trout (Oncorhynchus mykiss) characterized by large sex-specific differences in recombination rates. Genetics 155: 1331–1345.

Singer, A., H. Perlman, Y. Yan, C. Walker, G. Corley-Smith, B. Brandhorst & J. Postlethwait, 2002. Sex-specific recombination rates in zebrafish (Danio rerio). Genetics 160: 649–657.

Trivers, R.L., 1988. Sex differences in rates of recombination and sexual selection. pp. 270–286. in: The Evolution of Sex, edited by R.E. Michod & B.R. Levin. Sinauer Associates Inc., Sunderlan, Massachusetts.

van Oorschot, R.A.H., P.A. Porter, C.M. Kammerer & J.L. VandeBerg, 1992. Severely reduced recombination in females of the South American marsupial Monodelphis domestica. Cytogenet. Cell Genet. 60: 64–67.

Watson, I.D. & H.G. Callan, 1963. The form of bivalent chromosomes in newt oocytes at first meaphase of meiosis. Q. J. Microsc. Sci. 104: 281–295.

Zeng, Z.-B., 1994. Precision mapping of quantitative trait loci. Genetics 136: 1457–1466.

Genetica (2005) 123: 49–62 © Springer 2005

Genetics and adaptation in structured populations: sex ratio evolution in *Silene vulgaris*

Matthew S. Olson[1,*], David E. McCauley[2] & Douglas Taylor[3]
[1]*Institute of Arctic Biology and Department of Biology and Wildlife, University of Alaska Fairbanks, Fairbanks, AK;* [2]*Department of Biological Science, Vanderbilt University, Nashville, TN;* [3]*Department of Biology, University of Virginia, Charlottesville, VA, USA;* **Author for correspondence: Institute of Arctic Biology and Department of Biology and Wildlife, University of Alaska Fairbanks, Fairbanks, AK, USA (Phone: +1-970-474-2766; Fax: +1-907-474-6967; E-mail: matt.olson@uaf.edu)*

Received 21 May 2002 Accepted 9 December 2003

Key words: cytoplasmic male sterility, gynodioecy, mitochondrial DNA, subjective fitness

Abstract

Theoretical models suggest that population structure can interact with frequency dependent selection to affect fitness in such a way that adaptation is dependent not only on the genotype of an individual and the genotypes with which it co-occurs within populations (demes), but also the distribution of genotypes among populations. A canonical example is the evolution of altruistic behavior, where the costs and benefits of cooperation depend on the local frequency of other altruists, and can vary from one population to another. Here we review research on sex ratio evolution that we have conducted over the past several years on the gynodioecious herb *Silene vulgaris* in which we combine studies of negative frequency dependent fitness on female phenotypes with studies of the population structure of cytoplasmic genes affecting sex expression. This is presented as a contrast to a hypothetical example of selection on similar genotypes and phenotypes, but in the absence of population structure. Sex ratio evolution in *Silene vulgaris* provides one of the clearest examples of how selection occurs at multiple levels and how population structure, *per se*, can influence adaptive evolution.

Abbreviations: CMS – cytoplasmic male sterility.

Introduction

The role of population structuring (the degree of subdivision of individuals or genes in a metapopulation into discrete local breeding units) and its importance for adaptive evolution is a contentious issue (Wright, 1931; Fisher, 1958; Coyne, Barton & Turelli, 1997, 2000; Wade & Goodnight, 1998; Goodnight & Wade, 2000). However, there are some cases where, in the short term, the role of population structuring is likely to have emergent effects that cannot be determined through just understanding fitness at the level of the individual. For instance, the presence of population structure implies the restriction of gene flow and one corollary of this restriction is that genotypes or phenotypes associated within demes are more similar to one another than they are to genotypes or phenotypes picked at random from all demes (Wilson, 1979). When these associations influence fitness, the fitness of individuals in demes cannot be predicted by averaging across populations. Such situations arise when fitness is frequency dependent and individual fitness is influenced by the presence of individuals with the same phenotype. In these cases, the population structure provides the context for an emergent property that affects fitness in such a manner that individual fitness cannot be predicted without incorporating the effects of structure.

The most well known theoretical example of how population structure can alter the outcome of evolution is in the evolution of cooperative or altruistic behavior (Goodnight, Scwartz & Stevens, 1992). Selection on cooperative versus selfish behavior is inherently frequency dependent, and cooperation is increasingly favored in structured populations because altruists are clustered into a subset of demes where they benefit from being the recipients of altruism. Although both frequency dependent selection and population structure are common in nature, there are few empirical examples of how population structure, *per se*, can influence evolution in this way.

Recently, the effects of population structuring on the relative fitnesses of the two sexes in gynodioecious species have drawn considerable interest in this context (McCauley & Taylor, 1997; Pannell, 1997; Couvet, Ronce & Gliddon, 1998; Hatcher, 2000; Frank & Barr, 2001). Gynodioecy is a breeding system characterized by the co-occurrence of females and hermaphrodites. Theoretical models suggest that population structure may contribute to at least two aspects of population sex ratio. First, it may create the conditions that allow cytoplasmic genes effecting male sterility (CMS or cytoplasmic male sterility genes) to evade nuclear male fertility restorer genes (Frank, 1989; Gouyon, Vichot & Van Damme, 1991). Second, population structure in the presence of pollen limitation may alter the fitness of CMS types relative to the case of no population structure (McCauley & Taylor, 1997).

A complex web of selection at different levels

Sex ratio evolution in gynodioecious species is known to involve selection at different levels of organization (Cosmides & Tooby, 1981; Saumitou-LaPrade, Cuguen & Vernet, 1994; Hurst, Atlan & Bentsson, 1996). In many gynodioecious species, gender is genetically determined by an interaction between CMS factors and nuclear male fertility restorers (Schnabel & Wise, 1998). The CMS factors block pollen production and are maternally inherited. Male fertility restorers, located in the nuclear genome, are biparentally inherited and reinstate viable pollen production. Individuals with CMS genes and lack nuclear restorers express a female phenotype, whereas those with CMS and nuclear restorers express hermaphroditic phenotypes. Within a species, multiple CMS/restorer systems further complicate the association between genotype and phenotype (Schnabel & Wise, 1998). These CMS/restorer systems are generally thought to interact in a gene-for-gene manner whereby only one type of restorer will reinstate male fertility for a given CMS type (Frank, 1989; Schnabel & Wise, 1998), though this has not been studied extensively in natural systems.

A consideration of selection at the level of the gene is necessary to understand the spread of CMS genes. From the vantage of CMS genes, fitness is increased only via increasing seed production and the complete loss of male fertility does not directly affect fitness, or for that matter, the fitness of any maternally inherited element. In contrast, the nuclear male fertility restorer genes are biparentally inherited and their fitness is maximized through balancing allocation to both male and female reproductive modes (Fisher, 1958; Frank, 1989). What makes this system so compelling is that the CMS and restorer genes directly affect the genetic transmission system; thus, their expression affects the selective environment (Jacobs & Wade, 2003). Moreover, since gender is epistatically determined the fitness of each component of the genetic determination system is dependent on the frequencies of other components (Jacobs & Wade, 2003). The commonness of CMS/restorer systems in plants and their importance to agriculture (Levings, 1993; Frank & Barr, 2001) contribute to making this one of the most celebrated examples of the conflict of interest between cytoplasmic and nuclear genomes (Cosmides & Tooby, 1981; Hurst, Atlan & Bengtsson, 1996; Werren & Beukeboom, 1998).

Obviously, CMS genes will spread when they are over-represented relative to other cytoplasmic types in future generations. Such over-representation results from the production of more seeds or higher quality seeds by females than hermaphrodites and is a common attribute of gynodioecious plants (Gouyon & Couvet, 1987). This 'reproductive compensation' may result from reallocation of resources that would otherwise be used for pollen production (Ashman, 1999) and is affirmed by the observation that many gynodioecious species exhibit negative genetic tradeoffs between male and female reproductive allocation (Olson &

Antonovics, 2000). Because females produce more seeds than hermaphrodites, CMS factors have higher fitness when they reside in females than hermaphrodites in the absence of pollen limitation. The opposite is not necessarily true for restorers; nuclear fitness depends both on the gender of the individual as well as the population sex ratio.

Theoretically, the interactive effects of variation in population sex ratio and pollen limitation can slow the spread of selfish CMS genes (McCauley & Taylor, 1997; Hatcher, 2000). In particular, pollen limitation may inhibit the seed production of females when sex ratios become sufficiently female biased (Lewis, 1941; Lloyd, 1974). Such negative frequency dependent fitness has been observed in both field experiments (McCauley & Brock, 1998) and natural populations (Graff, 1999; McCauley et al., 2001) of gynodioecious species. Variation in local population sex ratio created by sampling effects coupled with restricted pollen flow between populations will result in lower global female fecundity because most females are located in female-biased populations where they have low relative fitness because of pollen limitation (McCauley & Taylor, 1997). The higher the level of population structure is, the greater the expected fecundity reduction due to the effects of population structure. Additionally, if specific CMS factors are non-randomly associated with females, their fitness may also be influenced by the population sex ratio.

Clearly, the evolution of the sex ratio in gynodioecious species with cytonuclear sex determination is complex, being affected by selection at many levels of organization. In such complex systems, it would seem prudent to identify simple elements of the system that can be understood and then add complexity onto this foundation. In this vain, here we focus on the selective pressures potentially driving changes in the frequencies of CMS factors and cytoplasmic haplotypes through field, molecular, and crossing studies in the gynodioecious plant *Silene vulgaris*. In *S. vulgaris*, all studies of inheritance of mtDNA and cpDNA markers to date have found only maternal inheritance (McCauley, 1998; Olson & McCauley, 2000); thus, the fitness of cytoplasmic factors can be summarized by assessing components of fitness through seed. Here we summarize and integrate the studies to date that have contributed to our understanding of the evolution of sex ratio in subdivided populations of a gynodioecious species with cytonuclear sex determination.

Case study: sex ratio evolution in subdivided populations of *Silene vulgaris*

To demonstrate that population structure, *per se*, influences the *selection* of CMS genes, we would need to show that there is variation in sex ratio among demes and that this sex ratio variance influences the relative fitnesses of different CMS cytotypes. Specifically, cytotypes that are associated with more females than the average cytotype may suffer reduced fitness via pollen limitation when the sexes are increasingly segregated into different demes. For population structure to influence the *evolution* of CMS genes, we would need to demonstrate further that among-deme variation in the sex ratio is due to underlying population structure at the genes that control sex expression. If this were true, then the effect of population structure on fitness has genetic consequences. Finally, we would need to show that the individual demes contribute to some larger group of demes – a metapopulation (Couvet, Ronce & Gliddon, 1998). If this were true, then demes with clusters of less fit female genotypes would contribute relatively few progeny to the global pool, resulting in an overall change in gene frequency. Over the past several years, we have demonstrated all but the very last of these conditions, making sex ratio evolution in *Silene vulgaris* one of the clearest examples of how natural selection can be influenced by the fact that it occurs in a spatially explicit context.

The model system

Silene vulgaris is a gynodioecious short-lived perennial native to Europe that became naturalized throughout much of northeastern and north central North America sometime after Europeans colonized the New World. Migration of seeds to North America probably occurred several times in the past and may continue today. Individuals are capable of reproducing annually and it is not known how many generations may have passed since colonization. We can estimate a range of 40–400 generations might have passed since this time (assuming 400 years since colonization and decadal to annual generation times). Individuals are weakly clonal, so genetically different individuals are not difficult to distinguish in the field.

Gender in *Silene vulgaris* is either female or hermaphrodite and its inheritance is consistent with genetic determination through an interaction between mitochondrial CMS genes and nuclear male fertility restorers (Charlesworth & LaPorte, 1998; Taylor, Olson & McCauley, 2001). Hermaphrodites produce both pollen and ovules, are capable of self-fertilization, and produce flowers with slightly larger petals and sepals than females produce (Dulberger & Horovitz, 1984; Olson, unpublished data). Female flowers are easily identified by the absence of developed stamens. Generalist pollinators including small bees and moths frequent flowers on both sexes. When averaged across populations, females produce 22.5% more seeds per fruit than hermaphrodites; however, this ratio varies depending on local population sex ratio (see below). Seeds from females have slightly higher germination rates than seeds from hermaphrodites, but this difference is not statistically significant (McCauley et al., 2000a). Seeds generated from self-fertilization in hermaphrodites suffer fitness effects from inbreeding depression (Jolls, 1984; Jolls & Chenier, 1989; Petterson, 1992; Emery, 2001) that can vary among populations (Emery, 2001). Seeds are passively dispersed.

Natural populations (demes) of *Silene vulgaris* show a large amount of variation in sex ratio. A series of naturally occurring populations in the valleys of Giles and Craig counties in the Allegheny Mountains of Virginia, USA have been the focus of studies by several researchers over the past few years. These populations range in size from < 10 to >>1000 individuals and are scattered along roadsides and agricultural fields; thus, their ecology is likely to be affected by anthropogenic factors associated with roadside maintenance as well as natural processes. In these valleys the global sex ratio is about 28% female but population sex ratios vary from 0 to 75% females; population sex ratios are significantly over dispersed compared to the random expectation from a binomial model (Figure 1; $G_{\text{ind}} = 94.1$, $\text{df} = 19$, $p < 0.001$).

Sex ratio and fitness variation in natural populations

Several authors have implicated pollen limitation as an important factor in the evolution of population sex ratio in gynodioecious species (Lewis, 1941; Lloyd, 1974; McCauley & Taylor, 1997). With strong effects of pollen-limited fecundity, seed or fruit production of females will decrease with decreasing pollen availability. Hermaphrodite fecundity may not decrease, however, if self-fertilization is possible.

One criteria that must be met for pollen limitation to be important within populations (demes) is that pollen flow between populations must be minimal. Genetic marker studies estimate that in *S. vulgaris* gene flow through pollen may be three times as high as through seed (McCauley, 1998), but detailed measures of gene flow among populations have not yet been conducted. It is clear, however, that seed fitness of females decreases dramatically with increased distance from pollen sources (Taylor, McCauley & Trimble, 1999). Taylor, McCauley & Trimble, (1999) experimentally placed single females at varying distances,

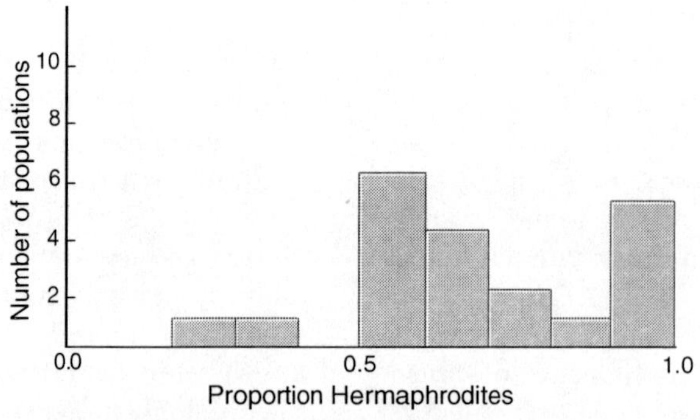

Figure 1. The distribution of sex ratios (proportion of hermaphrodites) in 20 natural populations of *Silene vulgaris* near Mountain Lake, Virginia.

from 20 to 160 m from source populations with included nine hermaphrodites and three females. Compared to females in the source populations, females at 20 m suffered fitness decreases of 95% and females at 160 m produced no seeds. Thus, it is reasonable to surmise that seed set in populations separated by > 500 m is only weakly affected by pollen flow from sources outside the deme.

In *Silene vulgaris*, McCauley and Brock (1998) experimentally assessed the effects of population sex ratio on fruit set and seeds per fruit by manipulating sex ratio in experimental populations. This study clearly showed that fruit to flower ratios increase and females produce more seeds as the frequency of hermaphrodites in the population increases. Although explicit pollen limitation studies were not conducted by testing for increased fitness with pollen addition (*sensu* Bierzychudek, 1981), the experimental nature of the study accounted for environmental effects and implicates pollen limitation via female-biased population sex ratios as a major source of population variation in female fecundity.

Patterns consistent with pollen limitation also have been observed in the natural roadside populations in Giles and Craig Counties, Virginia (McCauley et al., 2000a; McCauley, Olson & Taylor, 2000b). In this study, each population was visited twice, the first time to assess population specific sex ratios and the second time to collect mature fruits from females and hermaphrodites in each population. Fruits were transported to the laboratory where the seeds were counted and germinated. Because both population sex ratio and the local environment might have affected seed set, the ratio of the average seed production by females compared with hermaphrodites was computed. In these populations, the ratio of seed production of females compared to that of hermaphrodites decreased with increasing frequencies of females (Figure 2). The weight of evidence from the combination of the manipulative and observational studies indicates that female fecundity in *S. vulgaris* is influenced strongly by the frequency of hermaphrodites.

Clearly, seed production is only one component of fitness, but it appears to be a good metric to determine an effect of pollen limitation. Although seed production accounts for less than half of the fitness of hermaphrodites (Lloyd, 1974), it accounts for the entire reproductive

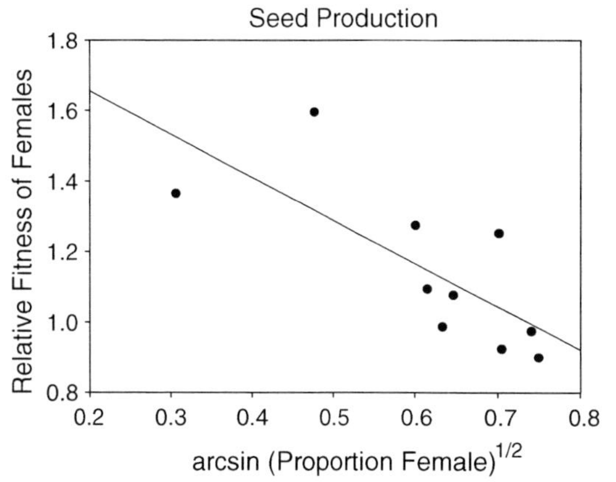

Figure 2. Seed production of females relative to hermaphrodites (square root of seeds per capsule for females divided by the square root of seeds per capsule for hermaphrodites) as a function of the arcsin square root transformed field sex ratio of those populations ($Y = 1.90 - 1.22X$; $p < 0.02$).

output of females and from the perspective of the genes controlling gender, seed production accounts for the entire fitness of cytoplasmically inherited elements even if they reside in hermaphrodites. Taken together, therefore, local variation in the sex ratio appears to influence the relative fitness of the two genders by reducing the fitness of females when they are clustered into a subset of demes.

Population structure of genes controlling gender expression

Two methods have been used to quantify the level of population-to-population variation in the cytoplasmic and nuclear factors controlling gender expression in *S. vulgaris*: (1) experimental crosses of individuals within and among populations and (2) studies of the variation in maternally inherited genetic markers among populations.

Factorial crossing studies

Experimental crosses can be designed to partition the relative contributions of sire and dam genotypes to offspring sex ratios and assess whether these contributions vary more within or among populations. Nuclear factors affecting offspring gender expression (i.e. male fertility restorers) are assumed to be inherited in a strictly Mendelian fashion and thus can be contributed through both the dam and sire. The dam additionally contributes cytoplasmic factors affecting family sex ratio (i.e. CMS factors). Within a CMS type, dominant restorers contribute only to the effect of the sire because dams are homozygous for non-restorer alleles. In contrast, recessive restorers contribute only to dam effects because sires are homozygous for restoration alleles. Co-dominant restorers contribute to both sire and dam effects. Studies in *S. vulgaris* and other species indicate that restoration is due to a mix of mainly dominant restorers and, less often, recessive restorers (Van Damme, 1983; Charlesworth & Laporte, 1998; Taylor, Olson & McCauley, 2001). The variance in cytoplasmic and nuclear genes controlling gender can be further partitioned into within- and among-population differences by comparing progeny sex ratios from crosses between parents from the same and from different populations.

Taylor, Olson and McCauley (2001) conducted two series of crosses, termed within- and among-population, using parents derived from the roadside populations in Giles and Craig counties. For the within-population crosses, each of 3–4 females was crossed with each of 3–4 hermaphrodites from within the same population in a factorial design. These within-population crosses were replicated across nine different populations. For the among-population crosses, one female and one hermaphrodite were randomly selected from each of the within-population designs; each of these hermaphrodites was crossed to each female in a full factorial design. For both sets of crosses, up to 50 progeny from each cross were grown to flowering and their gender was determined.

Progeny sex ratio from the within-population crosses was influenced by the different dams within populations, the sire × dam interaction, and the block effect of population but not sire effects (Figure 3A; Taylor, Olson & McCauley, 2001). Progeny sex ratio from the among-population crosses was strongly influenced by different dams and the interaction between particular sires and dams (Figure 3B; Taylor, Olson & McCauley, 2001). The variance components associated with the two sets of crosses showed that the dam effect was nearly four times stronger in the among- than within-population crossing design and this difference was significant when the variance components were compared using a jackknife procedure. The variance components did not differ between the two sets of crosses for any other treatment. The far stronger dam effect in the among-population crosses is a quantitative genetic analogue to Wright's F_{ST}. It shows that the among-population variance in male sterility elements is greater than the within-population variance, and gives direct evidence for population-to-population differences in the maternally inherited genes controlling gender. Dams chosen from different populations had very strong and different effects on progeny sex ratio suggesting that different populations may harbor different CMS factors. The strong population structure was somewhat surprising given the small spatial scale over which these roadside populations are dispersed and the similarity in the grassy disturbance-prone environments in which they are found. Population variation in the frequency of CMS factors has also been found in other gynodioecious species (de Haan, 1997) and

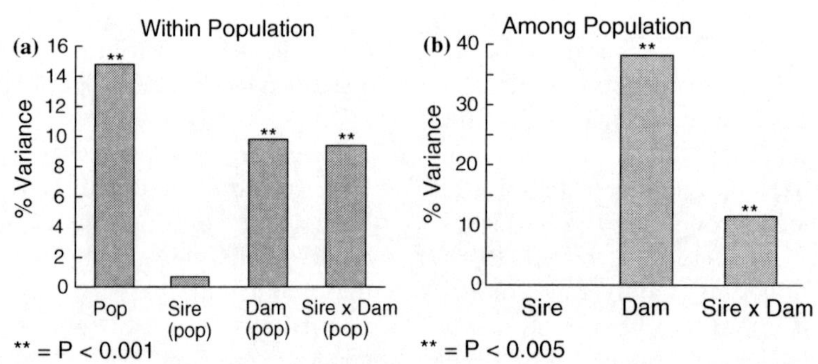

Figure 3. The proportion of variance accounted for by the treatments in the within and among-population crosses between female and hermaphroditic *Silene vulgaris* plants. See text and Taylor, Olson and McCauley (2001) for details. **Significant treatment effect in both ANOVA and logistic regression analyses.

may be a common phenomenon at ecological time scales (Frank & Barr, 2001).

Our results also point to two other conclusions. First, the strong dam effects in both analyses are indicative that there are multiple CMS genes contributing to variation in progeny sex ratio. Since restorers tend toward dominance, it is unlikely that the dam effect can be accounted for solely through the action of recessive male fertility restorers (Taylor, Olson & McCauley, 2001). Second, epistatic cytonuclear effects of sex expression were indicated by the strong sire × dam interaction influence on progeny sex ratios. This is a clear example of non-additive gene interactions (epistasis) being an important component of genetic variance in nature (Galloway & Fenster, 2000). Although the main effect of sire was inconsequential in both sets of crosses, the sire × dam component suggests that there is some variation among different sires in their nuclear contributions to gender (i.e. restorer genotypes) both within and among populations and this nuclear genetic variance tends is expressed statistically as an epistatic interaction with cytoplasmic loci.

The population structuring of cytoplasmic elements affecting progeny gender was underscored by the presence of an association between mtDNA haplotype markers and the progeny sex ratio in the within-population crosses. Using Southern blotting techniques, 12 RFLP haplotypes associated with the region surrounding the *cytochrome oxidase I* gene were identified from the dams used in the within-population crossing design. *A posteriori* tests revealed an association between these markers and progeny gender expression (Taylor, Olson & McCauley, 2001). Although this is suggestive that these markers are in some way associated with the CMS genes, this conclusion should be approached cautiously. The strong population structuring in the mtDNA haplotypes ($F_{ST} = 0.42$) prevented mtDNA haplotype being tested independently from the population effect, and thus population effects were confounded with mtDNA haplotype effects. Nonetheless, it is not unreasonable to expect some sort of relationship between genes and markers in the mitochondrial genome since the entire genome is inherited as a unit. In essence, these mtDNA haplotypes can be considered putative *qualitative trait loci* for gender. We refer to them as *qualitative trait loci* because unlike the more traditional quantitative trait locus, the trait is qualitative and environmental factors have little effect on its expression. Dissociations between CMS factors and the RFLP haplotypes may arise from different mutation and/or fixation rates of the two elements (see below and Olson & McCauley, 2000).

Population structure of mtDNA qualitative trait loci in natural populations

The potential population structuring of cytoplasmic factors affecting gender expression might also be inferred from the patterns of molecular markers associated with these genes in natural populations. As stated in the previous section, there is some evidence that mtDNA RFLP haplotypes are linked to different CMS types, but these associations currently are not defined. Since the mtDNA is inherited as a single unit, RFLP polymorphisms must be in linkage disequilibrium with CMS genes, but the extent of this linkage depends on the relative rates of evolution of CMS factors and RFLP haplotypes (Olson & McCauley, 2000). If CMS factors persist for long periods relative to the evolution of new RFLP alleles, there may be sufficient time for many RFLP haplotypes to become associated with a single CMS factor. On the other hand, if the mutation rate of new CMS factors is rapid relative to that of mtDNA RFLP haplotypes, more than one CMS factor may be associated with the same mtDNA RFLP haplotype. Nonetheless, the strong pattern of co-inheritance of the entire mitochondrial genome and the inability for mitochondria from different individuals to recombine will limit the development of random associations between CMS factors and mtDNA haplotypes.

Associations between mtDNA haplotypes and CMS may also be eroded by parallel evolution of the same mtDNA RFLP haplotype in two different lineages. Although, this effect does not appear to be pervasive in the haplotypes screened in *S. vulgaris*, there is some homoplasy in phylogenetic trees constructed with mtDNA RFLP haplotypes (Olson & McCauley, 2000). It is not yet known however, whether this homoplasy results from parallel evolution of the exact same RFLP haplotype, or if it results from the inability to differentiate very similar RFLP haplotypes using Southern blotting techniques.

Finally, detection of the association between mtDNA haplotypes and gender is also made difficult by the epistatic interactions (shown in Taylor, Olson & McCauley, 2001) that cloud the association between CMS genotypes and phenotypes.

With this in mind, mtDNA RFLP haplotypes are currently the best way to detect large amounts of molecular variation in the mitochondrial genome at local spatial scales in *S. vulgaris* (Olson & McCauley, 2002), and their population structure is likely to reflect the structure of CMS factors. In a recent study, Olson and McCauley (2002) assessed the population structure of mtDNA RFLP haplotypes in the Giles and Craig county populations. These mtDNA RFLP polymorphisms were screened in 250 individuals from 18 natural populations in the Giles and Craig counties. Thirteen haplotypes were recognized (Figure 4). Their population distribution was highly structured ($F_{ST} = 0.574 \pm 0.066$ s.e.) with four populations containing single haplotypes and eight other populations containing only two haplotypes.

At the same time as the leaf tissue was collected for genetic analysis, the gender of the plant was recorded. A strong statistical association between haplotypes and gender was apparent when individuals were pooled across populations ($p < 0.005$; Figure 5). For instance, 63% of individuals with haplotype *g* were females, while >80% of individuals with haplotypes *a* and *d* were hermaphroditic. A pattern of variation in the sex ratios of individuals carrying different mtDNA haplotypes might reflect that different mtDNA haplotypes are associated with different CMS factors.

Strong population structure of cytoplasmic genomes has been observed in other ruderal plants where it has been hypothesized to arise from the combination of limited seed flow among populations and relatively dynamic extinction and colonization ecology (McCauley, 1998). In gynodioecious species, such patterns have also been theorized to result from selection on genes controlling gender (Lewis, 1941; Frank, 1989; Gouyon, Vichot & Van Damme, 1991; Olson & McCauley, 2002). For instance, low within-population haplotype diversity can result from selective sweeps of unrestored CMS haplotypes through the local population. Among-population haplotype diversity can be generated when CMS

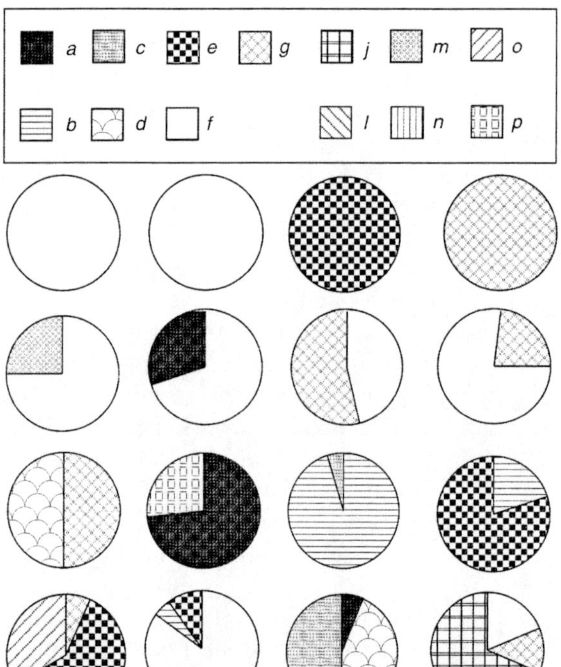

Figure 4. Distribution of the 13 mtDNA haplotypes among the 18 studied populations in Giles County, Virginia. Each pie chart represents the frequency of different mtDNA haplotypes found in each population. See Olson and McCauley (2002) for further details.

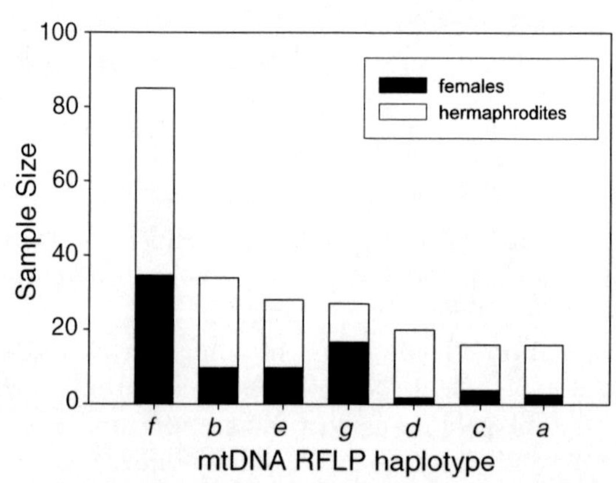

Figure 5. Numbers of females (black fill) and hermaphrodites (white fill) associated with each of the eight common mtDNA haplotypes near Mountain Lake, Virginia. Individuals were pooled across populations. See Olson and McCauley (2002) for further details.

types 'escape' from their restorers into populations where they are generally not restored and are at a selective advantage (Frank, 1989). Such dynamics, however, imply that populations also differ in their frequencies of restorers for different CMS types and that there are fitness costs to harboring restorer alleles (Frank, 1989; Gouyon, Vichot & Van Damme, 1991). Both of these patterns have been somewhat elusive.

Three empirical studies suggest that frequencies of nuclear restorers may differ among populations. First, the frequencies of nuclear allozyme polymorphisms in the roadside populations of Giles and Craig counties vary among populations ($F_{ST} = 0.22$; McCauley, 1998), but to a lesser extent than cytoplasmic polymorphisms (McCauley, 1998; Olson & McCauley, 2002). Second, the significant sire × dam interaction from the among-population crossing studies in Taylor, Olson and McCauley (2001) indicates that hermaphrodites drawn randomly from different populations vary in their abilities to restore male sterility in the same female; however, a sire × dam interaction effect was also present in the within-population crosses and thus the interaction in the among population crosses may simply reflect sampling variance within different populations. Finally, significantly different associations between the gender of individuals with the same mtDNA RFLP haplotypes were found in different roadside populations in western Virginia (Figure 6; Olson & McCauley, 2002). Such patterns in allozyme frequencies, crossing studies, and associations among-gender and mtDNA haplotypes support the presence of some population structuring of nuclear male fertility restorers, but not to the same extend as the structuring of cytoplasmic factors.

Estimating the magnitude of the effect of population structure

Both molecular and crossing evidence suggest that there is substantial population structuring of cytoplasmic genes affecting gender expression in *Silene vulgaris*, but how important is this structuring for the evolution of sex ratio or the fitness of different cytotypes? Answering this question is a goal of current and future empirical studies of the roadside populations, but it can also be addressed with regard to theoretical models concerning the effect of population structure on the relative fitness of females and hermaphrodites (McCauley & Taylor, 1997).

Our marker studies indicate that mtDNA haplotypes are in some way associated with CMS factors across the entire Giles and Craig county metapopulation. Clearly, differential association with females and hermaphrodites will affect the fitness of cytoplasmic elements (both CMS factors and mtDNA haplotypes) since cytoplasmic elements are transmitted only through seed and female seed production relative to hermaphrodites is dependent on local population sex ratio. Here

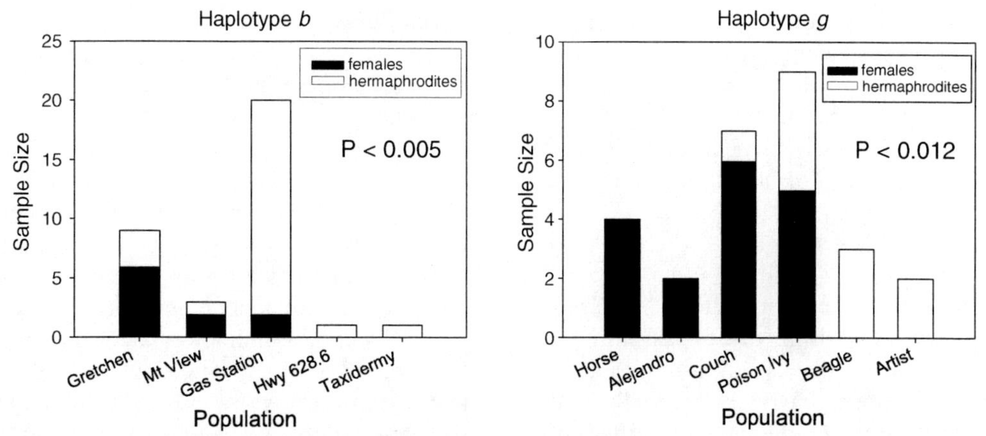

Figure 6. The numbers of females (black fill) and hermaphrodites (white fill) associated haplotypes *b* and *g* across the populations in which they were found. *p* values refer to the results of Fisher's exact tests. See Olson and McCauley (2002) for further details.

we ask whether population structure can differentially affect cytoplasmic haplotypes. This can be assessed through comparing the potential fitness of cytoplasmic haplotypes in a single large panmictic population to their fitness in several small subdivided populations using the theoretical calculations of the expected frequencies with which the haplotype (independent of gender) will encounter hermaphrodites in panmictic and subdivided populations (McCauley & Taylor, 1997; McCauley, Olson & Taylor, 2000b).

Given random pollen movement in a panmictic population, the expected frequency at which a haplotype will encounter hermaphrodites is simply the global frequency of hermaphrodites. In a subdivided population, this theoretical frequency can be calculated by applying the concept of subjective frequencies (Wilson, 1980) and can be adjusted according to the variance in hermaphrodite frequency among populations. These calculations require the unrealistic assumption that CMS factors are randomly associated with nuclear backgrounds both within and among populations and the effects of violation of this assumption will be discussed later. We refer the reader to McCauley and Taylor (1997) and McCauley, Olson & Taylor (2000b) for detailed presentations of the theoretical underpinnings and assumptions underlying the following analysis.

Let us assume that the CMS factor or mtDNA haplotype α produces hermaphrodites with the probability X_α. All other haplotypes combined produce hermaphrodites with the probability of $X_{\text{not } \alpha}$. Let the global frequency of haplotype α be p_α and the global frequency of all other factors be $1 - p_\alpha$ and the among-population variance in the relative frequencies of haplotype α (relative to all other factors in all populations) be $V_{\alpha,\text{not } \alpha}$. Theoretically, in a large panmictic population, individuals with CMS_α experience hermaphrodites at the global the frequency of hermaphrodites:

$$H_{\text{no structure}} = p_\alpha X_\alpha + p_{\text{not } \alpha}(X_{\text{not } \alpha}) \qquad (1)$$

but in subdivided populations they experience hermaphrodites only within demes (assuming no pollen flow among demes). With population structuring, individuals carrying a specific CMS type experience other individuals with the same CMS type more often than they do in a panmictic population. A CMS type that produces more females than others will therefore experience more females in a subdivided metapopulation than in a single panmictic population and its seed production will be diminished. Individuals with CMS_α experience hermaphrodites in a subdivided, structured population at the frequency of

$$H_{\text{structure}} = p_\alpha X_\alpha + p_{\text{not}\,\alpha}(X_{\text{not}\,\alpha}) + (V_{\alpha,\text{not}\,\alpha}/p_\alpha)(X_\alpha - X_{\text{not}\,\alpha}) \qquad (2)$$

(Equation (6) in McCauley, Olson & Taylor, 2000b). The observant reader will note that Equations (1) and (2) differ only by the second term. This term penalizes CMS types that produce high frequencies of females in proportion to the degree at which CMS types are segregated into different demes (= population structure).

Application to the Virginia metapopulation

In this simplified model, the parameters necessary for estimating the effect of population structure are the population-to-population variation in the frequency of haplotype α and the frequency at which haplotype α produces females relative to the other haplotypes in the metapopulation. Both of these parameters can be estimated from the Olson and McCauley (2002) mtDNA data. The population variation in the frequency of mtDNA haplotype α can be estimated via Wright's F_{ST} and by recognizing that the general equation for F_{ST} of a single allele is

$$F_{\text{ST}} = V_{\alpha,\text{not}\,\alpha}/p_\alpha p_{\text{not}\,\alpha}. \qquad (3)$$

The frequency at which haplotype α produces hermaphrodites can be estimated from the association between haplotype and gender from the roadside populations (Figure 5).

Let us consider the potential effects of population structure on the mtDNA haplotypes that are associated with the highest proportions of females (haplotype *g*) and hermaphrodites (haplotype *d*) in the Virginia populations (Figures 4 and 5). In this sample, the global frequency of hermaphrodites was 64% and in a panmictic population we would theoretically expect all haplotypes (including *g* and *d*) to experience hermaphrodites at this frequency. (Note that this sex ratio is female biased relative to the actual global sex ratio of the roadside populations because haplotypes in large populations were proportionally underrepresented and small populations tended to be more female biased than large populations. Nonetheless, this ratio will serve to illustrate the theoretical effect of population structuring on fitness.)

To calculate the subjective frequency at which each haplotype experiences hermaphrodites, first assume that the frequency at which haplotype *g* produces hermaphrodites is $X_g = 10/27 = 0.370$ (see Figure 5). Accordingly, all other haplotypes produce hermaphrodites at a frequency of $X_{\text{not } g} = 150/223 = 0.673$ (Olson & McCauley, 2002). The F_{ST} of haplotype *g* in the roadside populations was estimated at 0.523. Thus, by Equations (2) and (3) the theoretical frequency at which haplotype *g* experiences hermaphrodites in the roadside populations is 0.499. This value is 22% less than the global frequency of hermaphrodites in the mtDNA sample. If fitness were proportional to the local population sex ratio in the same manner as in Figure 2, population structure would decrease the fitness of haplotype *g* by 15% because carriers of this haplotype are usually female and the haplotype is rare within the metapopulation.

In contrast, the fitness of haplotypes that have a high probability of being carried by hermaphrodites may be increased by population structure. Assume that the frequency at which haplotype *d* produces hermaphrodites is $X_d = 18/20 = 0.900$. Accordingly, all other haplotypes produce hermaphrodites at a frequency of $X_{\text{not } d} = 142/230 = 0.617$ (Olson & McCauley, 2002). The F_{ST} of haplotype *d* in the roadside populations was estimated at 0.474. Thus, by Equations (2) and (3) the theoretical frequency at which haplotype *d* experiences hermaphrodites in the roadside populations is 0.711. Thus if fitness were proportional to the local population sex ratio in the same manner as in Figure 2, the fitness of haplotype *d* increases 8% in structured compared to panmictic population structure. The above analysis indicates that population structuring is likely to have strong affects on the spread of mtDNA haplotypes and CMS factors in natural populations. This effect may be particularly acute for CMS haplotypes because when conditions allow CMS factors to spread via female advantage within populations (e.g. population frequencies of male fertility restorers are low), female-biased sex ratios are likely to develop, eventually restricting the fitness of the CMS due to pollen limitation. Thus, the effect of local population sex ratio and its interplay with pollen limitation may limit the spread of CMS factors compared to the absence of population structure and cannot be overlooked in an analysis of the microevolutionary dynamics of sex ratio evolution in cytonuclear gynodioecious species.

Limitations of the model

To generate these theoretical predictions it was necessary to make assumptions that may not be true for natural populations. Here we review three of these assumptions and their potential effects. The reader is referred to McCauley and Taylor (1997) and McCauley, Olson & Taylor (2000b) for further discussion. First, we assumed that population structure in the frequencies of nuclear male fertility restorers was absent, so that each haplotype produced similar frequencies of hermaphrodites in every deme. However, there is compelling evidence that suggests that this is not the case. Nuclear allozyme polymorphisms show significant levels of population structure in these same populations (McCauley, 1998) and the same mtDNA haplotype is associated with different proportions of hermaphrodites in different demes, suggesting population-to-population variation in the frequencies of male fertility restorers (Olson & McCauley, 2002). Population structure in restorers affects both our estimate of the sex ratio associated with each CMS type from field data (X_α) and the ability of the model to correctly predict the sex ratio associated with each CMS type within demes. One might suspect violation of this assumption to be particularly menacing because restorer genotypes frequencies may be driven by CMS frequencies within populations and vice versa (Frank, 1989; Gouyon et al., 1991). However, as long as the frequencies of all CMS types and their restorers are correlated in the same manner across demes, violation of this assumption should have a reduced effect on the estimate of the 'true' effect of population structure because the predictions are calculated relative to all other CMS types.

Second, this analysis assumes that individuals of the same gender but bearing different CMS types have the same fitness through seed. However, recent data suggests that that fitness is dependent both on an individual's sex and mtDNA haplotype in *Silene vulgaris* (McCauley & Olson, 2003). Such patterns are consistent with the 'cost of restoration' or fitness decrease associated with harboring additional or incompatible male fertility restorer alleles (Charlesworth, 1981; Delannay, Gouyon & Valdeyron, 1981; Gouyon, Vichot & Van Damme, 1991). Gregorius and Ross (1984) have shown that joint CMS and restorer polymorphism can be maintained when there is a tradeoff between seed fitness of hermaphrodites and females within a CMS type. When this is the case, some CMS types gain more fitness via seed output through females relative to hermaphrodites than do others. Although the interactive effects between different costs of restoration for different CMS types and population structure have not yet been formally modeled, one might expect that the spread of haplotypes that rely on high female seed production for their maintenance will be more strongly suppressed by population structure than those that rely more equally on seed production from both females and hermaphrodites. In a larger sense, our analysis ignores all interactive and coevolutionary influences of male fertility restorers on the spread of CMS factors. Understanding these interactions is currently one of the most pressing concerns for advancing our understanding of the evolution of the CMS elements.

Finally, McCauley and Taylor (1997) assume annual population turnover whereas roadside populations persist for decades. A decadal time scale may be necessary for several generations of adaptation to occur within demes. Within deme adaptation between the CMS and male fertility restorer loci will result in heterogeneity of the genetic environment across populations. For this reason, future investigations should aim to understand the spatial and temporal scales across which this adaptation occurs.

Conclusions

We have been using sex ratio evolution in the gynodioecious plant, *Silene vulgaris*, as a model system for studying evolution in spatially structured populations. In our series of ongoing studies of roadside populations in western Virginia, we have shown that there is sex ratio variation among local populations, and since the fitness of females and hermaphrodites is frequency dependent, this sex ratio affects the relative fitness of the two phenotypes. We have also shown that the frequency of CMS elements (as associated mtDNA markers) varies across populations, and that there is some evidence that nuclear genes affecting gender are structured. Our empirical results and theoretical investigations indicate that population subdivision has the potential to have emergent effects on the evolution of sex ratio in gynodioecious species that cannot be predicted from studies that do not take into account the effects of subdivision. The one observation we are missing is whether demes with different sex ratios (and therefore different fitnesses) differentially export propagules to the metapopulation at large and affect a change in gene frequency across generations, simply as a result of how those genotypes are distributed in space. Further progress in understanding sex ratio evolution in gynodioecious populations must also focus on how population structure of nuclear male fertility restorers interacts with that of the CMS factors. The population structure of these interacting sets of genes is probably affected by both stochastic and selective factors, but the relative importance of these factors is currently unknown.

We can draw an analogy between the way population structure can favor the evolution of cooperative versus selfish behaviors, and the way population structure can favor the evolution of hermaphroditism over male sterility. In our view, this analogy runs very deep. A cytoplasmic gene that makes pollen is 'altruistic' because it contributes to the seed production (fitness) of other cytoplasmic genomes, and to the extent that pollen is costly to produce, it does so at its own expense. CMS factors are appropriately called 'selfish' genes because they opt out of pollen production while benefiting from the pollen production by other cytoplasms. Accordingly, the same type of frequency dependent selection influences both selfish genes and selfish behaviors. Both are favored when are rare because they receive the benefits of altruism without paying the cost. Population structure influences both systems in the same way, clustering 'altruists' into a subset of demes where those that pay the costs of cooperative behaviors also receive its benefits. It is interesting, though perhaps not surprising, that such a clear empirical example of selection acting in this way comes from plants. In gynodioecious plants, the mechanism causing frequency dependent selection (pollen limitation) is relatively to easy measure, the genetic basis of the 'behavior' is never in question, and the distribution of plant populations in space makes them amenable to an experimental approach.

Acknowledgements

The authors would like to thank Rodney Mauricio and three additional reviewers for their helpful comments. Logistical support was provided through Mountain Lake Biological Station and financial support was provided through the Institute of Arctic Biology at the University of Alaska, NSF Alaska EPSCoR, a National Science Foundation/Alfred P. Sloan Postdoctoral Fellowship in Molecular Evolution (DBI-9750033), and a grant from the National Science Foundation (DEB-0078531).

References

Ashman, T.-L., 1999. Determinants of sex allocation in a gynodioecious wild strawberry: implications for the evolution of dioecy and sexual dimorphism. J. Evol. Biol. 12: 648–661.

Bierzychudek, P., 1981. Pollinator limitation of plant reproductive effort. Am. Nat. 117: 838–840.

Charlesworth, D., 1981. A further study of the problem of the maintenance of females in gynodioecious species. Heredity 46: 27–39.

Charlesworth, D. & V. Laporte, 1998. The male-sterility polymorphism of *Silene vulgaris:* analysis of genetic data from two populations and comparison with *Thymus vulgaris*. Genetics 150: 1267–1282.

Cosmides, L.M. & J. Tooby, 1981. Cytoplasmic inheritance and intragenomic conflict. J. Theor. Biol. 89: 83–129.

Couvet, D.O. Ronce, & C. Gliddon, 1998. The maintenance of nucleocytoplasmic polymorphism in a metapopulation: the case of gynodioecy. Am. Nat. 152: 59–70.

Coyne, J.A., N.H. Barton & M. Turelli, 1997. Perspective: a critique of Sewall Wright's shifting balance theory of evolution. Evolution 51: 643–671.

Coyne, J.A., N.H. Barton & M. Turelli, 2000. Is Wright's shifting balance process important in evolution? Evolution 54: 306–317.

de Haan, A.A., H.P. Koelewijn, M.P.J. Hundscheid & J.M.M Van Damme, 1997. The dynamics of gynodioecy in *Plantago lanceolata* L. II. Mode of action and frequencies of restorer alleles. Genetics 147: 1317–1328.

Delannay, X., P.H. Gouyon & G. Valdeyron, 1981. Mathematical study of the evolution of gynodioecy with cytoplasmic inheritance under the effect of a nuclear restorer gene. Genetics 99: 169–181.

Dulberger, R. & A. Horovitz, 1984. Gender polymorphism in flowers of *Silene vulgaris* (Moench) Garcke (Caryophyllaceae). Bot. J. Linn. Soc. 89: 101–117.

Emery, S.N., 2001. Inbreeding depression and its consequences in *Silene vulgaris*. Masters Thesis, Vanderbilt University, Nashville, TN, USA.

Fisher, R.A., 1958. The Genetical Theory of Natural Selection, 2nd revised edn. Oxford University Press, Oxford.

Frank, S.A., 1989. The evolutionary dynamics of cytoplasmic male sterility. Am. Nat. 133: 345–376.

Frank, S.A. & C.M. Barr, 2001. Spatial dynamics of cytoplasmic male sterility, pp. 219-243 in Integrating Ecology and Evolution in a Spatial Context, edited by J. Silvertown & J. Antonovics. Blackwell Science, Oxford, UK.

Galloway, L.F. & C.B. Fenster, 2000. Population differentiation in an annual legume: local adaptation. Evolution 54: 1173–1181.

Goodnight, C.J., J.M. Scwartz & L. Stevens, 1992. Contextual analysis of models of group selection, soft selection, hard selection, and the evolution of altruism. Am. Nat. 140: 743–761.

Goodnight, C.J. & M.J. Wade, 2000. The ongoing synthesis: a reply to Coyne, Barton, and Turelli. Evolution 54: 317–324.

Gouyon, P.H. & D. Couvet, 1987, A conflict between two sexes, females and hermaphrodites, pp. 245–260 in The Evolution of Sex and its Consequences, edited by S.C. Stearns. Birkauser Verlag, Basel.

Gouyon, P.-H., F. Vichot & J.M.M. Van Damme, 1991. Nuclear-cytoplasmic male sterility: single-point equilibria versus limit cycles. Am. Nat. 137: 198–514.

Graff, A., 1999. Population sex structure and reproductive fitness in gynodioecious *Sidalaceea malviflora malviflora* (Malvaceae). Evolution 53: 1714–1722.

Gregorius, H.-R. & M.D. Ross, 1984. Selection with gene-cytoplasm interactions. I. Maintenance of cytoplasm polymorphisms. Genetics 107: 165–178.

Hatcher, M.J., 2000. Persistence of selfish genetic elements: population structure and conflict. Trends Ecol. Evol. 15: 271–277.

Hurst, L.D., A. Atlan & B.O. Bengtsson, 1996. Genetic conflicts. Q. Rev. Biol. 71: 317–364.

Jacobs, M.S. & M.J. Wade, 2003. A synthetic review of the theory of gynodioecy. Am. Nat. 161: 837–851.

Jolls, C.L., 1984. The maintenance of hermaphrodites and females in populations of *Silene vulgaris* (Moench) Garcke (Caryoplyllaceae. Am. J. Bot. 71: 80.

Jolls, C.L. & T.C. Chenier, 1989. Gynodioecy in *Silene vulgaris* (Caryophyllaceae): progeny success, experimental design, and maternal effects. Am. J. Bot. 76: 1360–1367.

Levings, C.S. III., 1993. Thoughts on cytoplasmic male sterility in *cms-T* maize. Plant Cell 5: 1285–1290.

Lewis, D., 1941. Male sterility in natural populations of hermaphroditic plants. New Phytol. 40: 56–63.

Lloyd, D.G., 1974. Theoretical sex ratios of dioecious and gynodioecious angiosperms. Heredity 32: 11–31.

McCauley, D.E., 1998. The genetic structure of a gynodioecious plant: nuclear and cytoplasmic genes. Evolution 52: 255–260.

McCauley, D.E. & M.T. Brock, 1998. Frequency-dependent fitness in *Silene vulgaris*, a gynodioecious plant. Evolution 52: 30–36.

McCauley, D.E., M.S. Olson, S.N. Emery & D.L. Taylor, 2000a. Sex ratio variation in a gynodioecious plant: spatial scale and fitness consequences. Am. Nat. 155: 814–819.

McCauley, D.E., M.S. Olson & D.R. Taylor, 2000b. The influence of metapopulation structure on genotypic fitness in a gynodioecious plant. Evol. Ecol. 14: 181–194.

McCauley, D.E. & D.R. Taylor, 1997. Local population structure and sex ratio: evolution in gynodioecious plants. Am. Nat. 150: 406–419.

Olson, M.S. & J.A. Antonovics, 2000. Correlations between male and female reproduction in the near-dioecious herb *Astilbe biternata*. Am. J. Bot. 87: 837–844.

Olson, M.S. & D.E. McCauley, 2000. Linkage disequilibrium and phylogenetic congruence between chloroplast and mitochondrial haplotypes in *Silene vulgaris*. Proc. R. Soc. Lond. B Biol. Sci. 267: 1801–1808.

Olson, M.S. & D.E. McCauley, 2002. Mitochondrial DNA diversity, population structure, and gender association in the gynodioecious plant *Silene vulgaris*. Evolution 56: 253–262.

Pannell, J., 1997. The maintenance of gynodioecy and androdioecy in a metapopulation. Evolution 51: 10–20.

Petterson, M.W., 1992. Advantages of being a specialist female in nondioecious *Silene vulgaris* S.L. (Caryophyllaceae). Am. J. Bot. 79: 1389–1395.

Ross, M.D. & H.-R. Gregorius, 1985. Selection with genecytoplasm interactions. II. Maintenance of gynodioecy. Genetics 109: 427–439.

Saumitou-Laprade, P., J. Cuguen & P. Vernet, 1994. Cytoplasmic male sterility in plants: molecular evidence and the nucleocytoplasmic conflict. Trends Ecol. Evol. 9: 431–435.

Schnable, P.S. & R.P. Wise, 1998. The molecular basis of cytoplasmic male sterility and fertility restoration. Trends Plant Sci. 3: 175–180.

Slatkin, M., 1977. Gene flow and genetic drift in a species subject to frequent local extinctions. Theor. Popul. Biol. 12: 253–262.

Taylor, D.R., D. McCauley & S. Trimble, 1999. Colonization success of females and hermaphrodites in the gynodioecious plant, *Silene vulgaris*. Evolution 53: 745–751.

Taylor, D.R., M.S. Olson & D.E. McCauley, 2001. A quantitative genetic analysis of nuclear-cytoplasmic male sterility in structured populations of *Silene vulgaris*. Genetics 158: 833–841.

Van Damme, J.M.M., 1983. Gynodioecy in *Plantago lanceolata* L. II. Inheritance of three male sterility types. Heredity 50: 253–273.

Wade, M.J. & C.J. Goodnight, 1998. Perspective: The theories of Fisher and Wright in the context of metapopulations: when nature does many small experiments. Evolution 52: 1537–1553.

Werren, J.H. & L.W. Beukeboom, 1998. Sex determination, sex ratios, and genetic conflict. Annu. Rev. Ecol. Syst. 29: 233–261.

Wilson, D.S., 1979. Structured demes and trait-group variation. Am. Nat. 113: 606–610.

Wilson, D.S., 1980. The Natural Selection of Populations and Communities. Benjamin/Cummings, Menlo Park, CA, USA.

Wright, S., 1931. Evolution in Mendelian populations. Genetics 16: 97–159.

Genetica (2005) 123: 63–74 © Springer 2005

Studying genetics of adaptive variation in model organisms: flowering time variation in *Arabidopsis lyrata*

Mona Riihimäki[1], Robert Podolsky[1,2], Helmi Kuittinen[1], Hans Koelewijn[1,3] & Outi Savolainen[1,*]
[1]*Department of Biology, University of Oulu, FIN-90014 University of Oulu, Finland;* [2] *Present address: Office of Biostatistics and Bioinformatics, Medical College of Georgia, Augusta, GA 30912-4900, USA;* [3]*Present address: ALTERRA, Center for Ecosystem Studies, P.O. Box 47, 6700 AA Wageningen, The Netherlands;* **Author for correspondence (Phone +358-8-553 1782; Fax +358-8-5531061; E-mail: outi.savolainen@oulu.fi)*

Received 3 March 2003 Accepted 26 November 2003

Key words: adaptation, *Arabidopsis lyrata, A. thaliana,* flowering time, genetic differentiation

Abstract

Arabidopsis thaliana has emerged as a model organism for plant developmental genetics, but it is also now being widely used for population genetic studies. Outcrossing relatives of *A. thaliana* are likely to provide suitable additional or alternative species for studies of evolutionary and population genetics. We have examined patterns of adaptive flowering time variation in the outcrossing, perennial *A. lyrata*. In addition, we examine the distribution of variation at marker genes in populations form North America and Europe. The probability of flowering in this species differs between southern and northern populations. Northern populations are much less likely to flower in short than in long days. A significant daylength by region interaction shows that the northern and southern populations respond differently to the daylength. The timing of flowering also differs between populations, and is made shorter by long days, and in some populations, by vernalization. North American and European populations show consistent genetic differentiation over microsatellite and isozyme loci and alcohol dehydrogenase sequences. Thus, the patterns of variation are quite different from those in *A. thaliana*, where flowering time differences show little relationship to latitude of origin and the genealogical trees of accessions vary depending on the genomic region studied. The genetic architecture of adaptation can be compared in these species with different life histories.

Introduction

Arabidopsis thaliana is the best known plant species in terms of its genome and molecular biology (Arabidopsis Genome Initiative, 2000). Its small genome and readily available mutants have made it a favorite organism for developmental and molecular genetic studies. Recently, the interest in the population genetics of *A. thaliana* has increased (Hanfstingl et al., 1994; Innan et al., 1996; Mitchell-Olds, 2001). At the same time, related species have begun to be seen also as potential model organisms. These relatives offer possibilities to study species with different life histories and the molecular genetic tools of *A. thaliana* can be often readily applied in the relatives (e.g., Kuittinen et al., 2002a). *A. lyrata* is a self-incompatible outcrossing species (Schierup, 1998; Kärkkäinen et al., 1999), to which the extensive population genetics theory of random mating populations can

Table 1. Comparison of *A. lyrata* and *A. thaliana* features

Trait	*A. lyrata*	*A. thaliana*	Reference
Outcrossing rate	1.0	0.02	Abbot and Gomes (1989) Kärkkäinen et al. (1999) and Schierup (1998)
Life cycle	Perennial	Annual	
Diploid genome size	0.46–0.51 pg	0.23–0.29	*Arabidopsis* Genome Initiative (2000), Earle (unpublished)
Chromosome #	8	5	Jones (1963)
Distribution	Palearctic, nearctic	Worldwide	

be applied. In outcrossing species, the different genes evolve more independently than in selfing species, where extensive linkage disequilibrium (LD) of genomes is maintained (Nordborg et al., 2002). The more independent variation of genes may make it easier to examine the evolution and its causes of individual genes. Further, *A. thaliana* is a weedy species, and outcrossing relatives may offer a possibility of studying populations where the effects of recent population expansions are not as much confounding in analyses of sequence variation. Third, for studies of local adaption, it may well be profitable to also use species that are not global generalist weeds.

In this paper, we examine the patterns of variation in one potentially adaptive trait, flowering time. Based on the life history differences between *A. thaliana* and *A. lyrata,* we can ask several questions. First, do the more stable, less weedy populations of *A. lyrata* show signs of local adaptation e.g. in flowering time, related to the environmental conditions. Do the populations of the outcrossing species have much variation within the populations, in comparison to the selfing *A. thaliana.* (e.g. Charlesworth & Charlesworth, 1995). Third, is the current distribution reflected in the genetic structure of *A. lyrata* populations? Do we find consistent patterns of genetic relationships between populations, using data from different parts of the genome. We address these questions with new data on the variation of flowering time, and with some new data and new analysis of earlier genetic markers and sequences. We discuss the implications of the differences between the species for the study of genetics of adaptation.

Materials and methods

Natural history of Arabidopsis lyrata

Arabidopsis lyrata is among the closest relatives to *A. thaliana* based on restriction fragment length polymorphism (RFLP) studies of cpDNA, and sequences of rbcL {Price, Palmer & Al-Shehbaz, 1994). Until recently, the two subspecies of *A. lyrata* (ssp. *lyrata* and ssp. *petraea)* were called *Arabis lyrata* and *Cardaminopsis petreaea,* but O'Kane and Al-Shebaz (1997) placed the species (and several others) in the genus *Arabidopsis*. This view of the systematics has been confirmed in many later studies of the Brassicaceae, using both cpDNA and nuclear sequences (Koch, Bishop & Mitchell-Olds, 1999; Koch, Haubold & Mitchell-Olds, 2000, 2001). The proportion of synonymous substitutions between the two species ranges between 10 and 15%, and for aminoacid changing nonsynonymous substitutions the divergence level is about 1–2%. Koch, Haubold and Mitchell-Olds (2000) have estimated a divergence time of about 5 MY for these two species based on *Adh* and *Chs* sequences.

The diploid genome size of *A. lyrata* (Swedish Mjällom and US Michigan populations) measured with flow cytometry is 0.46–0.51 pg, compared with the estimates for *A. thaliana* of 0.23–0.29 pg in the same set of measurements (Earle, pers. comm.). *A. lyrata* and other close relatives have eight chromosomes, against the five of *A. thaliana* (Jones, 1963). The two species can be crossed (Mesicek, 1967; Redei, 1974). Nasrallah et al. (2000) produced viable vigorous offspring from the hybrid seeds after embryo rescue. In the backcross offspring of the hybrids, there was no evidence of crossing over between homeologous segments of the genomes of the two species (Nasrallah et al., 2000).

There are important life history differences. *A. thaliana* is annual, *A. lyrata* perennial (See Table 1). There is a well developed self-incompatibility system in *A. lyrata* (Kusaba et al., 2001), which gives rise to a fully outcrossing mating system (Schierup, 1998; Kärkkäinen et al., 1999). This difference is reflected also in the relatively large, pollinator-attracting petals of *A. lyrata* (see figures in Nasrallah et al., 2000). The species *A. lyrata* has a fragmented distribution in Europe, Japan and North America, with largely unknown distribution in Russia (Figure 1), references in Savolainen et al. (2000), whereas *A. thaliana* is a

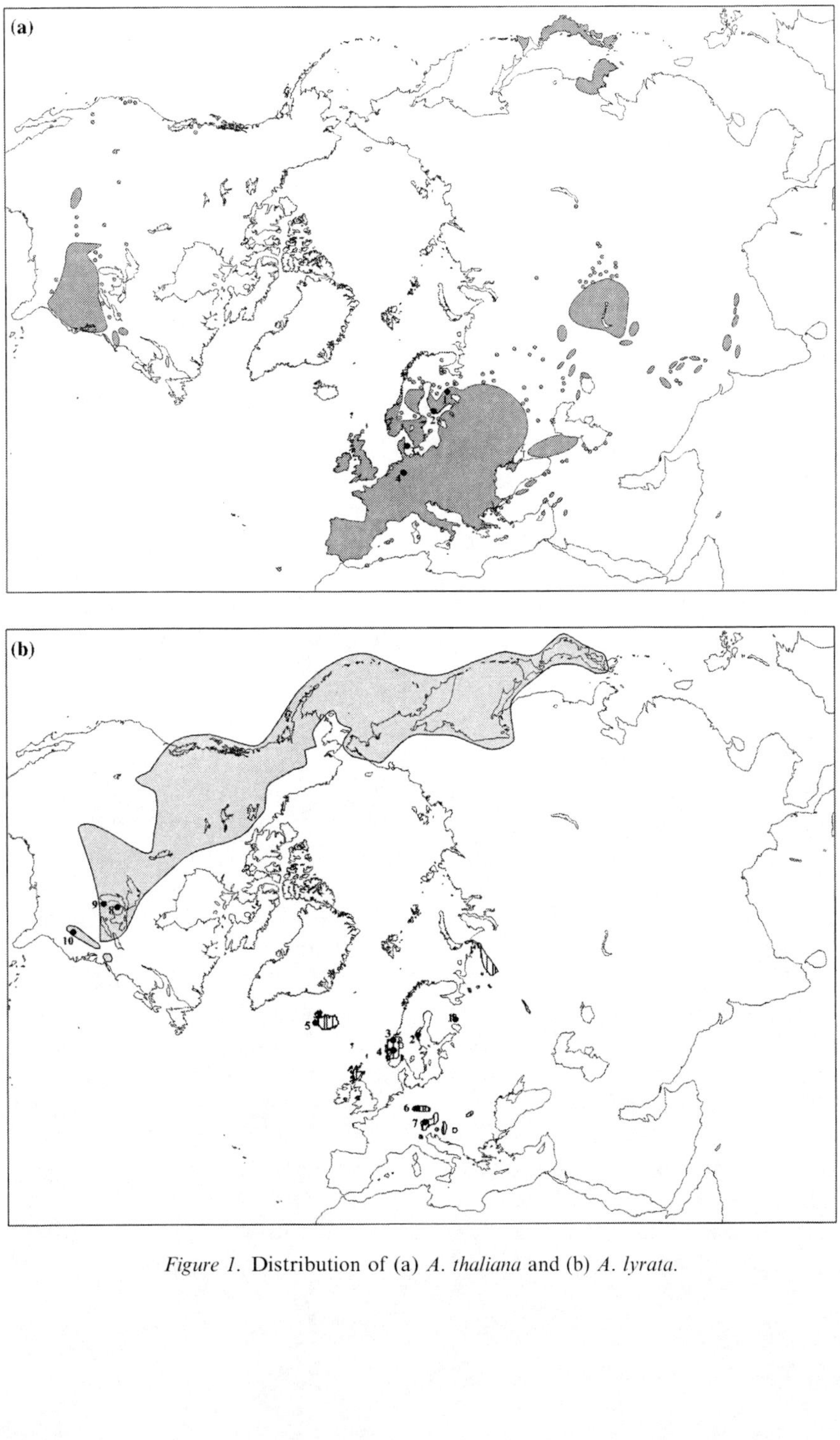

Figure 1. Distribution of (a) *A. thaliana* and (b) *A. lyrata*.

widespread weed. It has its origins in Asia and has spread to Europe (Price, Palmer & Al-Shehbaz, 1994), and has been introduced to other parts of the world, such as to the USA.

Measuring flowering time variation

We have examined the flowering variation in the perennial *Arabidopsis lyrata*. Six populations were chosen for the study (Plech, Germany 49°39′N. Bohemia, The Czech Republic 50°03′N, Spiterstulen, Norway 61°38′N, alt. 1100 m, Litldalen, Norway 62°32′N, Karhumäki, Russia 62°55′N, and Mjällom, Sweden 62°55′N). The seed samples were germinated and grown in long (LD, 20 h) and short (SD, 14 h) daylengths (+20°C). After 6 weeks of growth half of the plants from both daylengths were vernalized in +4°C, for 4 weeks. The nonvernalized plants were kept in +15°C to reduce growth. Both sets of plants received 8 h of light. After vernalization the plants were moved back to LD and SD at +20°C. In each of the four treatments, each population was represented by 12 plants.

We also grew a small set of crosses (12 females crossed each with four males) from the population of Karhumäki, Russia. The plants were not vernalized. They were grown under natural light conditions in the spring time in a greenhouse. The date when the first plant flowered was designated 1.

Statistical analyses

The flowering time data in the different environments were analyzed using the linear mixed effects model of R, after logarithmic transformation (Pinheiro & Bates, 2000; Team R Development Core Group, 2002). For the purposes of the analysis, the data from the four northern populations were combined to form a northern region, and the two southern populations were likewise combined to form a southern region. Region, daylength and vernalization were treated as fixed factors. The plants were randomized within daylengths on six trays. The tray was regarded as a random factor. The within population family data were also analyzed with ANOVA in R. Mothers and fathers were both treated as fixed effects.

The proportions of flowering could not be transformed to have normal distributions. Hence, we used a Bayesian generalized linear mixed model (GLMM) analysis for this kind of data. The analysis is implemented in the program WIN-BUGS. Rather than testing significance, the method results in an estimate of the probability that the factor in question has an effect (Clayton, 1996; Spiegelhalter, Thomas & Best, 2000).

Genetic markers and sequencing of A. lyrata

The methods for sequencing the alcoholdehydrogenase gene (*Adh*) of *A. lyrata* have been described by Savolainen et al. (2000). We obtained additional sequences from plants from Mayodan, North Carolina (seeds kindly provided by C.H. Langley) and from Mjällom, Sweden (see Van Treuren et al., 1997 for description of the locations). The earlier data of nine polymorphic enzyme and five microsatellite loci of Van Treuren et al. (Saitou & Nei, 1987) were also used for making genealogical trees of the populations. Neighbor-joining trees (Saitou & Nei, 1987) were constructed with the MEGA program version 1.3 (Kumar et al., 2001).

Results

Probability of flowering

We characterize the flowering of the populations in two ways, first the probability of flowering, and second, the time to flower formation. The measurements were made in four different environmental conditions, long and short days with and without vernalization. The Bayesian analysis of the probabilities showed that the northern and southern (regions) populations differed (Figure 2, Table 2). In short days, the southern populations of Plech and Bohemia were more likely to flower than any of the four northern populations. The northern populations were more likely to flower in long days than in short days. These different reactions to the conditions showed up as a significant interaction between region and daylength. Vernalization effects varied across daylengths and regions. It increased the probability of flowering in the northern populations in both short and long days, but did not have a consistent effect in the southern populations. This resulted in a significant interaction between vernalization and daylength. It should be noted that the results are based on

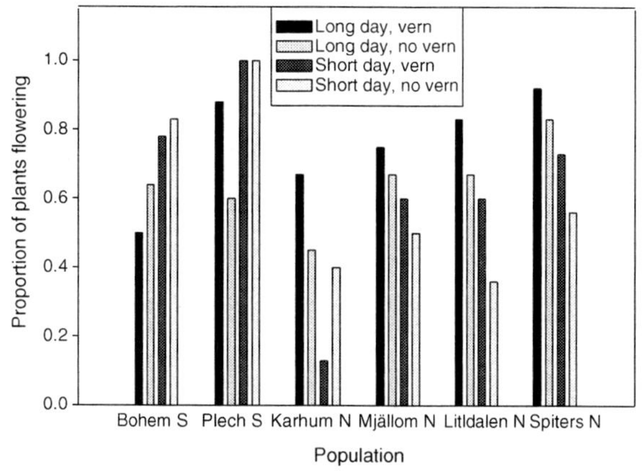

Figure 2. Proportion of plants from six different populations (N – northern, S – southern) flowering in the different day length – vernalization treatments. Long days (LD 20:4), short days (LD 14:10), vernalization – rosette cold treatment during 4 weeks.

Table 2. Bayesian generalized linear mixed model analysis of flowering probability of *A. lyrata* using WinBUGS 3.1

Node	Probability
Region	0.97
Daylength	0.75
Vernalization	0.89
Reg × Dayl	0.98
Reg × Vern	0.67
Dayl × Vern	0.99
Reg × Dayl × Vern	0.79

Names of factors and the probability that the factor has an effect on probability of flowering.

rather small samples, and need to be confirmed in later studies.

Timing of flowering

The shortest flowering times were for southern populations in long days, less than a hundred days, while northern populations in short days could take more than 150 days to flower (Figure 3). In all environmental conditions, the two southern populations (Plech, Bohemia) flowered earlier than the northern populations (for region, $p < 0.001$). All populations also flowered more rapidly in the long days than in the short days. This effect was similar in all populations, with no interaction for region and daylength. Vernalization had an overall effect of speeding up flowering (Table 3), but this effect was strongest in the northern populations of Spiterstulen and Litldalen, resulting in a region by vernalization interaction.

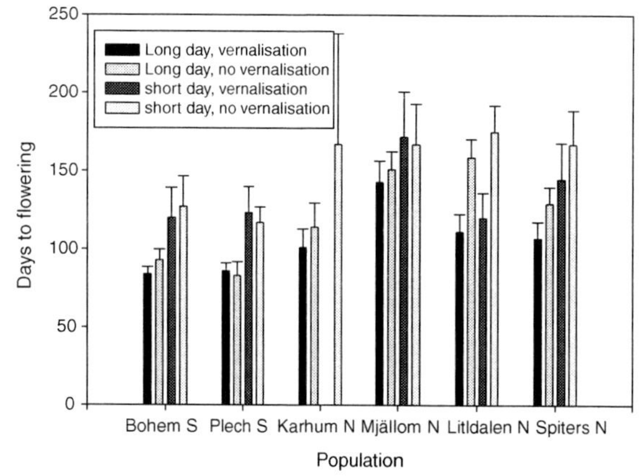

Figure 3. Flowering time of six populations of *A. lyrata* in different environmental conditions. Long days (LD 20:4), short days (LD 14:10), vernalization – rosette cold treatment during 4 weeks. Days to flowering (means and standard errors of the mean). (Too few plants flowered in Karhumäki, short days, vernalization – no result presented).

Table 3. Analysis of variance of flowering time of *A. lyrata* of the grouped northern and southern populations in four different environments

Effect	df	F	p
Region	1	25.78	0.001
Daylength	1	18.49	0.002
Vernalization	1	10.84	0.013
Reg × Dayl	1	0.98	0.320
Reg × Vern	1	3.66	0.050
Dayl × Vern	1	0.058	0.810
Reg × Dayl × Vern	1	0.002	0.966

Variation in flowering time within the population

The average flowering time of individual families of Karhumäki, in long days, with no vernalization had a range of 25 days (the date when the first plant flowered was designated 1). In this pilot study, there were significant differences in flowering time both between the maternal ($F_{11,316} = 7.29$, $p < 0.001$) and paternal ($F_{3,\,324} = 3.45$, $p < 0.02$) families. Figure 4 shows large maternal family influences, probably partly due to maternal effects,

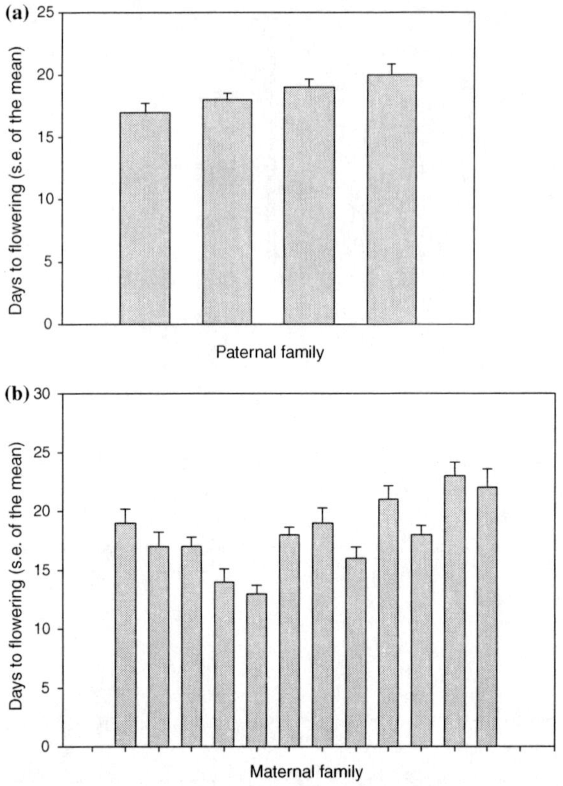

Figure 4. Flowering times for families of *A. lyrata* from Karhumäki, Russia in greenhouse conditions. (a) Means of maternal families (in days after first plant to flower) (± standard error of the mean); (b) paternal families (± standard error of the mean).

Table 4. ANOVA for flowering time variation within *A. lyrata* population of Karhumäki

Factor	df	Mean square	F	p
Mothers	11	229.3	7.29	0.001
Residuals	316	31.4		
Fathers	3	128.6	3.45	0.017
Residuals	324	37.2		

but the paternal family differences are evidence for genetic variation within the population.

Phylogeographic relationships between populations

Isozyme and microsatellite allele frequencies were available from four different populations (Van Treuren et al., 1997). In addition, we used the alcoholdehydrogenase (*Adh*) sequences of (Savolainen et al., 2000), and some additional sequences obtained for the current purpose from Sweden and North Carolina. From these data, we constructed neighbor-joining trees shown in Figure 5. All data sets give a similar picture of the grouping of the North American and European populations. There is very high bootstrap support for this with the Adh sequences. The two north American populations Michigan and Indiana are rather close to each other based on microsatellites and allozymes, and the Adh sequences show that North Carolina also is not much diverged from Indiana.

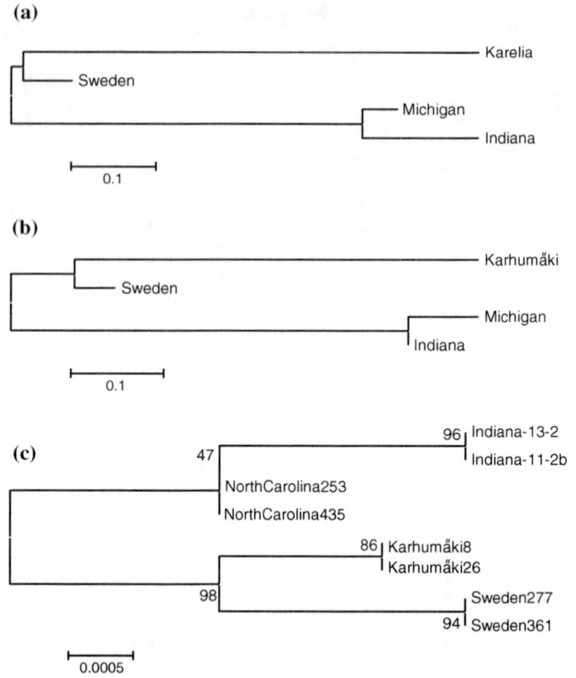

Figure 5. Neighbor-joining trees of *A. lyrata* based on (a) isozyme loci and (b) microsatellites (data of Van Treuren et al., 1997) and (c) *ADH* sequences from Savolainen et al. (2000) and additional sequences from North Carolina and Sweden (with bootstrap support).

Discussion

We have above described patterns of mainly between population variation in the outcrossing *Arabidopsis lyrata*. In comparing the patterns to *Arabidopsis thaliana*, we can test for effects of the outcrossing mating system, but these are counfounded with the effects of demographic differences between the species.

Patterns of flowering time variation between populations

The set of six populations of *Arabidopsis lyrata* showed consistent differences between populations for both the probability to flower in different conditions and the time to flowering. Southern populations were more likely to flower and flowered more rapidly than the northern ones. Latitudinal clines in timing of reproduction or growth are common in many plant species (Mikola, 1982; Thomas & Vince-Prue, 1999). These patterns are interpreted as adaptations due to natural selection by climatic factors. The flowering time of *Arabidopsis thaliana* accessions has also been extensively studied. In these studies, it is rare that the plants would not flower at all, rather the lack of flowering of *A. lyrata* may correspond to very late flowering in *A. thaliana*. Based on the data of Karlson, Sills and Nienhuis (1993) we have plotted the flowering time (recorded as leaf number at flowering) of accessions against the latitude of origin (Figure 6), which shows that there is no clinal variation. The data of Nordborg and Bergelson (1999) showed a similar lack of clinal variation. Johanson et al. (2000) also did not find a strong relationship of flowering time with latitude. Stenøien et al. (2002) also failed to find clinal variation in populations collected in a south–north transect along the Norwegian coast, even if the same populations did show a cline in hypocotyl responses to red and far-red light. Thus, the early and later flowering of *A. thaliana* seems to be a reflection of whether the plants are winter annuals requiring vernalization or summer annuals without such a requirement. The quantitative variation among the genotypes requiring vernalization does not seem to be directly related to the length of the growing season (latitude of origin).

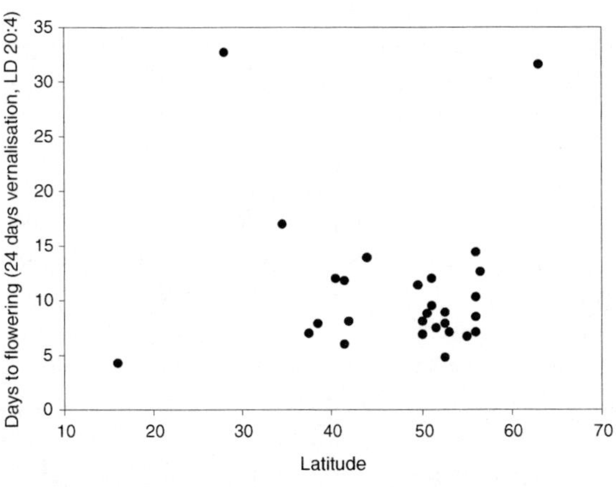

Figure 6. Variation of flowering time (measured as leaf number at flowering) in *Arabidopsis thaliana* in relation to latitude, based on data of Karlsson et al. (1993).

Environmental factors influencing flowering

We also gained some understanding on the factors controlling the probability to flower by growing the plants in several environments. The region by daylength interaction suggests that the southern and northern populations respond differentially to daylength, with northern populations more likely to flower in long days. In this experiment, vernalization had a stronger effect on the time to flowering rather than the probability to flower. In *A. thaliana,* the different accessions or ecotypes differ considerably with respect to vernalization response. It is well known that there are winter annual ecotypes requiring vernalization (e.g. Stockholm), late flowering summer annuals that flower faster after a cold treatment (such as Gr) and early flowering summer annuals which are not influenced by cold treatment (such as Li-5) (Zenker, 1955; Napp-Zinn, 1957). Napp-Zin (1957) already identified the locus *FRI*. This gene has been recently cloned and its role in determining flowering time differences in the wild between winter and summer annual ecotypes has been examined in detail (Johanson et al., 2000).

However, as mentioned, the distribution of these ecotypes is not related to latitudinal climatic variation. All populations do eventually flower even in the absence of vernalization. Interestingly, a third close relative, *A. hirsuta,* seems not to flower at all without a vernalization treatment (Zenker, 1955). Thus, in the related species the relative importance of the different pathways may vary.

The *A. thaliana* ecotypes also have variable responses to photoperiod (Karlsson, Sills & Nienhuis 1993), and in their G × E interactions (Pigliucci, Pollard & Cruzan, 2003). The photoperiodic pathway of *Arabidopsis thaliana* and its relationship to flowering time control has been well described (e.g. Koornneef, Hanhart & van der Veen, 1991; Suarez-López et al., 2001). Developmental studies of the gene CONSTANS have shown that it has an important role (Putterill et al., 1995; Yanovsky & McKay, 2002). Further, El-Assal et al. (2001) recently demonstrated that the *CRY2* cryptochrome locus is largely responsible for a flowering time difference between two early flowering accessions. QTL studies have identified other loci in crosses between summer annuals (Jansen et al., 1995). In addition, phytochrome *A* has been shown to influence flowering time differences between natural populations of *A. thaliana* (Maloof et al., 2001). The initial results on *A. lyrata*, in combination with the well known pathways of *A. thaliana*, suggest further studies on the genetic mechanisms governing these differences.

Variation within populations

We also demonstrated that there are quantitative genetic differences between families in the Russian Karhumäki population, when plants were grown under long days without vernalization. These findings are consistent with the existence of considerable within population genetic variation, as has been found earlier for marker genes (Van Treuren et al., 1997; Schierup, 1998; Clauss, Cobban & Mitchell-Olds, 2002) and for sequence variation at the *Adh* gene (Savolainen et al., 2000). *Arabidopsis thaliana* populations have been examined only rarely for quantitative genetic variation. Early British studies found evidence of segregating major gene variation (Westerman & Lawrence, 1970; Jones, 1971b, a), presumably due to the FRl gene (Johanson et al., 2000). Kuittinen, Mattila and Savolainen (1997) found that many marginal populations had no variation for flowering time. Likewise, the within population variation in microsatellites or isozymes has been found to be low (Abbott & Gomes, 1989; Todokoro, Terauchi & Kawano, 1996), as well as in restriction fragment length polymorphism (RFLP) studies (Bergelson et al., 1998) and studies of sequence variation (Stahl et al., 1999; Kuittinen, Salguero & Aguadé, 2002b).

The reduced level of genetic variation in flowering time and other traits found in at least some populations could be due to the effects of the mating system and a reduction of effective population size, due to background selection or hitchhiking (Kaplan, Hudson & Langley, 1989; Charlesworth, Morgan & Charlesworth, 1993; Charlesworth & Charlesworth, 1995). In addition to this, the weedy life history of *Arabidopsis thaliana* may also give rise to extinctions and recolonizations. The metapopulation structure is expected to lead to much reduced variation within populations, beyond the mere effects of selfing (Ingvarsson, 2002). *A. lyrata* in turn is a perennial, and is less likely to suffer frequent population extinctions.

Population history in the light of distribution of marker and sequence genetic variation

The small set of populations that was studied in *A. lyrata* demonstrated that isozymes (nine loci), microsatellites (five loci) and sequence variation at the alcoholdehydrogenase (*Adh*) locus (1700 nt) all result in a clear separation of the North American and European populations. The *Adh* trees also show that the variation between populations is high relative to within population variation, as was found earlier for isozymes and microsatellites (Van Treuren et al., 1997). These sets of populations have evidently been isolated for considerable time. When we use all our available *Adh* data (34 sequences from North America, 15 from Europe), we obtain a net divergence d_A of 0.0033 (Nei & Kumar, 2000). If we use the rate of synonymous substitution at the *Adh* locus suggested by Koch, Haubold and Mitchell-Olds (2000) of 1.5×10^{-8}/bp/year, we obtain a rough estimate of separation of at least 100,000 years for the North American and European populations. The European populations are more similar to each other, as are the two North American ones.

This pattern is in strong contrast to the situation found in *A. thaliana*. Most studies of molecular genetic variation in *A. thaliana* have been based on examining a set of accessions collected from around the world (Hanfstingl et al., 1994; Innan, Terauchi & Miyashita, 1997; Miyashita, Kawabe & Innan, 1999; Sharbel, Haubold & Mitchell-Olds, 2000). Several loci have shown a pattern of strong dimorphism, with two divergent haplotypes (e.g. Aguadé, 2001), whereas others show no such pattern. Evidence of recombination has been found in all the genes examined to date (Innan et al., 1996). Gene genealogies of

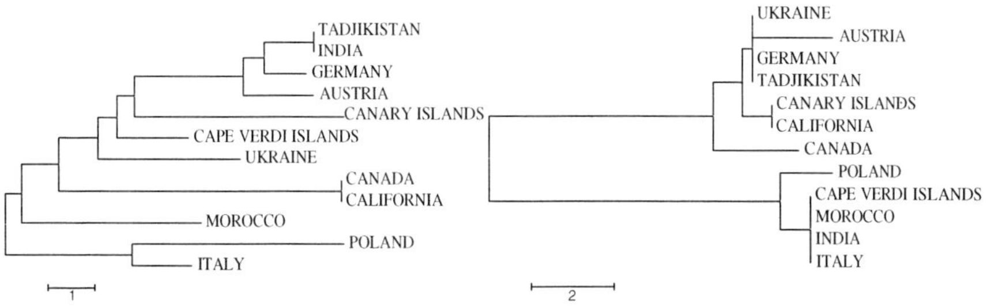

Figure 7. Neighbor-joining trees of *A. thaliana* accessions, on top *CHI,* based on data of Kuittinen et al. (2000) and on the bottom, *FAH,* based on the data of Aguadé (2001). The trees show the geographical areas, the accession names can be found in the original papers. The scale shows numbers of nucleotide substitutions.

accessions from different geographical areas based on variation at the different loci do not show geographic consistency. Figure 7 shows examples of the genealogies for the dimorphic *FAH1* and the nondimorphic *CHI* based on data of Kuittinen and Aguadé (2000) and Aguadé (2001). Several studies (e.g. Sharbel, Haubold & Mitchell-Olds, 2000) suggest that the population has expanded recently. Thus, there seems to be no one genealogical tree of the accessions or populations, an important feature of *A. thaliana.*

Implications for studying the molecular basis of adaptation

Most studies on the genetic basis of quantitative variation in plants have been on cultivated species, such as *Brassicas* (Lagergrantz et al., 1996; Lagercrantz, 1998), where domestication may have influenced patterns of variation. *A. lyrata* and other relatives of *A. thaliana* offer many opportunities to the study of adaptation in natural population, with variable mating systems and life histories.

The mating system is one of the key determinants of plant population genetics (Hamrick & Godt, 1996), and potentially modes of adaptation. Population genetics theory has several specific predictions about the expected levels of neutral variation within and between populations (Charlesworth, Morgan & Charlesworth, 1993). A comparison of the closely related species *A. thaliana* and *A.lyrata* allows investigation of the effects of the mating system of patterns of sequence evolution (Savolainen et al., 2000; Wright, Lauga & Charlesworth, 2002). The mating system effects, however, are also confounded with other life history traits, for instance as perenniality and other demographic aspects, such as the level of migration or the occurrence of extinction/colonization cycles (Pannell & Charlesworth, 1999). Variable selfing and a possible metapopulation structure add complexity to the models (Nordborg & Donnelly, 1997; Pannell & Charlesworth, 2000; Wakeley & Aliacar, 2001). Interpreting the effects of natural selection against a background of other evolutionary forces, such as effects of history, genetic drift, selection at other linked loci may be easier in random mating species as the population genetical theory for random mating populations with reasonable constant size is well developed (e.g. Hudson, 1990).

The mating system also influences patterns of linkage disequilibrium, i.e. statistical association between alleles at different loci or nucleotide sites. LD has become an important tool in genetic mapping of human diseases (Nordborg & Tavaré, 2002; Weiss & Clark, 2002) or loci responsible for quantitative genetic variation in plants (Thornsberry et al., 2001). This technique relies on examining the association of densely situated single nucleotide polymorhisms (SNPs) and phenotypic traits. SNPs close or at the disease/phenotype causing nucleotide site will be in disequilibrium, while those further away will show less association. Selfing species such as *A. thaliana* are expected to have high LD because of little effective recombination in mostly homozygous individuals (Allard, Jain & Workman, 1968). Recently, Nordborg et al. (2002) found that extent in a global sample *A. thaliana,* LD decayed over 250 kb, indicating that recombination has occurred over the long time span represented by this sample. In a local sample, LD extended over whole chromosomes, as there had been little breakdown in disequilibrium over the short time span represented by these collections. Association mapping cannot be used within local populations, as the linkage disequilibrium will be uniformly high across large parts of chromosomes. Worldwide samples will have the necessary structure of declining disequilibrium, but in such a sample the quantitative traits may be genetically heterogeneous (Nordborg et al., 2002). Disequilibrium within populations of *A. lyrata* will decline much more rapidly than in *A. thaliana*. Then associations of nucleotide variation with the phenotypic variation could be studied at a smaller scale, utilizing also the within population phenotypic variation that has been demonstrated above.

Acknowledgments

We acknowledge the financial support of the Research council for Environment and Biosciences of Finland, and the Graduate School for Forest Sciences for financial support. We thank Mark McNair, Maria Clauss, Katri Kärkkäinen and C. H. Langley for help with seeds.

References

Abbott, R.J. & M.F. Gomes, 1989. Population genetic structure and outcrossing rate of *Arabidopsis thaliana* (L.) Heynh. Heredity 62: 411–418.

Aguadé, M., 2001. Nucleotide sequence variation at two genes of the phenylpropanoid pathway, the *FAHl* and *F3H* genes, in *Arabidopsis thaliana*. Mol. BioI. Evol. 18: 1–9.

Allard, R.W., S.K. Jain & P. Workman, 1968. The genetics of inbreeding species. Adv. Genetics 14: 55–131.

Arabidopsis Genome Initiative, 2000. Analysis of the genome sequence of the flowering plant *Arabidopsis thaliana*. Nature 408: 796–815.

Bergelson, J., E. Stahl, S. Dudek & M. Kreitman, 1998. Genetic variation within and among populations of *Arabidopsis thaliana*. Genetics 148: 1311–1323.

Charlesworth, B., M.T. Morgan & D. Charlesworth, 1993. The effect of deleterious mutations on neutral molecular variation. Genetics 134: 1289–1303.

Charlesworth, D. & B. Charlesworth, 1995. Quantitative genetics of plants: the effect of the breeding system on genetic variability. Evolution 49: 911–920.

Clauss, M.J., H. Cobban & T. Mitchell-Olds, 2002. Cross-species microsatellite markers for elucidating population genetic structure in *Arabidopsis* and *Arabis* (Brassicaceae). Mol. Ecol. 11: 591–601.

Clayton, D., 1996. Generalized linear models, pp. 275–301 in Markov Chain Monte Carlo Methods in Practice, edited by W.R. Gilks, S. Richardson & D.J. Spiegelhalter. Chapman and Hall, London.

El-Assal, S.E.D., A.-B.C., A.J.M. Peeters, V. Raz & M. Koorneef, 2001. A QTL for flowering time in Arabidopsis reveals a novel allele of CRY2. Nature Genetics 29: 435–440.

Hamrick, J.L. & M.J.W. Godt, 1996. Effects of life history traits on genetic diversity in plant species. Phil. Trans. R. Soc. Lond. B 351: 1291–1298.

Hanfstingl, U., A. Berry, E.A. Kellogg, J.T. Costa, W. Ruediger & F.M. Ausubel, 1994. Haplotypic divergence coupled with lack of diversity at the *Arabidopsis thaliana* alcohol dehydrogenase locus: roles for both balancing and directional selection? Genetics 138: 811–828.

Hudson, R.R., 1990. Gene genealogies and the coalescent process, pp. 1–44 in Oxford Surveys in Evolutionary Biology, edited by D. Futuyma & J. Antonovics. Oxford.

Ingvarsson, P.K., 2002. A metapopulation perspective on genetic diversity and differentiation in partially self-fertilizing plants. Evolution 56: 2368–2373.

Innan, H., F. Tajima, R. Terauchi & N.T. Miyashita, 1996. Intragenic recombination in the *Adh* locus of the wild plant *Arabidopsis thaliana*. Genetics 143: 1761–1770.

Innan, H., R. Terauchi & N. Miyashita, 1997. Microsatellite polymorphism in natural populations of the wild plant *Arabidopsis thaliana*. Genetics 146: 1441–1731.

Jansen, R., J. Van Ooijen, P. Starn, C. Lister & C. Dean, 1995. Genotype-by-environment interaction in genetic mapping of multiple quantitative trait loci. Theor. Appl. Genet. 91: 33–37.

Johanson, U., J. West, C. Lister, S. Michaels, R. Amasino & C. Dean, 2000. Molecular analysis of FRIGIDA, a major determinant of natural variation in *Arabidopsis* flowering time. Science 290: 344–347.

Jones, B.M.G., 1963 Experimental Taxonomy of the Genus *Arabis*. University of Leicester, UK, Leicester.

Jones, M.E., 1971a. The population genetics of *Arabidopsis thaliana*. II. Population structure. Heredity 27: 51–58.

Jones, M.E., 1971b. The population genetics of *Arabidopsis thaliana*. I. The breeding system. Heredity 27: 39–50.

Kaplan, N.L., R.R. Hudson & C.H. Langley, 1989. The 'hitchhiking' effect revisited. Genetics 123: 887–899.

Kärkkäinen, K., H. Kuittinen, R. van Treuren, C. Vogl, S. Oikarinen & O. Savolainen, 1999. Genetic basis of inbreeding depression in *Arabis petraea*. Evolution 53: 1354–1365.

Karlsson, B.H., G.R. Sills & *J.* Nienhuis, 1993. Effects of photoperiod and vernalization on the number of leaves at flowering in 32 *Arabidopsis thaliana* (Brassicaceae) ecotypes. Amer. *J.* Bot. 80: 646–648.

Koch, M., J. Bishop & T. Mitchell-Olds, 1999. Molecular systematics and evolution of *Arabidopsis* and *Arabis*. Plant Biol. 1: 529–537.

Koch, M., R Haubold & T. Mitchell-Olds, 2001. Molecular systematics of the Brassicaceae: evidence from coding platidic *MatK* and nuclear *CHS*. Amer. J. Bot. 88: 534–544.

Koch, M.A., B. Haubold & T. Mitchell-Olds, 2000. Comparative evolutionary analysis of chalcone synthase and alcohol dehydrogenase loci in *Arabidopsis, Arabis,* and related genera (Brassicaceae). Mol. Biol. Evol. 17: 1483–1498.

Koornneef, M., C.J. Hanhart & J.H. van der Veen, 1991. A genetic and physiological analysis of late flowering mutants in *Arabidopsis thaliana*. Mol. Gen. Genet. 229: 57–66.

Kuittinen, H. & M. Aguadé, 2000. Nucleotide variation at the *Chalcone Isomerase* locus in *Arabidopsis thaliana*. Genetics 155: 863–872.

Kuittinen, H., M. Aguadé, D. Charlesworth, A. De Haan, B. Lauga, T. Mitchell-Olds, S. Oikarinen, S. Ramos-Onsins,

B. Stranger, P. van Tienderen & O. Savolainen, 2002a. Primers for 22 candidate genes for ecological adaptations in Brassicaceae. Mol. Ecol. Notes 2: 258–262.

Kuittinen, H., A. Mattila & O. Savolainen, 1997. Genetic variation at marker loci and in quantitative traits in natural populations of *Arabidopsis thaliana*. Heredity 79: 144–152.

Kuittinen, H., D. Salguero & M. Aguadé, 2002b. Parallel patterns of sequence variation within and between populations at three loci of *Arabidopsis thaliana*. Mol. Biol. Evol. 19: 2030–2034.

Kumar, S., K. Tamura, LB. Jakobsen & M. Nei, 2001. MEGA2: Molecular Evolutionary Genetics Analysis Software. Arizona State University, Tempe, Arizona.

Kusaba, M., K. Dwyer, J. Hendershot, J. Vrebalov, J.B. Nasrallah & M.E. Nasrallah, 2001. Self-incompatibility in the genus *Arabidopsis*: characterization of the S locus in the outcrossing *A. lyrata* and its autogamous relative *A. thaliana*. Plant Cell 13: 627–643.

Lagercrantz, U., 1998. Comparative mapping between *Arabidopsis thaliana* and *Brassica nigra* indicates that *Brassica* genomes have evolved through extensive genome replication accompanied by chromosome fusions and frequent rearrangements. Genetics 150: 1217–1228.

Lagercrantz, U., J. Putterill, G. Coupland & D. Lydiate, 1996. Comparative mapping in *Arabidopsis* and *Brassica*, fine scale genome collinearity and congruence of genes controlling flowering time. Plant J. 19: 13–20.

Maloof, J., J. Borevitz, T. Dabi, J. Lutes, R. Nehring, J. Redfern, G. Trainer, J. Wilson, T. Asami, C. Berry, D. Weigel & J. Chory, 2001. Natural variation in light sensitivity of *Arabidopsis*. Nat. Genet. 29: 441–446.

Mesicek, J., 1967. The chromosome morphology of *Arabidopsis thaliana* (L.) Heynh. and some remarks on the problem of *Hylandra suecica* (Fr.) Love. Folia Geobot. Phytotaxonom. 2: 433–436.

Mikola, J., 1982. Bud-set phenology as an indicator of climatic adaptation of Scots pine in Finland. Silva Fennica 16: 178–184.

Mitchell-Olds, T., 2001. *Arabidopsis thaliana* and its wild relatives: a model system for ecology and evolution. Trends Ecol. Evol. 16: 693–700.

Miyashita, N.T., A. Kawabe & H. Innan, 1999. DNA variation in the wild plant *Arabidopsis thaliana* revealed by amplified fragment length polymorphism analysis. Genetics 152: 1723–1731.

Napp-Zinn, K., 1957. Untersuchungen zur Genetik des Kältebedurfnisses bei *Arabidopsis thaliana*. Zeitschrift fur indukt. Abstammungs- und Vererbungslehre 88: 253–285.

Nasrallah, M.E., K. Yogeeswaran, S. Snyder & J.B. Nasrallah, 2000. *Arabidopsis* species hybrids in the study of species differences and evolution of amphiploidy in plants. Plant Physiol. 124: 1605–1614.

Nei, M. & S. Kumar, 2000. Molecular Evolutionary Phylogenetics. Oxford University Press, Oxford.

Nordborg, M. & J. Bergelson, 1999. The effect of seed and rosette cold treatment on germination and flowering time in some *Arabidopsis thaliana* (Brassicaceae) ecotypes. Am. J. Bot. 86: 470–475.

Nordborg, M., J.O. Borewitz, J. Bergelson, C. Berry, J. Chory, J. Hagenblad, M. Kreitman, J. Maloof, T. Noyes, P.J. Oefner, E.A. Stahl & D. Weigel, 2002. The extent of linkage disequilibrium in *Arabidopsis thaliana*. Nat. Genet. 30: 190–193.

Nordborg, M. & P. Donnelly, 1997. The coalescent process with selfing. Genetics 146: 1185–1195.

Nordborg, M. & S. Tavaré, 2002. Linkage disequilibrium: what history has to tell us. Trends Genet. 18: 83–90.

O'Kane, S.L. & LA. Al-Shehbaz, 1997. A synopsis of *Arabidopsis* (Brassicaceae). Novon 7: 323–327.

Pannell, J.R. & B. Charlesworth, 1999. Neutral genetic diversity in a metapopulation with recurrent local extinction and recolonization. Evolution 53: 664–676.

Pannell, J.R. & B. Charlesworth, 2000. Effects of metapopulation processes on measures of genetic diversity. Phil. Trans. R. Soc. Lond. B 355: 1851–1864.

Pigliucci, M., H. Pollard & M.B. Cruzan, 2003. Comparative studies of evolutionary responses to light environments in *Arabidopsis*. Amer. Natur 161: 68–82.

Pinheiro, J.C. & D.M. Bates, 2000. Mixed Effects Models in Sand S-PLUS. Springer Verlag, New York.

Price, R.A., J.D. Palmer & I.A. Al-Shehbaz, 1994. Systematic relationships of *Arabidopsis*, pp. 7–19 in *Arabidopsis*, edited by E.M. Meyerowitz & C. Somerville. Cold Spring Harbor Press, Cold Spring Harbor, NY.

Putterill, J., F. Robson, K. Lee, R. Simon & G. Coupland, 1995. The CONSTANS gene of *Arabidopsis* promotes flowering and encodes a protein showing similarities to zinc finger transcription factors. Cell 80: 847–857.

R Development Core Team. 2002. R: a language and environment for statistical computing. Vienna, Austria.

Redei, G.P., 1974. Is *Hylandra* an amphidiploid of *Arabidopsis* and *Cardaminopsis arenosa? Arabidopsis* Inf. Servo 11: 5.

Saitou & Nei, 1987. The nieghbor-joining method: a new method for constructing phylogenetic trees. Mol. Biol. Evol. 4: 406–425.

Savolainen, O., C.H. Langley, B. Lazzaro & H. Freville, 2000. Contrasting patterns of nucleotide variation at the alcohol dehydrogenase locus in the outcrossing *Arabidopsis lyrata* and the selfing *Arabidopsis thaliana*. Mol. Biol. Evol. 17: 645–655.

Schierup, M.H., 1998. The effect of enzyme heterozygosity on growth in a strictly outcrossing species, the self-incompatible *Arabis petraea* (Brassicaceae). Hereditas 128: 21–31.

Sharbel, T.F., B. Haubold & T. Mitchell-Olds, 2000. Genetic isolation by distance in *Arabidopsis thaliana:* biogeography and postglacial colonization of Europe. Mol. Ecol. 9: 2109–2118.

Spiegelhalter, D.J., A. Thomas & N.G. Best, 2000. WinBUGS Version 1.3. User Manual. MRC Biostatistics Unit, Cambridge, UK.

Stahl, E.A., G. Dwyer, R. Mauricio, R. Kreitman & J. Bergelson, 1999. Dynamics of disease resistance polymorphism at the *Rpml* locus of *Arabidopsis*. Nature 400: 667–671.

Stenøien, H., C. Fenster, H. Kuittinen & O. Savolainen, 2002. Quantifying latitudinal clines to light responses in natural populations of *Arabidopsis thaliana* (Brassicaceae). Am. J. Bot.

Suarez-López, P., K. Wheatley, F. Robson, H. Onouchi, F. Valverde & G. Coupland, 2001. CONSTANS mediates between the cirdacian clock and the control of flowering time in *Arabidopsis*. Nature 410: 1116–1120.

Thomas, B. & D. Vince-Prue, 1999. Photoperiodism in Plants. Academic Press, San Diego.

Thornsberry, J.M., M.M. Goodman, J. Doebley , S. Kresovich, D. Nielsen & E.S. Buckler, 2001. Dwarf8 polymorphisms associate with variation in flowering time. Nature Genet. 28: 286–289.

Todokoro, S., R. Terauchi & S. Kawano, 1996. Microsatellite polymorphism in natural populations of *Arabidopsis thaliana* in Japan. Jpn. J. Genet. 70: 543–554.

Van Treuren, R., H. Kuittinen, K. Kärkkäinen, E. Baena-Gonzalez & O. Savolainen, 1997. Evolution of microsatellites in *Arabis petraea* and *A. lyrata,* outcrossing relatives of *Arabidopsis thaliana*. Mol. Biol. Evol. 14: 220–229.

Wakeley, J. & N. Aliacar, 2001. Gene genealogies in a metapopulation. Genetics 159: 893–905.

Weiss, K.M. & A.G. Clark, 2002. Linkage disequilibrium and the mapping of complex human traits. Trends Genet. 18: 19–24.

Westerman, J.M. & M.J. Lawrence, 1970. Genotype-environment interaction and developmental regulation in *Arabidopsis thaliana*. I. Inbred lines; description. Heredity 25: 609–627.

Wright, S.I., B. Lauga & D. Charlesworth, 2002. Rates and patterns of molecular evolution in inbred and outbred *Arabidopsis*. Mol. Biol. Evol. 19: 1407–1420.

Yanovsky, M.J. & S.A. McKay, 2002. Molecular basis of seasonal time measurement in *Arabidopsis*. Nature 419: 308–312.

Zenker, A.M., 1955. Jarowisationsuntersuchungen an sommerannuellen *Arabidopsis* Rassen. Beitr. Biol. Pflantz. 32: 135–170.

Genetica (2005) 123: 75–85 © Springer 2005

Ontogenetics of QTL: the genetic architecture of trichome density over time in *Arabidopsis thaliana*

Rodney Mauricio
Department of Genetics, University of Georgia, Athens, GA 30602, USA (Phone: +1-(706)-542-1417; Fax: +1-(706)-542-3910; E-mail: mauricio@uga.edu)

Received 15 January 2002 Accepted 20 July 2002

Key words: adaptation, adult, *Arabidopsis*, development, juvenile, ontogeny, phase change, quantitative trait loci, QTL, trichome

Abstract

Although much is known about the molecular genetic basis of trichome development in *Arabidopsis thaliana*, less is known about the underlying genetic basis of continuous variation in a trait known to be of adaptive importance: trichome density. The density of leaf trichomes is known to be a major determinant of herbivore damage in natural populations of *A. thaliana* and herbivores are a significant selective force on genetic variation for trichome density. A number of developmental changes occur during ontogeny in *A. thaliana*, including changes in trichome density. I used multiple interval mapping (MIM) analysis to identify QTL responsible for trichome density on both juvenile leaves and adult leaves in replicate, independent trials and asked whether those QTL changed with ontogeny. In both juvenile and adult leaves, I detected a single major QTL on chromosome 2 that explained much of the genetic variance. Although additional QTL were detected, there were no consistent differences in the genetic architecture of trichome density measured on juvenile and adult leaves. The finding of a single QTL of major effect for a trait of known adaptive importance suggests that genes of major effect may play an important role in adaptation.

Abbreviations: cM – centiMorgans; LOD – logarithm of the odds; MIM – multiple interval mapping; n – sample size; QTL – quantitative trait locus; RI – recombinant inbred; SE – standard error.

Introduction

The density of leaf hairs, or trichomes, is a trait of considerable ecological importance for many plants. One of the primary adaptive hypotheses commonly proposed for the presence and density of plant hairs involves their role in defense against herbivores (Levin, 1973; Johnson, 1975; Ågren and Schemske, 1994; Elle et al., 1999). For example, in natural populations of *Arabidopsis thaliana*, genotypes with higher trichome densities suffer significantly less herbivore damage than genotypes with lower trichome densities (Mauricio, 1998). Furthermore, herbivores have been shown to be a significant selective agent acting on genetic variation for trichome density in *A. thaliana* (Mauricio and Rausher, 1997).

In many plants, trichomes differ on leaves of different age (Poethig, 1997, 2000, 2003). Leaf age has long been recognized as having an important effect on plant resistance to herbivores – herbivores often have strong preferences for tissue of a particular age (Janzen, 1979; Coley 1980; Krischik and Denno, 1983; Karban and Thaler, 1999; Lawrence et al., 2003). Damage to leaves of different ages can have different effects on plant fitness (Stinchcombe, 2002). Therefore, herbivores can impose very different selective pressures on plants depending on their pattern of feeding (Mauricio et al., 1993). Difference in trichome density on juvenile and adult leaves might mediate such selection.

The vegetative phase change from juvenile to adult rosette leaves in *A. thaliana* is well-described, particularly with respect to trichomes (Telfer et al., 1997). The distribution and density of trichomes varies during vegetative development and has been used in *A. thaliana* to distinguish the juvenile and adult rosette (Lawson and Poethig, 1995; Telfer et al., 1997). Leaves produced early in development have no trichomes on the abaxial (lower) surface and rosette leaves produced later have trichomes on both adaxial (upper) and abaxial surfaces. There are differences in the density of trichomes between juvenile and adult leaves in *A. thaliana*, although the change in trichome density between these vegetative phases occurs gradually through development (Telfer et al., 1997). In particular, total trichome number in *A. thaliana* has been reported to increase with rosette age (Martínez-Zapater et al., 1995; Payne et al., 2000).

Since the magnitude of selection on plants by herbivores may differ depending on the age of the leaves eaten and the density of trichomes on those leaves, the ability to predict the evolutionary response of the plants to that selection will depend on an understanding of the genetic architecture of the traits under selection. Our ability to predict the potential response to selection is directly predicated on knowledge of the number of genes and their effects on the expression of the phenotype (Lande, 1983; Lynch and Walsh, 1998; Barton and Keightley, 2002). Although much is known about the molecular genetics of trichome development in plants (Hülskamp and Schnittger, 1998; Hülskamp and Kirik, 2000; Szymanski et al., 2000; Walker and Marks, 2000), less is known about the genetic basis of trichome density (Larkin et al., 1996) and very little is known about whether the genetic architecture of trichome density changes with ontogeny.

There is a strong genotypic component to variation in trichome density in *A. thaliana*. Considerable among- and within-population variation for trichome density exists in natural populations of *A. thaliana* (Mauricio, 1998, 2001a). The segregation of trichome density in *A. thaliana* strongly suggests that multiple genetic factors and the environment affect the inheritance of this trait (Larkin et al., 1996; Mauricio, 1998). Trichome density is, therefore, a quantitative trait and the appropriate tool for genetic analysis is QTL (quantitative trait loci) mapping (Mackay, 2001; Mauricio, 2001b).

A QTL mapping approach is likely to be a fruitful one in a completely sequenced model organism, such as *A. thaliana*. Many genetic markers are available, as are several sets of mapping populations. Genome scans for QTL have the potential to identify chromosomal segments containing genes that contribute to variation in a trait of interest (e.g., Doebley et al., 1997; Frary et al., 2000; Johanson et al., 2000).

Despite the fact that QTL mapping has been used extensively in the past decade, some caveats have been raised as to its use (Beavis, 1994, 1998; Mauricio, 2001b). In at least one study, replicate crosses were made from the same parents and QTL analyses were completed on each of the replicates – although the same QTL were detected across studies, some of the QTL detected were unique to each cross (Beavis, 1994, 1998). Environmental conditions have also been shown to play a significant role in the outcome of QTL mapping experiments (Paterson et al., 1991). Obviously, the ability to replicate QTL experiments is of paramount interest, but few studies have specifically addressed this question. In this study, we take advantage of replicate experiments to examine the repeatability of QTL studies.

In addition to providing information about the genetic basis of complex traits, genome scans for quantitative traits provide an empirical basis for testing one of the more enduring controversies in evolutionary biology: the genetic basis of adaptation. Fisher (1930) suggested that mutations of very small effect were responsible for adaptive evolution. Orr and Coyne (1992) reexamined the evidence for this Fisherian view and argued that both the theoretical and empirical basis for it were weak and that adaptive traits may well be controlled by genes of major effect. They encouraged evolutionary biologists to reexamine this research question by the genetic analysis of adaptive differences in natural populations.

In the present study, I investigate three questions addressing the genetic architecture of quantitative variation in trichome density in the plant, *A. thaliana*. First, using QTL analysis, what chromosomal segments in the *A. thaliana* genome contribute to trichome density variation in juvenile leaves and in adult leaves? Second, do QTL for trichome density change with ontogeny? Third, how variable are QTL analyses completed on a similar trait but performed at different times and in different labs?

Materials and methods

All seeds were obtained from the Arabidopsis Biological Resource Center (ABRC, Columbus, OH, USA). I used a mapping population of 100 recombinant inbred (RI) lines (ABRC stock number CS1899) that had been generated from a cross between the "Columbia" (Col-4; ABRC stock number CS-933) and "Landsberg erecta" (Ler-0; ABRC stock number CS-20) accessions of *A. thaliana* (L.) Heynh. (Lister and Dean, 1993). Progeny from the initial cross were taken through eight generations of selfing *via* single seed descent to produce nearly homozygous lines with an estimated heterozygosity of 0.42% (Lister and Dean, 1993; Juenger et al., 2000). I constructed a linkage map using a total of 228 markers (chromosome I, 54 markers; chromosome II, 33 markers; chromosome III, 37 markers; chromosome IV, 50 markers; and chromosome V, 54 markers). The map position of each marker was estimated from the observed recombination frequencies using the Kosambi mapping function as implemented by the software MapMaker 3.0 (Lander et. al., 1987). This analysis provided unique positions for each marker and a map spanning 592 centiMorgans (cM) of the *A. thaliana* genome (99% of the 597 cM estimated size of the *A. thaliana* genome based both on the *Arabidopsis* Genome Initiative sequence map and the Lister and Dean RI genetic map; www.arabidopsis.org/servlets/mapper). The mean intermarker distance was 2.8 cM. The map did not differ in marker order from the published linkage map of *A. thaliana* (www.arabidopsis.org).

Plants were grown from seed sowed singly in an approximately 5 × 5 × 6 cm plastic pot filled with a soilless mix of peat moss, perlite, pine bark and vermiculite (Fafard #3B, Agawam, MA). All replicates of each RI line were randomly assigned to an individual pot in a flat. The seeds were cold stratified at 4°C for three days and then transferred to a single growth chamber with control for both daylength (14 hours) and temperature (18°C). Five replicate plants were grown for each of the RI lines and trichome density was measured on leaves of the same age. Trichome density was estimated as the total number of trichomes within a 2.4 mm^2 area (using a micrometer in a dissecting microscope) of the upper central area of the adaxial leaf surface. In the first experiment (trial 1), I measured adult leaf trichome density on three fully expanded leaves from each replicate. In the second experiment (trial 2), I measured juvenile trichome density on the first two true leaves (the first two leaves of *A. thaliana* are initiated simultaneously) and adult trichome density on three fully expanded leaves of the same whorl. Larkin et al. (1996) counted the total number of trichomes (not density) on the first leaf of ten replicate plants from the same RI lines used here. J. C. Larkin kindly provided me with the original data from his experiment, which I have reanalyzed using this map and statistical approach.

Genome scans for QTL were done using the multiple interval mapping (MIM) procedure described by Kao and Zeng (1997), Kao et al. (1999) and Zeng et al. (1999) and implemented by the software package, QTL Cartographer, version 2.0 (Basten et al., 1994, 2004). Like other QTL approaches, this procedure tests the hypothesis that an interval flanked by two adjacent markers contains a QTL affecting the trait. Multiple interval mapping statistically accounts for the effects of additional segregating QTL outside the tested interval by using multiple marker intervals rather than individual markers. The procedure can specifically condition the statistical model on all putative QTL identified rather than markers alone. Kao et al. (1999) have shown that MIM tends to more powerful and precise in detecting QTL as compared to such techniques as interval mapping (Lander and Botstein, 1989) or composite interval mapping (Zeng, 1993, 1994).

The MIM procedure tests each parameter (putative QTL) in an initial model for significance using a backward elimination procedure and those parameters that do not lead to a significant improvement in fit are dropped (Basten et al., 2004). For the refinement of QTL position, for each QTL, the position is moved within the QTL interval from one end to the other and an information criterion is calculated for each position (Basten et al., 2004). The information criterion is a function that gives an indication of how good the model fits the data and that depends upon the likelihood ratio and the number of parameters in the model. The function is

$$I(L_k, k, n) = -2(\ln(L_k) - kc(n)/2),$$

where L_k is the likelihood for a k parameter model, $c(n)$ is a penalty function and ln is the natural log (Basten et al., 2004). For a model with k QTL, MIM searches for k + 1st QTL over all intervals that do not presently have a QTL in them. For each of these intervals, the program walks along the interval and calculates the information criterion for the presence of a QTL. The MIM protocol keeps track of the minimum information criterion (equivalent to the maximum likelihood) within each interval. When all intervals have been tested, the minimum over all intervals is determined and compared to the information criterion of the k QTL model. If $I(L_k, k, n) - I(L_{k+1}, k + 1, n)$ is greater than the threshold, the QTL at that site is retained in the model. The process repeats until no new QTL are retained (Basten et al., 2004).

I began analysis in the MIM module of QTL Cartographer using the MIM default parameters to search for an initial model. I used a walking speed of 1 cM and an initial penalty function, $c(n)$, equal to the $\ln(n) = 4.6$, with a threshold value of 0.0. After this initial run of the analysis, I iteratively reran the model in phases. In the first phase, QTL were located. In the second phase, the positions of those QTL were refined. In the third phase, I searched for additional QTLs. In order to obtain a more conservative estimate of additional QTL, I doubled the penalty function to 2 $\ln(n)$ in this phase of the analysis. In the final phase, I tested for significance of all the QTLs. I calculated conservative confidence intervals (CI) around each QTL by estimating a drop of approximately two LOD scores around the likelihood peak (van Ooijen, 1992; Juenger et al., 2000). The markers located closest to these likelihood cutoffs were considered the two LOD CI flanking markers (Juenger et al., 2000). For some QTL of small effect, I could not detect a drop off of two LOD scores. In those cases, the confidence interval effectively extends across the entire linkage group.

The MIM procedure also estimates such quantitative genetic parameters as variance components, heritabilities, and additive effects. I used the estimates of phenotypic variance, genetic variance, additive effect, and percentage of variance explained that were directly calculated by QTL Cartographer for each trait and QTL. I calculated the coefficient of genetic variation, CV_G, as $(\sqrt{V_G}/\bar{x})$ in order to facilitate comparisons of evolvability between trials (Houle, 1992). The biological interpretation of the additive (or average) effect of an allele is the difference between the mean genotypic value of individuals carrying at least one copy of that allele and the mean genotypic value of a random individual from the entire population. Statistically, the additive effect is a least squares regression coefficient of genotypic value on the gene content (Lynch and Walsh, 1998). The expected population mean value of the additive effect is zero. In these RI lines, a positive additive effect indicates the action of the "Columbia" allele and a negative additive effect indicates the effect of the "Landsberg" allele. In other words, the "Columbia" allele acts to increase trichome density and the "Landsberg" allele decreases trichome density.

Results

The "Columbia" and "Landsberg" accessions of *A. thaliana* differ significantly in their trichomes densities for both adult (Figure 1) and juvenile leaves. The average trichome density on adult leaves from a sample of the Col-4 accession was

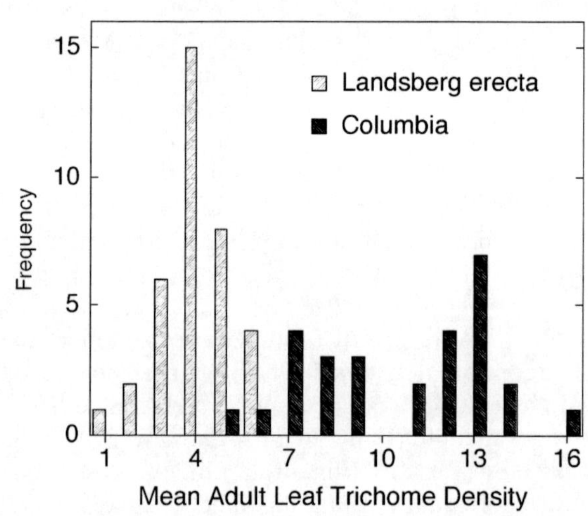

Figure 1. Frequency distribution of trichome density measured on adult leaves of the "Columbia" and "Landsberg" accessions of *A. thaliana.*

10.6 ($n = 28$ individuals; 140 leaves examined; SE = 0.6) but only 4.2 on the Ler-0 accession ($n = 36$ individuals; 180 leaves examined; SE = 0.2). The difference between these 2 accessions for juvenile leaf trichome number was of even greater magnitude. Larkin et al. (1996) reported that the mean number of trichomes per juvenile leaf in the "Columbia" accession was 30.5 ($n = 50$ individuals; 50 leaves examined; SE = 0.9) and 8.9 for the "Landsberg" accession ($n = 50$ individuals; 50 leaves examined; SE = 0.3).

Measurements of trichome density on the set of RI lines showed that juvenile leaves had double the trichome density compared to the adult leaves (Table 1). The mean trichome density on juvenile leaves in trial 2 was 18.8 while the mean trichome density on adult leaves from the two trials was 9.3 (Table 1).

The environment had a significant influence on trichome density, although that effect was more pronounced for trichome density on adult leaves than for juvenile leaves (Table 1). The environmental variance can be estimated from subtracting the genetic variance from the phenotypic variance (since $V_P = V_G + V_E$). An alternative expression of this phenomenon can be seen by comparing the proportion of the total phenotypic variance explained by the among RI line variance (the genetic variance). This "heritability" and coefficient of genetic variation were higher for juvenile leaf trichome density than for adult leaf estimates (Table 1).

The significant differences in trichome densities among the RI lines and between juvenile and adult leaves allowed me to correlate those traits with specific segments of the *A. thaliana* genome using QTL mapping techniques. The most striking result from these four analyses was that a single QTL of major effect was detected for trichome density on both juvenile and adult leaves (Table 2; Figure 2). That QTL, located on chromosome 2, was localized to an interval between 6 and 23 cM in size (depending on the trial) and explained 68.1–70.% of the variance in juvenile leaf trichome density and 28.4–27.6% of the variance in adult leaf trichome density. Comparing across the two trials for each leaf age, the magnitude of the additive effect and the variance explained for this major QTL were similar (Table 2). The additive effect of the "Columbia" allele of this QTL was uniformly positive (Table 2), meaning that the substitution of the "Columbia" allele for the "Landsberg" allele would result in a significant increase in the trichome density of that individual. An additional QTL of major effect, explaining 13.6% of the variance, was found for adult trichome density (Table 2), but only in 1 trial. This QTL has a negative additive effect and is located on chromosome 1 in a region of approximately 19 cM in size.

The QTL analysis revealed several other QTL, but most of them were of minor effect (explaining less than 10% of the variance) (Table 2; Figure 2). In most cases, it was impossible to accurately estimate a confidence interval for these minor QTL: effectively, the confidence interval extends over the entire linkage group. Despite this, in two cases the best estimates for the region associated with a minor QTL for juvenile trichome density did co-localize (Figure 2). In the first case, at approximately 48 cM on chromosome 3 (Table 2) I identified a QTL for juvenile leaf trichome density in both trial 2 and in my re-analysis of the Larkin et al. (1996) data. That QTL explained a similar amount of the variation and had a similar additive effect in the two trials (Table 2). The other case of co-localization also involved juvenile leaf trichome density and was found between position 10.9 and 18.3 cM on chromosome 4 in both trial 2

Table 1. Quantitative genetic parameters for trichome density measured in the RI lines

Trichome density measured on	$\bar{x} \pm (SE)$	Phenotypic variance (V_P)	Genetic variance (V_G)	V_G/V_P	CV_G
Juvenile leaves (Trial 2)	18.8 (0.8)	57.68	44.31	0.77	0.35
Juvenile leaves (Larkin)	20.2 (0.9)	81.64	69.60	0.85	0.41
Adult leaves (Trial 2)	11.5 (0.4)	12.40	7.24	0.58	0.23
Adult leaves (Trial 1)	7.1 (0.4)	12.50	5.44	0.44	0.33

Table 2. Trichome density QTL identified using multiple interval mapping analysis

Trichome density measured on	Linkage group	Position (cM)	2-LOD confidence interval (cM)	Nearest marker	2-LOD confidence interval markers	Additive effect	% variance explained
Juvenile leaves (Trial 2)	2	46.03	41–49	er	GPA1 – mi54	+6.52	68.1
Juvenile leaves (Trial 2)	3	49.61	NE	mi178	NE	+1.08	2.5
Juvenile leaves (Trial 2)	4	10.90	NE	mi390	NE	+1.89	6.3
Juvenile leaves (Larkin)	2	46.04	43–49	er	er – mi54	+7.72	70.5
Juvenile leaves (Larkin)	3	47.70	NE	mi178	NE	+1.57	3.3
Juvenile leaves (Larkin)	4	18.30	6–27	app	g3843 – HY4	+2.30	5.8
Juvenile leaves (Larkin)	4	55.30	23–113	m226	mi167 – ve031	−1.57	1.6
Juvenile leaves (Larkin)	5	60.20	NE	mi125	NE	−1.45	4.0
Adult leaves (Trial 2)	1	83.55	NE	mi72	NE	+0.93	6.5
Adult leaves (Trial 2)	1	150.10	138–157	g17311	PAB5 – pAtT32CX	−1.28	13.6
Adult leaves (Trial 2)	2	53.93	45–57	m220	er – ve096	+1.84	28.4
Adult leaves (Trial 2)	4	78.70	33–113	O6455	pCITf3 – ve031	+1.08	9.9
Adult leaves (Trial 1)	2	40.95	35–58	GPA1	O802F – mi277	+1.93	27.6
Adult leaves (Trial 1)	3	67.00	NE	g4117	NE	+0.94	8.4
Adult leaves (Trial 1)	4	115.61	NE	g3713	NE	−0.85	7.6

NE = not estimable.

and in my re-analysis of the Larkin et al. (1996) data (Table 2; Figure 2). Again, that minor QTL explained a similar amount of variation and had a similar additive effect in the two trials (Table 2).

The four different trials did yield different results for the remaining minor QTL (Table 2; Figure 2). Given the inability of the analysis to accurately localize those QTL, it is impossible to conclude that, for example, 4 of the 5 QTL identified on chromosome 4 or the 3 QTL on chromosome 3 are different (Figure 2). Clearly, two independent QTL were detected on chromosome 4 in the Larkin et al. (1996) trial (Figure 2). In another case, the QTL for adult leaf trichome density located on chromosome 4, has an additive effect that differs in sign in the two trials, suggesting that these are, in fact, different QTL (Table 2).

Discussion

Although much is known about the molecular genetic basis of trichome development in *A. thaliana*, less is known about the underlying genetic basis of continuous variation in trichome density: a trait known to be of adaptive importance. The density of leaf trichomes is a major determinant of herbivore damage in natural populations of *A. thaliana* (Mauricio, 1998). Herbivores have been shown to be a significant selective force on genetic variation for trichome density in natural populations of *A. thaliana* (Mauricio and Rausher, 1997).

In the present study, I investigated three questions related to understanding the genetic architecture of quantitative variation in trichome density in *A. thaliana*. The first and second questions focused on identifying QTL responsible for trichome density on juvenile leaves and adult leaves and asked whether those QTL changed with ontogeny. A considerable literature has demonstrated that a number of developmental changes occur during vegetative phase change in *A. thaliana* (e.g., Telfer et al., 1997), including changes in trichomes. I found dramatic differences in the mean trichome density in both juvenile and adult leaves between two parental lines of *A. thaliana* that

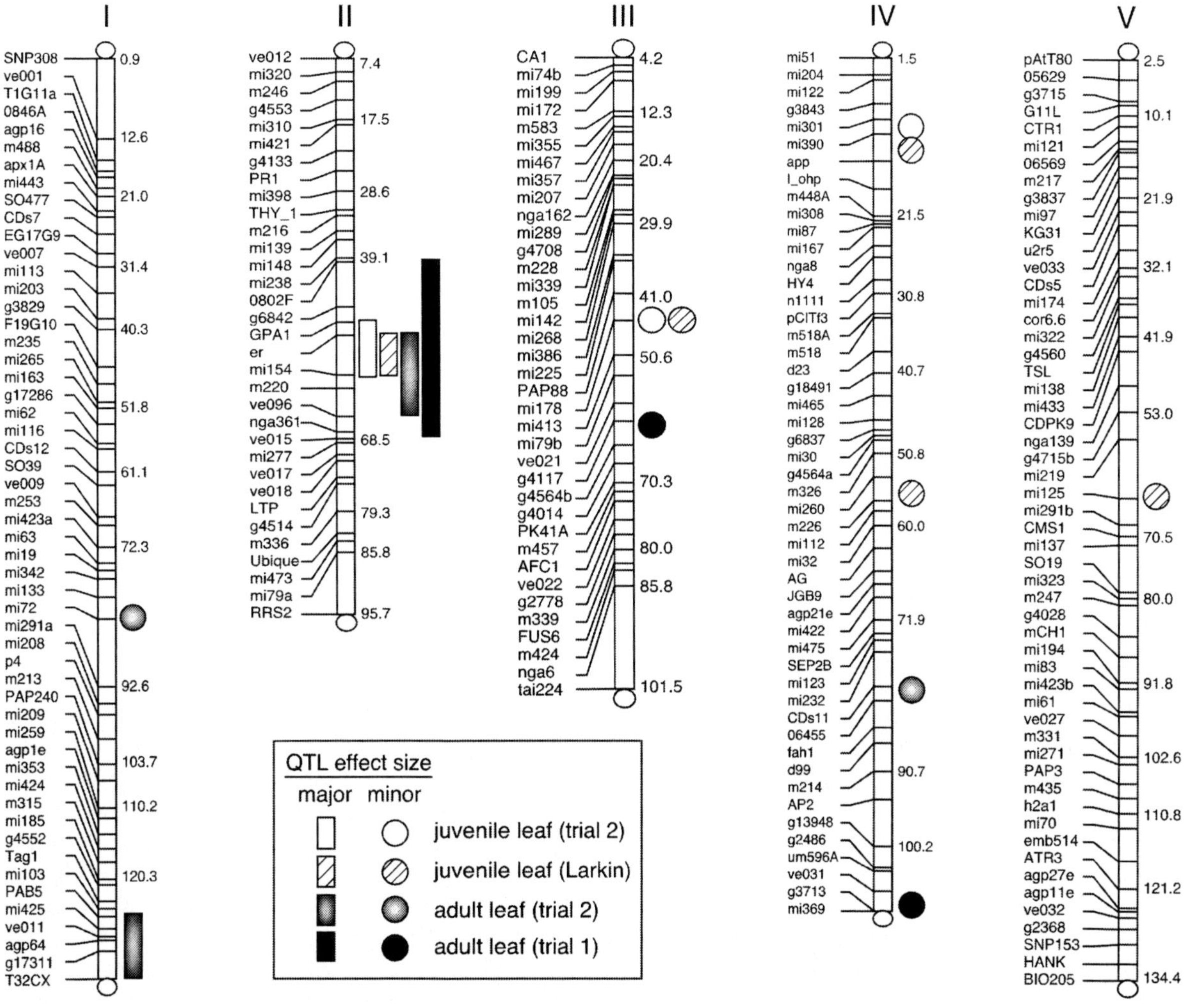

Figure 2. The five chromosomes of *A.thaliana* showing all QTL identified using multiple interval mapping. Markers used in the analysis are listed to the left of the chromosome and genetic distances in Kosambi centiMorgans are listed to the right. QTLs of major effect (explaining > 10% of the variance) are identified by bars. The length of the bar spans the markers included in the 2-LOD confidence interval. Minor QTL are indicated by circles to the right of the marker identified as being linked to the QTL. The analysis was unable to establish confidence intervals for most minor QTL and the entire linkage group on which the minor QTL is located should be considered as the confidence interval. The shading of the bar/circle indicates the trait and experimental trial from which the data were obtained.

allowed for mapping of QTL. In addition, I found that trichome density differed between juvenile and adult leaves with juvenile leaves tending to have higher trichome densities than adult leaves. On the surface, this finding contradicts the results of Martínez-Zapater et al. (1995) and Payne et al. (2000) who found that trichome number increased with age. Because they measured total trichome number and I measured trichome density (number of trichomes per unit area), our measures are not directly comparable.

Despite these ontogenetic differences, the most striking result of this study is that there were no consistent differences in the genetic architecture of trichome density measured on juvenile and adult leaves. In all cases, a single QTL on chromosome 2 explained much of the genetic variance. In juvenile leaves, this QTL explained approximately 70% of the variation. In adult leaves, the proportion of genetic variation explained was approximately 28%, although that is twice the variance of any other single QTL identified. A

QTL on chromosome 2 is clearly a major determinant of trichome density variation in both juvenile and adult leaves.

The QTL on chromosome 2 maps in the same rough location as another QTL first identified by Larkin et al. (1996) in juvenile leaves, which they called the *RTN* locus. Larkin et al. (1996) were able to specifically localize *RTN* to the interval on chromosome 2 between the *er* and the *m220* markers. Larkin et al. (1996) observed that the difference in trichome density between the "Columbia" and "Landsberg" parents was related to the duration of trichome development in the leaf primordia. In "Landsberg", trichome development ceases when the leaf primordia are about 500 μm long while in "Columbia", trichome production continues until even after the leaf primordia reach 700 μm in length (Larkin et al., 1996).

A number of QTL of minor effect seemed to be detected in leaves of all ages. Since I was generally unable to establish a confidence interval smaller than the entire length of the chromosome for minor QTL, the QTLs detected on chromosomes 3 and 4 are possibly located in the same region. Those minor QTL were identified from both juvenile and adult leaves.

Although the chromosome 2 QTL was the most significant QTL identified, there were some differences in the QTL detected for leaves of different ages. A major QTL, explaining almost 14% of the genetic variation for trichome density on adult leaves, was detected on the end of chromosome 1. This QTL was not detected in either of the trials on juvenile leaves. However, the inability to detect that same QTL in the adult leaves in trial 1 suggests that the identification of that QTL be considered tentative. Similarly, a minor QTL unique to juvenile leaf trichome density was detected on chromosome 5, but was not found in the juvenile leaf trial 2.

There were clear differences in the contribution of the environment to phenotypic variation in trichome density on leaves of different ages. The heritability of juvenile leaf trichome density was very high. In contrast, the heritability for adult leaf trichome density was much lower. This is not surprising considering the development of trichomes. Because trichome development ceases before the leaves are fully developed, a number of sources of environmental variation can be introduced in the time it takes for the leaves to fully develop and age.

Much is known of the molecular genetic basis of trichome development in *A. thaliana* since plant developmental biologists use trichomes as a model system for understanding pattern formation (Marks, 1997; Hülskamp and Schnittger, 1998; Hülskamp and Kirik, 2000; Szymanski et al., 2000). At least 24 distinct loci are required for normal trichome development and expression (Hülskamp et al., 1994; Marks, 1997). Seven loci, *GL1* (Marks and Feldmann, 1989; Herman and Marks, 1989; Larkin et al., 1993, 1994, 1999; Esch et al., 1994; Schnittger et al., 1998; Szymanski and Marks, 1998), *GL3* (Payne et al., 2000), *TTG* (Larkin et al., 1994, 1999), *GL2* (Rerie et al., 1994; Szymanski et al., 1998a), *TRY* (Schnittger et al., 1998; Szymanski and Marks, 1998), *CPC* (Wada et al., 1997) and *COT1* (Szymanski et al., 1998b) have been described that may play a role in the regulation of trichome density (Szymanski et al., 2000). The mutant alleles identified for *TTG* completely eliminate leaf trichomes, as do most of the alleles for *GL1*. However, at least one mutant allele of *GL1* (*gl1-2*) produces a plant with lower trichome density compared to the wild-type allele (Esch et al., 1994). The mutant alleles identified at the *GL3* locus produces plants with reduced trichome density (Payne et al., 2000). Mutant alleles identified at the four other loci have normal trichome densities, but have been functionally shown to play a role in trichome initiation.

Five of these loci have been genetically mapped (www.arabidopsis.org). The *GL1* locus has been definitively located on chromosome 3 between positions 48 and 49 cM. *GL1* appears on the sequence-based map as well as on genetic maps. The positions of *GL3*, *GL2*, *TTG* and *CPC* are less well localized (only listed on the classical map). *GL3* has been mapped to chromosome 5 at 53 cM. *GL2* is located on the bottom of chromosome 1. *TTG* is located on chromosome 5 at 28 cM. *CPC* has been mapped to chromosome 2 at 63 cM. Neither *TRY* nor *COT1* have been mapped. Given the positions, it is possible that *GL1* is the QTL I identified on chromosome 3 for both juvenile and adult trichome density. The QTL I identified for adult leaf trichome density on chromosome 1 may co-localize with *GL2*. Finally, the juvenile leaf trichome density QTL identified on chromosome 5 may co-localize with either *GL3* or *TTG*. The QTL located on chromosome 4 do not correspond to any known trichome density loci.

Obviously, given the resolution of QTL mapping, any attempt to identify a candidate gene from these data are preliminary and should be considered only as hypotheses for further investigation. Even in model organisms, the ability to move from QTL to gene is not trivial. In this study, the tightest confidence intervals around any major QTL extended between 6 and 23 cM. Even in the best of QTL studies, many QTL are defined by markers more than 10 cM apart. For example, the mean confidence interval around floral trait QTL in *A. thaliana* reported by Juenger et al. (2000) was 10.9 cM (range: 4–23). In *A. thaliana*, the estimated genetic map is 597 cM and the physical size is approximately 125 Megabases (Kaul et al., 2000). On average, there are 213 Kilobases of DNA and approximately 50 genes per cM in *A. thaliana* (Copenhaver et al., 1998). Thus, in a typical 10 cM interval, there are possibly 500 genes. Even if a genome project has identified each of the genes in that interval, proving that any particular gene is responsible for variation in a trait of interest is labor-intensive.

This study has relevance for the debate on the genetic basis of complex adaptive traits (Orr and Coyne 1992). Again, trichome density is known to be of significant adaptive value in natural populations of *A. thaliana* (Mauricio 1998; Mauricio and Rausher, 1997). Quantitative genetic studies of trichome density in *A. thaliana* support the hypothesis that this is a quantitative trait (Larkin et al., 1996). Fisher (1930) argued that many mutations of very small effect were responsible for adaptive evolution. Orr and Coyne (1992) argued that Fisher may have been premature in rejecting the hypothesis that genes of major phenotypic effect played a role in adaptation. My finding of a single QTL of major effect for a trait of known adaptive importance suggests that genes of major effect may play an important role in adaptation.

It has been argued that QTL of large phenotypic effect seen in studies of this kind are an artifact of the strong directional selection often used to create the phenotypically divergent parental lines that are used for mapping (Lande, 1983). Strong selection can fix alleles that normally segregate in the base population. In addition, artificial selection may create repeated bottlenecks through which only a sample of segregating alleles pass. Thus, fewer QTL will be able to be detected and the QTL that are eventually detected may explain an inflated portion of the phenotypic variance. As the parental lines used in this cross were not actively selected, at least not with respect to differences in trichome density, this criticism likely does not apply in this case.

The third question investigated in this study involved the variability in QTL analyses completed on a similar trait but performed at different times and in different labs. There has been some concern expressed in the literature about the repeatability of QTL studies (Mauricio, 2001b). Beavis (1994, 1998) summarized the results of a number of QTL mapping experiments on yield and height of maize, including replicate studies of the same crosses. Although the same QTL were detected across studies, some of the QTL detected were unique to each cross. Even the replicate studies did not detect the same QTL. In this paper, I measured adult trichome density on leaves from the same cross, but in independent experiments. I measured juvenile leaf trichome density and Larkin et al., (1996) measured the total number of trichomes on juvenile leaves. By and large, the similarities across the paired studies outweighed any differences. The means and heritabilities of both adult traits and both juvenile traits were very similar, even though the measures of juvenile leaf trichomes were distinctly different. And, both pairs of studies identified the same major QTL on chromosome 2. Certainly, there were differences detected within the paired trials. But, in all but one case (the QTL for adult leaf trichome density detected on chromosome 1 in only one trial) those involved QTL of minor effect.

A final caveat is that the QTL mapping approach is strictly limited to detecting the genetic variation segregating in the particular cross used. The cross I used in these experiments represents only a sample of the naturally segregating variation found in natural populations of *A. thaliana*. In order to better understand the nature of quantitative genetic variation, it would be extremely valuable to repeat these kinds of QTL studies using a much wider sample of parental accessions collected from natural populations.

Acknowledgements

I am indebted to Amy Bouck for many helpful discussions on the intricacies of QTL mapping. I thank Andy Tull and Michael Boyd in the Franklin College Plant Biology Greenhouses for their expert plant care. I greatly appreciate the considerable time Zhao–Bang Zeng spent with me answering questions relating to QTL mapping. John Larkin kindly provided the original data from his work on trichome number. I thank Michael Arnold, Amy Bouck and Russell Malmberg for their thoughtful comments on earlier drafts of the manuscript. This material is based upon work supported by the National Science Foundation under Grant No. 0129191.

References

Ågren, J. & D.W. Schemske, 1994. Evolution of trichome number in a naturalized population of *Brassica rapa*. Am. Nat. 143: 1–13.

Basten, C.J., B.S. Weir & Z.-B. Zeng, 1994. Zmap–a QTL cartographer, pp. 65–66 in Proceedings of the 5th World Congress on Genetics Applied to Livestock Production: Computing Strategies and Software, edited by C. Smith, J.S. Gavora, B. Benkel, J. Chesnais, W. Fairfull, J.P. Gibson, B.W. Kennedy and E.B. Burnside. Organizing Committee, 5th World Congress on Genetics Applied to Livestock Production, Guelph, Ontario, Canada.

Basten, C.J., B.S. Weir & Z.-B. Zeng, 2004. QTL Cartographer, Version 2.0. Department of Statistics, North Carolina State University, Raleigh, NC, USA.

Barton, N.H. & P.D. Keightley, 2002. Understanding quantitative genetic variation. Nat. Rev. Genet. 3: 11–21

Beavis, W.D., 1994. The power and deceit of QTL experiments: lessons from comparative QTL studies, pp. 255–266 in Proc. Corn and Sorghum Industry Research Conference, American Seed Trade Association, Washington, DC, USA.

Beavis, W.D., 1998. QTL analyses: power, precision, and accuracy, pp. 145–162 in Molecular Dissection of Complex Traits, edited by A.H. Paterson. CRC Press, Boca Raton, FL, USA.

Coley, P.D., 1980. Effects of leaf age and plant life history patterns on herbivory. Nature 284: 545–546.

Copenhaver, G.P., Browne, W.E. & D. Preuss, 1998. Assaying genome-wide recombination and centromere functions with *Arabidopsis* tetrads. Proc. Natl. Acad. Sci. USA 9: 247–252.

Doebley, J., A. Stec & L. Hubbard, 1997. The evolution of apical dominance in maize. Nature 386: 485–488.

Elle, E.,N.M. van Dam & J. D. Hare, 1999. Cost of glandular trichomes, a 'resistance' character in *Datura wrightii* Regel (Solanaceae). Evolution 53: 22–35.

Esch, J.J., D.G. Oppenheimer & M.D. Marks, 1994. Characterization of a weak allele of the *GL1* gene of *Arabidopsis thaliana*. Plant Mol. Biol. 24: 203–207.

Fisher, R.A., 1930. The Genetical Theory of Natural Selection. Clarendon Press, Oxford, UK.

Frary, A., T.C. Nesbitt, A. Frary, S. Grandillo, E. van der Knaap, B. Cong, J. Liu, J. Meller, R. Elber, K.B. Alpert & S.D. Tanksley, 2000. *fw2.2*: a quantitative trait locus key to the evolution of tomato fruit size. Science 289: 85–88.

Herman, P.L. & M.D. Marks, 1989. Trichome development in *Arabidopsis thaliana*. II. Isolation and complementation of the *GLABROUS1* gene. Plant Cell 1: 1051–1055.

Houle, D., 1992. Comparing evolvability and variability of quantitative traits. Genetics 130: 195–204

Hülskamp, M., S. Misera & G. Jurgens, 1994. Genetic dissection of trichome cell development in *Arabidopsis*. Cell 76: 555–566.

Hülskamp, M. & A. Schnittger, 1998. Spatial regulation of trichome formation in *Arabidopsis thaliana*. Semin. Cell Dev. Biol. 9: 213–220.

Hülskamp, M. & V. Kirik, 2000. Trichome differentiation and morphogenesis in *Arabidopsis*, pp. 237–260 in Advances in Botanical Research incorporating Advances in Plant Pathology, Vol. 31, Plant Trichomes edited by D.L. Hallahan & J.C. Gray. Academic Press, San Diego, CA, USA.

Janzen, D.H., 1979. New horizons in the biology of plant defenses, pp. 331–350 in Herbivores: Their Interaction with Secondary Plant Metabolites, edited by D.H. Janzen & G.A. Rosenthal. Academic Press, NY, USA.

Jiang, C. & Z.-B. Zeng, 1995. Multiple trait analysis of genetic mapping for quantitative trait loci. Genetics 140: 1111–1127.

Johanson, U., J. West, C. Lister, S. Michaels, R. Amasino & C. Dean, 2000. Molecular analysis of FRIGIDA, a major determinant of natural variation in A*rabidopsis* flowering time. Science 290: 344–347.

Johnson, H.B., 1975. Plant pubescence: an ecological perspective. Bot. Rev. 41: 233–258.

Juenger, T., M. Purugganan & T.F.C. Mackay, 2000. Quantitative trait loci for floral morphology in *Arabidopsis thaliana*. Genetics 156: 1379–1392.

Kao, C.-H. & Z.-B. Zeng, 1997. General formulae for obtaining the MLEs and the asymptotic variance-covariance matrix in mapping quantitative trait loci. Biometrics 53: 653–665.

Kao, C.-H., Z.-B. Zeng & R.D. Teasdale, 1999. Multiple interval mapping for quantitative trait loci. Genetics 152: 1203–1216.

Karban R. & J.S. Thaler, 1999. Plant phase change and resistance to herbivory. Ecology 80: 510–517.

Kaul, S. et al., 2000. Analysis of the genome sequence of the flowering plant *Arabidopsis thaliana*. Nature 408: 796–815.

Krischik, V.A. & R.F. Denno, 1983. Individual, population, and geographic patterns in plant defense, pp. 463–512 in Variable Plants and Herbivores in Natural and Managed Systems, edited by R.F. Denno & M.S. McClure. Academic Press, NY, USA.

Lande, R., 1983. The response to selection on major and minor mutations affecting a metrical trait. Heredity 50: 47–65.

Lander, E.S. & D. Botstein, 1989. Mapping Mendelian factors underlying quantitative traits using RFLP linkage maps. Genetics 121: 185–199.

Lander, E., P. Green, J. Abrahamson, A. Barlow, M. Daley, S. Lincoln, L. Newbury, 1987. MAPMAKER: an interactive computer package for constructing primary genetic linkage maps of experimental and natural populations. Genomics 1: 174–181.

Larkin, J.C., D.G. Oppenheimer, S. Pollock & M.D. Marks, 1993. *Arabidopsis GLABROUS1* gene requires downstream sequences for function. Plant Cell 5: 1739–1748.

Larkin, J.C., D.G. Oppenheimer, A.M. Lloyd, E.T. Paparozzi, & M.D. Marks, 1994. Roles of the GLABROUS1 and TRANSPARENT TESTA GLABRA genes in *Arabidopsis* trichome development. Plant Cell 6: 1065–1076.

Larkin, J.C., N. Young, M. Prigge & M.D. Marks, 1996. The control of trichome spacing and number in *Arabidopsis*. Development 122: 997–1005.

Larkin, J.C., J.D. Walker, A.C. Bolognesi-Winfield, J.C. Gray, & A.R. Walker, 1999. Allele-specific interactions between *ttg* and *gl1* during trichome development in *Arabidopsis thaliana*. Genetics 151: 1591–1604.

Lawrence R., B.M. Potts & T.G. Whitham. 2003. Relative importance of plant ontogeny, host genetic variation, and leaf age for a common herbivore. Ecology 84: 1171–1178.

Lawson, E. & R.S. Poethig, 1995. Shoot development in plants: time for a change. Trends Genet. 11: 263–268.

Levin, D.A., 1973. The role of trichomes in plant defenses. Q. Rev. Biol. 48: 3–15.

Lister, C. & C. Dean, 1993. Recombinant inbred lines for mapping RFLP and phenotypic markers in *Arabidopsis thaliana*. Plant J. 4: 745–750.

Lynch, M. & B. Walsh, 1998. Genetics and Analysis of Quantitative Traits. Sinauer Associates, Sunderland, MA, USA.

Mackay, T.F.C., 2001. Quantitative trait loci in *Drosophila*. Nat. Rev. Genet. 2: 11–20.

Marks, M.D., 1997. Molecular genetic analysis of trichome development in *Arabidopsis*. Annu. Rev. Plant Physiol. Plant Mol. Biol. 48: 137–163.

Marks, M.D. & K.A. Feldmann, 1989. Trichome development in *Arabidopsis thaliana*. I. T-DNA tagging of the *GLABROUS1* gene. Plant Cell 1: 1043–1050.

Martínez-Zapater, J.M., J.A. Jarillo, M. Cruz-Alvarez, M. Roldán & J. Salinas, 1995. *Arabidopsis* late-flowering *fve* mutants are affected in both vegetative and reproductive development. Plant J. 7: 543–551.

Mauricio, R., 1998. Costs of resistance to natural enemies in field populations of the annual plant, *Arabidopsis thaliana*. Am. Nat. 151: 20–28.

Mauricio, R., 2001a. An ecological genetic approach to the study of coevolution. Am. Zool. 41: 916–927.

Mauricio, R., 2001b. Mapping quantitative trait loci in plants: uses and caveats for evolutionary biology. Nat. Rev. Genet. 2: 370–381.

Mauricio, R., M.D. Bowers & F.A. Bazzaz, 1993. Pattern of leaf damage affects fitness of the annual plant, *Raphanus sativus* (Brassicaceae). Ecology 74: 2066–2071.

Mauricio, R. & M.D. Rausher, 1997. Experimental manipulation of putative selective agents provides evidence for the role of natural enemies in the evolution of plant defense. Evolution 51: 1435–1444.

Orr, H.A. & J.A. Coyne, 1992. The genetics of adaptation: a reassessment. Am. Nat. 140: 725–742.

Paterson, A.H., S. Damon, J.D. Hewitt, D. Zamir, H.D. Rabinowitch, S.E. Lincoln, E.S. Lander & S.D. Tanksley, 1991. Mendelian factors underlying quantitative traits in tomato: comparison across species, generations, and environments. Genetics 127: 181–197.

Payne, C.T., F. Zhang, & A.M. Lloyd, 2000. *GL3* encodes a bHLH protein that regulates trichome development in *Arabidopsis* through interaction with *GL1* and *TTG1*. Genetics 156: 1349–1362.

Poethig, R.S., 1990. Phase change and the regulation of shoot morphogenesis in plants. Science 250: 923–930

Poethig, R.S., 1997. Leaf morphogenesis in flowering plants. Plant Cell 9: 1077–1087.

Poethig R.S., 2003. Phase change and the regulation of developmental timing in plants. Science 301: 334–336.

Rerie, W.G., K.A. Feldmann, & M.D. Marks, 1994. The *GLABRA2* gene encodes a homeo domain protein required for normal trichome development in *Arabidopsis*. Genes Dev. 8: 1388–1399.

Schnittger, A., G. Jurgens, & M. Hülskamp, 1998. Tissue layer and organ specificity of trichome formation are regulated by *GLABRA1* and *TRIPTYCHON* in *Arabidopsis*. Development 125: 2283–2289.

Stinchcombe, J.R., 2002. Fitness consequences of cotyledon and mature-leaf damage in the ivyleaf morning glory. Oecologia 131: 220–226.

Szymanski, D.B., R.A. Jilk, S.M. Pollock, & M.D. Marks, 1998a. Control of *GL2* expression in *Arabidopsis* leaves and trichomes. Development 125: 1161–1171.

Szymanski, D.B., D.A. Klis, J.C. Larkin, & M.D. Marks, 1998b. *cot1*: a regulator of *Arabidopsis* trichome initiation. Genetics 149: 565–577.

Szymanski, D.B., & M.D. Marks, 1998. *GLABROUS1* overexpression and *TRIPTYCHON* alter the cell cycle and trichome cell fate in *Arabidopsis*. Plant Cell 10: 2047–2062.

Szymanski, D.B., A.M. Lloyd, & M.D. Marks, 2000. Progress in the molecular genetic analysis of trichome initiation and morphogenesis in *Arabidopsis*. Trends Plant Sci. 5: 214–219.

van Ooijen, J., 1992. Accuracy of mapping quantitative trait loci in autogamous species. Theor. Appl. Genet. 84: 803–811.

Wada, T., T. Tachibana, Y. Shimura, & K. Okada, 1997. Epidermal cell differentiation in *Arabidopsis* determined by a *Myb* homolog, *CPC*. Science 277: 1113–1116.

Zeng, Z.-B., 1993. Theoretical basis for separation of multiple linked gene effects in mapping quantitative trait loci. Proc. Natl. Acad. Sci. USA 90: 10972–10976.

Zeng, Z.-B., 1994. Precision mapping of quantitative trait loci. Genetics 136: 1457–1468.

Zeng, Z.-B., C. Kao, & C.J. Basten, 1999. Estimating the genetic architecture of quantitative traits. Genet. Res. 74: 279–289.

Genetica (2005) 123: 87–105 © Springer 2005

Epistasis and genotype-environment interaction for quantitative trait loci affecting flowering time in *Arabidopsis thaliana*

Thomas E. Juenger[1], Sáunak Sen[2], Kirk A. Stowe[3] & Ellen L. Simms[3]
[1]*University of Texas at Austin, Section of Integrative Biology, Austin, TX 78712, USA (Phone: +512-232-5751; Fax: +512-471-3878; E-mail: tjuenger@mail.utexas.edu);* [2]*University of California, San Francisco, Department of Epidemiology and Biostatistics, San Francisco, CA 94143-0560, USA;* [3]*University of California, Berkeley, Department of Integrative Biology, Berkeley, CA 94720, USA*

Received 22 October 2002 Accepted 14 May 2003

Key words: Arabidopsis thaliana, epistasis, flowering time, gene–environment interaction, non-additive gene action, phenology, quantitative genetics

Abstract

A major goal of evolutionary biology is to understand the genetic architecture of the complex quantitative traits that may lead to adaptations in natural populations. Of particular relevance is the evaluation of the frequency and magnitude of epistasis (gene–gene and gene–environment interaction) as it plays a controversial role in models of adaptation within and among populations. Here, we explore the genetic basis of flowering time in *Arabidopsis thaliana* using a series of quantitative trait loci (QTL) mapping experiments with two recombinant inbred line (RIL) mapping populations [Columbia (Col) x Landsberg *erecta* (L*er*), L*er* x Cape Verde Islands (Cvi)]. We focus on the response of RILs to a series of environmental conditions including drought stress, leaf damage, and apical damage. These data were explicitly evaluated for the presence of epistasis using Bayesian based multiple-QTL genome scans. Overall, we mapped fourteen QTL affecting flowering time. We detected two significant QTL–QTL interactions and several QTL–environment interactions for flowering time in the L*er* x Cvi population. QTL–environment interactions were due to environmentally induced changes in the magnitude of QTL effects and their interactions across environments – we did not detect antagonistic pleiotropy. We found no evidence for QTL interactions in the L*er* x Col population. We evaluate these results in the context of several other studies of flowering time in *Arabidopsis thaliana* and adaptive evolution in natural populations.

Introduction

A central goal of evolutionary biology is to elucidate processes that constrain or facilitate adaptive phenotypic change. Evolutionary biologists have traditionally used either single locus population genetic or quantitative genetic theory to understand the importance of selection, genetic architecture, mutation, recombination, and drift on phenotypic evolution (Lynch & Walsh, 1998). While great theoretical progress has been made in this regard (Barton & Turelli, 1989), many empirical questions remain concerning the details underlying the genetics of adaptation (Barton & Turelli, 1989; Orr & Coyne, 1992; Orr, 1998). In particular, accurate reconstructions or predictions of adaptive evolution based on theory will ultimately require a more detailed understanding of both the function and genetic basis of variation in traits within nature (Mitchell-Olds & Rutledge, 1989). Consequently, a current empirical challenge is to elucidate the genetic architecture, including the number, magnitude of effect, and mode of gene action of the loci controlling ecologically important traits.

Epistasis or gene interaction is of particular interest as it plays a controversial role in the theory of adaptive evolution within and among populations (Wade, 2000). Epistasis occurs when differences in the phenotypic values of an allele at one locus are dependent on differences in specific alleles at other loci (gene–gene interaction) or across environmental heterogeneity (gene–environment interaction). These differences manifest as changes in the magnitude or order of allelic values contingent on the genetic or environmental background. Epistasis is thought to be important in several areas of evolutionary biology including speciation, developmental canalization, phenotypic plasticity, inbreeding depression, the evolution of sex, genome evolution, the maintenance of genetic diversity, and adaptive evolution via Wright's shifting balance theory (Fenster et al., 1998; Wolf et al., 2000; Wade et al., 2001). Given the broad interest in the role of epistasis in the evolutionary process (Wolf et al., 2000) its evaluation is a critical aspect of modern quantitative genetics (Lynch & Walsh, 1998; Zeng et al., 1999).

Gene interactions are commonly detected in molecular genetic studies that utilize loss-of-function mutants to resolve molecular pathways. Much less is known about interactions among naturally occurring alleles and how these interactions contribute to the partitioning of overall phenotypic variation. Historically, epistasis has been studied in a quantitative genetics framework using inbred line crosses aimed at detecting departures from the predictions of strictly linear additive models. Unfortunately, these tools are of limited value as they are restricted to the evaluation of composite directional non-additive effects summed across entire genomes (Lynch & Walsh, 1998). More recently, quantitative trait locus (QTL) mapping methods have been utilized to explore QTL–QTL and QTL–environment interactions in experimental populations (Mackay, 1995; Routman & Cheverud, 1997; Gurganus et al., 1998; Lynch & Walsh, 1998; Vieira et al., 2000).

In its simplest form, QTL mapping is a search for statistical associations, due to linkage disequilibrium, between quantitative phenotypic variation and genetic marker alleles segregating in an experimental population. Although this technique is not new, recent advances in genetic markers, high-throughput genotyping, and statistical techniques have greatly improved the power and resolution of the approach. Most QTL mapping efforts have sought phenotypic associations using single QTL models and have explicitly ignored interactions. Several QTL studies have progressed to the secondary testing of interactions between QTL after first locating them through their strictly additive effects. Although this method has revealed numerous QTL–QTL interactions, it is clearly limited in scope and will necessarily fail to detect interacting pairs of loci that lack strictly additive effects (Wade, 1992; Cheverud, 2000; Sen & Churchill, 2001). Finally, the accuracy with which the 'real' genomic positions of QTL can be located depends critically on the development of an accurate description of the genetic model (Zeng et al., 1999). QTL models failing to incorporate complex interactions when they occur can produce spurious or inappropriate QTL localization and confidence intervals. Here, we explore the genetic architecture of flowering time using multiple-QTL genome scans that incorporate pairwise interactions (Sen & Churchill, 2001).

Timing of reproduction is an important component of life–history variation in many plants and animals. For example, theory and empirical data suggest that the flowering phenology of annual plants can influence a variety of ecological factors including interactions with other species (e.g., competitors, pollinators, natural enemies), the matching of vegetative growth with seasonal pulses in soil nutrients and moisture, and the completion of fruit set by the close of the growing season. These factors can have dramatic impacts on plant fitness.

A. thaliana is a small crucifer with a vegetative growth period that produces a leafy rosette followed by the bolting of an indeterminate reproductive shoot. In nature, *A. thaliana* populations exhibit a winter annual life–history (with an over-wintering rosette stage), a spring annual life–history (with over-wintering seeds) or a mixed strategy (Donohue, 2002). Life-history variation and within-season flowering time are probably both important ecological traits in *Arabidopsis* populations. For instance, several studies have documented natural selection imposed on *A. thaliana* flowering time (or related traits such as bolting time) within a reproductive season due to variation in seedling density (Dorn et al., 2000), shading (Scheiner & Callahan, 1999; Dorn et al., 2000; Callahan & Pigliucci, 2002), timing of germination (Donohue, 2002), and season length or vernalization (Pigliucci & Marlow, 2001). Our focus is on the genetic architecture of within-season flowering time for ecotypes that exhibit spring annual life histories.

Flowering time in *Arabidopsis* is well studied by both mutant/molecular genetic methods and by quantitative genetic analyses of natural allelic variation (Napp-Zinn, 1985; Koornneef et al., 1998; Levy & Dean, 1998). Classic mutant screens and transgenic analyses have revealed at least 54 loci that affect flowering time (Levy & Dean, 1998) with many interactions among loci and with environmental cues (Sanda & Amasino, 1996). These loci have been organized into a flowering time scheme composed of independent vernalization and photoperiod induced pathways and an autonomous developmental program. Genes involved in the vernalization and photoperiod pathways are thought to ensure flowering under appropriate environmental conditions. Two such loci, *FRIGIDA* (*FRI*) and *CRYPTOCHROME-2* (*CRY2*), are polymorphic in natural populations and have been associated with natural variation in seasonal life-history and flowering time (Johanson et al., 2000; El-Assal et al., 2001). For example, the developmental switch between winter and spring annual life-histories is controlled to a large extent by the interaction of natural loss-of-function alleles at *FRI* with alleles at *Flowering Locus C* (*FLC*) (Johanson et al., 2000). In addition, several flowering time QTL have been mapped (Kowalski et al., 1994; Clark et al., 1995; Mitchell-Olds, 1996; Kuittinen et al., 1997; Stratton, 1998; Alonso-Blanco et al., 1998a; Ungerer et al., 2002). A number of these QTL interact with cold vernalization treatments and photoperiod and light quality conditions (Clarke et al., 1995; Stratton, 1998; Alonso-Blanco et al., 1998a). We extend this previous work with a detailed inspection of the role of epistasis and environmental stress in phenological variation.

In this paper, we use flowering time in *A. thaliana* to explore the genetic architecture of a classic complex trait. Two RIL mapping populations were screened to ask, (1) which genomic regions control phenotypic variation in flowering time among early flowering Col, L*er*, and Cvi ecotypes? (2) do these QTL interact with either environmental variation (drought stress, leaf damage, apical damage) or each other? (3) do these QTL overlap with known candidate genes? These results are discussed in terms of the role of genetic architecture in the adaptive evolution of flowering phenology.

Materials and methods

Recombinant inbred lines

In our experiments, we used 96 recombinant inbred lines (RILs) generated from a cross between Columbia (Col) and Landsberg *erecta* (L*er*) ecotypes (Lister & Dean, 1993) and 164 RILs generated from a cross between L*er* and Cape Verde Islands (Cvi) (Alonso-Blanco et al., 1998b) ecotypes to map QTL. In both populations, F_1 progeny from an initial cross were taken through eight generations of selfing via single seed descent to produce nearly homozygous lines. We constructed linkage maps for each cross using a subset of 169 and 111 markers in the L*er* x Col and L*er* x Cvi populations, respectively. The RIL genotype at each marker locus was obtained from the published data available from the *Arabidopsis* stock center (http://arabidopsis.org). In both cases, maps were constructed using markers that were genotyped in at least 80% of the sampled lines. The map position of each marker (*d* cM) was estimated from the observed recombination frequencies (*r*) using the Kosambi mapping function as implemented by the software MapMaker 3.0 (Lander et al., 1987). These analyses provided a unique position for each marker which did not differ in order from the published *Arabidopsis* linkage maps.

Experimental design

We utilized three independent factorial experiments to investigate the flowering time response of *Arabidopsis* RILs to drought stress, leaf damage, and apical damage. In each case, replicate plants were grown under standard greenhouse conditions using Promix BT potting soil^{TM} and 115 ml Conetainer^{TM} pots (Stuewe & Sons, www.stuewe.com). Individual Conetainers^{TM} were racked in 2-ft. x 1-ft trays at half the possible density (49 plants per tray – skipping every other position). Plants experienced long-day photoperiod conditions (16L/8D) provided by 1000 watt (HID) supplemental overhead lighting. Greenhouse temperature was maintained at approx. 65–70°F with a standard evaporative pad and fan cooling system during the day. Several seeds were initially planted in each Conetainer^{TM} and subsequently thinned to a single replicate individual at the first true leaf stage. Seeds and rosettes did not receive cold vernalization or photoperiod treatments to induce germination or flowering. Plants in each experiment received several applications of 0.5x concentration Hoagland's solution as a fertilizer supplement. Aphid pests were controlled using pesticide applications, although these treatments were rarely needed.

For the drought stress experiment, we used a split-plot experiment with whole-plots arranged in a completely randomized block design (CRBD) (Littel et al., 1996). Our split-plot design involved two experimental factors, irrigation and RIL. There were 98 RIL, all derived from the L*er* x Col mapping population. The two levels of irrigation treatment were (1) flooded (liberally watered) and (2) drought (restricted water), manipulated at the whole-plot level. The restricted water treatment was applied by allowing treatment plants to exhibit substantial wilting across all whole-plots before each watering. A whole-plot corresponded to two Container^{TM} racks (98 plants). Each block contained two whole-plots (196 plants), one for each level of the irrigation treatment. Each RIL was replicated once in each whole-plot. Overall, 3920 plants were evaluated for responses to the irrigation treatment (2 treatments (whole-plot units) x 98 RIL (subplot units) x 20 blocks = 3920). This design provides more precise information about variation among RILs than the effect of the irrigation treatment, but considerably simplified the application of the watering treatment. Four blocks were harvested prior to flowering (to evaluate patterns of resource allocation) and therefore data on flowering time is restricted to 16 whole-plots (~3136 plants). Date of first flowering was recorded by daily inspection of the experimental plants and scored upon the observation of a single open flower bud – flowering time was measured on a scale that set a value of one to represent the earliest flowering individuals in the population. This experiment was conducted from Nov. 1999 to Jan. 2000.

For the leaf damage experiment, we used a factorial randomized complete block design involving two experimental factors, leaf damage and RIL. Again, we used 98 RILs, all derived from the L*er* x Col mapping population. Four adjacently arranged Conetainer^{TM} racks were considered a spatial block (392 plants). Each block contained four replicate plants from each RIL randomly and evenly split into either a control treatment or a 50% rosette-leaf damage manipulation. Leaf damage was imposed on individual plants by randomly smashing half of the available rosette leaves using small-needle nose pliers on all treatment plants on a single arbitrarily chosen day (average number of rosette leaves on the day of treatment: L*er*, 7.5; Col, 10.4). We utilized artificial damage to simulate the insect herbivory experienced by *Arabidopsis* in natural populations (Mauricio & Rausher, 1997). Overall 3920 plants were evaluated for response to the leaf damage manipulation (98 RIL x 20 replicates x 2 treatment levels = 3,920 arrayed across 20 blocks). Date of flowering was recorded as described above. This experiment was conducted from Feb. to April 2000.

For the apical damage experiment, we used a factorial randomized incomplete block experimental design involving two experimental factors, an apical damage treatment and RIL. Here, we used 164 RILs, all derived from the L*er* x Cvi mapping population. We randomly and evenly assigned twelve replicate plants from each RIL to a control and twelve to the artificial clipping treatment. The clipping treatment was applied to individual plants by removing the bolting inflorescence stalk on the day of first flowering using small sharp scissors. We utilized experimental clipping as a proxy for the small mammal herbivory experienced by *Arabidopsis* in experimental populations grown under field conditions (C. Weinig, personal communication). Replicate plants were randomly arrayed across individual Conetainer^{TM} trays each containing 49 plants – we considered each tray an incomplete block. Overall, 3936 plants were evaluated for response to apical damage (164 RI lines x 12 replicates x 2 treatment levels = 3936 arrayed across 81 trays). Date of flowering was recorded as above for plants in the control treatment and as the date of the first flower produced from regrowth branches in the apical damage treatment. The experiment was conducted from June to August 2000.

Each experiment was analyzed using PROC MIXED (Little et al., 1996) with an appropriate linear mixed model considering RIL, RIL x treatment interaction, and spatial blocking as random factors and the experimental treatment as a fixed factor. Flowering time data were approximately normally distributed in both experiments using L*er* x Col RILs; in these experiments we analyzed the raw data scores. Flowering time was slightly skewed in the L*er* x Cvi population; in the experiment using these RILs, we performed a log(1 + flowering date) transformation and analyzed these values. Since the L*er* x Cvi population was constructed from reciprocal crosses (Alonso-Blanco et al., 1998b), we tested for cytoplasmic effects by nesting RIL within cytoplasm. We found no evidence of cytoplasmic effects on flowering time, so this term was removed from the analysis. In each analysis, the variance components associated with the random effects were estimated using restricted maximum likelihood (REML) and assessments of significance were based on likelihood ratio tests (Littel et al., 1996). We obtained empirical *B*est *L*inear *U*nbiased *P*redictors (BLUPS) (Littel et al., 1996) associated with the random effects from each analysis and considered these estimates to be breeding values for each RIL (Lynch & Walsh, 1998). All subsequent QTL analyses were preformed on BLUPS. In each analysis, the residuals were normally distributed and did not exhibit heteroscedasticity.

We estimated broad-sense heritability by computing the ratio V_G/V_P, where V_G equals the among-RIL variance component and V_P equals the total phenotypic variance for flowering time. We estimated this value in each environment by conducting the above statistical analysis separately for each fixed treatment factor in every experiment. In addition, we calculated the coefficient of genetic variation (CV_G) as $(100\sqrt{V_G})/\overline{X}$ for each trait, where $\overline{X}$ is the mean. We estimated genetic correlations (r_G) among flowering times measured in the different treatments as the standard Pearson product-moment pairwise correlation between the flowering time BLUPs estimated in each treatment. The significance of each genetic correlation was determined using a *t*-test after a *Z* transformation of the correlation coefficient.

QTL analyses

We mapped flowering time QTL using the multiple-QTL framework presented by Sen & Churchill (2001). This method relies on a Bayesian perspective and the use of a Monte Carlo imputation algorithm to simulate multiple versions of complete genotype information on a dense genome-wide grid. This grid is scanned using both one and two QTL models at each position across the genome and evidence for a QTL or a QTL–QTL interaction is determined using a robust 2-dimensional permutation test. Sen and Churchill (2001) describe the imputed genotypes from their analysis as 'pseudomarkers' and therefore named their analytical software *Pseudomarker*. Imputation and the generation of pseudomarkers is an alternative approach to commonly used interval mapping (Lander & Botstein, 1989) and expectation-maximization (EM) methods.

Initially, we used a simplified *Pseudomarker* mapping strategy appropriate to the characteristics of these mapping populations. Specifically, a marker regression approach was used after imputing missing marker genotypes with a single Monte Carlo imputation. Evidence for a QTL was quantified by the sum of squares of residuals from regressing the phenotype on the genotypes at each marker. Marker regression closely approximated analyses based on a more densely imputed marker grid due to the high density of markers genotyped in these populations (average intermarker distance: L*er* x Col, 2.9 ± 1.70 cM; L*er* x Cvi, 4.4 ± 2.23 cM) (Juenger, unpublished analysis) while substantially reducing the computational demands of the analyses. QTL analyses were performed on BLUPS for each RIL estimated in each environmental treatment (wet or dry; control or leaf damage; control or apical damage).

Using cofactors to control for linked and unlinked genetic variation has produced a remarkable improvement in the accuracy and precision of QTL mapping analyses (Lynch & Walsh, 1998). These methods involve the development of complex models that test for a QTL at a particular genomic location while simultaneously controlling for other existing QTL. We utilized a model building strategy incorporating initial genome scans followed by subsequent scans that included cofactors. The first step in model building was to perform one- and two-dimensional scans at the genotyped markers. These were used to suggest a small number of two-QTL models using the following steps:

1. *Marker pair detection.* We detected interesting pairs of loci by comparing a full 2-QTL model with interaction (H_{full}) to the null model of no QTL (H_{null}), for all pairs of loci across the genome:

$$H_{null;}\ y = u + error$$

versus

$$H_{full};y = u + QTL_1 + QTL_2 + QTL_1 * QTL_2 + error$$

We established the genome-wide significance of H_{full} by permutation and an empirically derived threshhold corresponding to $P = 0.05$ (average LOD score = 5.80). If a pair of loci was deemed significant, we conducted two subsequent tests.

2. *Test for interaction.* We compared the full model (H_{full}) to an additive model ($H_{additive}$):

$$H_{full};y = u + QTL_1 + QTL_2 + QTL_1 * QTL_2 + error$$

versus

$$H_{additive};y = u + QTL_1 + QTL_2 + error$$

Significance was determined using a genome-wide threshold for the interaction test by permutation with an empirically derived threshold corresponding to $P = 0.05$ (average LOD score = 4.60).

3. *Test for 'coat-tail' effect.* If we found no evidence of interaction in Step 2, we compared the two locus additive model to each of the single QTL models (H_1 & H_2):

$$H_{additive};y = u + QTL_1 + QTL_2 + error$$

Versus

$$H_{1;}y = u + QTL_1 + error,$$

$$H_{2;}y = u + QTL_2 + error.$$

This comparison was done at the significance level corresponding to the permutation threshold for a one-dimensional scan (average LOD score = 2.62). This step avoided "coat-tail" effects in which the significance of one QTL may carry along another locus to produce a significant pair.

4. *Model pruning.* The marker pairs that were selected using the steps outlined above were combined into a large multiple-QTL model that was pruned by backward selection using a Type III analysis with PROC MIXED in SAS. The marker-pair selection method was then repeated, this time conditioning on the loci that were found at the end of the most recent iteration of the model pruning step. We repeated the marker pair selection and model pruning steps until we could add no more loci to the model. This model construction involved two cycles in all of the analyses, except in the control treatment of the experiment involving the L*er* x Cvi population, which necessitated three cycles. Note that this model building strategy allows the detection of interacting QTL even in the absence of additive effects.

We also performed a secondary fine-scale analysis of three particularly interesting linkage groups (Chromosome I, II, and V) in the L*er* x Cvi population This focus was motivated by two observations: first, two QTL–QTL interactions detected in our initial scans were located near moderate-sized gaps in the linkage map (top of chromosome I, top of chromosome V) and, second, a relatively broad peak (with two adjacent significant markers) was located on chromosome II. We used a series of 100 Monte Carlo imputations to estimate the missing marker data at genotyped locations as well as to infer the genotype of 'pseudomarkers' at 3 cM intervals on these linkage groups. Here, the residual sum of squares corresponding to a particular model was calculated by averaging the residual sum of squares over the imputations. For technical reasons, the average is not a simple arithmetic mean (see Appendices C and F, Sen & Churchill, 2001).

Our search for epistasis is based solely on the linear additive model and contributions of gene interaction to the interaction variance. We acknowledge that there are alternative definitions of epistasis and alternative partitioning that could be utilized in a search for gene interaction (Cheverud & Routman, 1995; Routman & Cheverud, 1997; Cheverud, 2000). We leave these analyses to future explorations of the data. We estimated the additive effect of each QTL on flowering time as half the difference in the phenotypic means for the two homozygous genotypes at a locus. The sign of the additive effect corresponds to the direction of the effect of alleles from the Col or the Cvi parent: positive values indicate that alleles from these parents slowed flowering while negative values indicate that alleles accelerated flowering. We estimated the proportion of the total genetic variance (%V_G) explained by each QTL using two methods. In the first, we estimated the percent total genetic variance explained by a QTL by calculating %$V_G = 2p(1 - p)a^2$, where a corresponds to the additive effect and p is the marker frequency (Falconer & Mackay,1998). This statistic assumes additivity and tight linkage between the markers and the QTL. The second method estimated the %V_G explained by each QTL by dividing the sums of squares for each significant marker by the total corrected model sums of squares from additive QTL models in PROC GLM in SAS. Both methods gave similar results and so we present only the former. We calculated the epistatic effect of each significant interaction (4*i*) as (A + D – B – C), where A and D represent the means of the homotypic classes (AA, BB), and B and C represent the means of the heterotypic classes (AB, BA). We estimated the proportion of genetic variation explained by interacting QTL as the difference in the adjusted R^2 of additive GLM models versus those incorporating interaction. We plotted the posterior probability distribution of the QTL locations under the final model to locate the genomic positions of QTL. In the case of linked or interacting QTL, we plotted the 2-dimensional posterior probability distribution under a multiple-QTL model.

The observation of different QTL effects under different treatment conditions provides evidence for QTL–environment interactions. We further explored these interactions by incorporating marker x treatment terms in a full linear model using the PROC MIXED procedure of SAS. For each experiment, we fit a series of models including the main and interactive effects of all significant markers detected in the *Pseudomarker* analysis and the interaction of these markers and the experimental treatment (Lynch & Walsh, 1998). In this framework, marker–treatment interaction indicates gene–environment interaction, marker x marker interaction represents epistasis averaged over the environments, and marker x marker x treatment interaction indicates environment-specific epistasis.

We generated hypotheses concerning candidate genes underlying the observed QTL by reviewing the existing literature and utilizing a summary of flowering time genes maintained by the D. Weigel lab at the Salk Institute, La Jolla CA (http://www.salk.edu/LABS/pbio-w/flower_web.html).

The *Pseudomarker* programs implemented in this paper are available at (www.jax.org/research/churchill/software/pseudomarker).

Results

Quantitative genetic analysis

We found no effect of the drought stress treatment (F-value = 1.46, df = 1, 35.8, P-value = 0.2341) and only a marginally significant effect of the leaf damage treatment on flowering time (F-value = 3.32, df = 1, 3425, P-value = 0.07, flowering time difference < 1 day). The apical damage treatment was applied to each plant the day it first flowered and therefore could have no effect on flowering time. We detected significant genetic variation for flowering time in each experimental population (in all cases, $\chi^2 > 100$, df = 1, $P < 0.0001$). The broad-sense heritability of flowering time was ~0.29 in the drought stress experiment and ~0.38 in the leaf damage experiment (Table 1). We found no interaction between RIL and either the drought stress or leaf damage treatments, suggesting a lack of genotype-environment interaction at the trait level.

In the apical damage experiment, we detected a significant RIL x treatment interaction that was due primarily to changes in scale (genetic variance across the treatment, 76% of the interaction variance) and to a lesser extent changes in rank (crossing reaction norms, 24% of the interaction variance) ($\chi^2 = 2210$, df = 1, $P < 0.0001$) (Figure 1). A variety of transformations failed to substantially alter this interaction. In this experiment, the heritability of flowering time was 0.86 and 0.45 for control and apically damaged plants, respectively (Table 1).

In general, cross-treatment genetic correlations were positive and high (drought stress, $r_G = 0.87$; leaf damage, $r_G = 0.90$; apical damage, $r_G = 0.78$). The genetic correlation across the two independent *Ler* x Col experiments was positive and moderate ($r_G = 0.64$; using BLUPS from each experiment averaged across treatments).

Table 1. Summary statistics and variance component partitioning for flowering time in each experiment

Flowering date	Mean (SE)		$[V_L]$ [a]		$[V_R]$ [b]		$[H^2]$ [c]		$[CV_G]$ [d]	
Drought stress	Wet	Dry	Wet	Dry	Wet	Dry	Wet	Dry	Wet	Dry
	16.69 (0.07)	16.35 (0.07)	2.38	1.77	5.27	5.17	0.31	0.26	9.24	8.14
Leaf damage	Control	Damaged	Control	Damage	Control	Damage	Control	Damage	Control	Damage
	17.1 (0.06)	17.20 (0.06)	2.29	2.25	3.61	4.05	0.39	0.36	8.84	8.72
Simulated browsing	Control	Clipped	Control	Clipped	Control	Clipped	Control	Clipped	Control	Clipped
	7.35 (0.12)	6.68 (0.14)	0.231	0.023	0.038	0.029	0.86	0.45	70.26	19.89

[a] Among-line variance component from PROC Mixed analysis split by treatment.
[b] Residual variance component from PROC Mixed analysis – the summation of Residual and Block variance components.
[c] Broad-sense heritability calculated as $V_L/(V_L + V_R)$.
[d] Coefficient of genetic variation calculated as $(100 \times \sqrt{V_L})/X$ using untransformed values for the estimation of parameters.

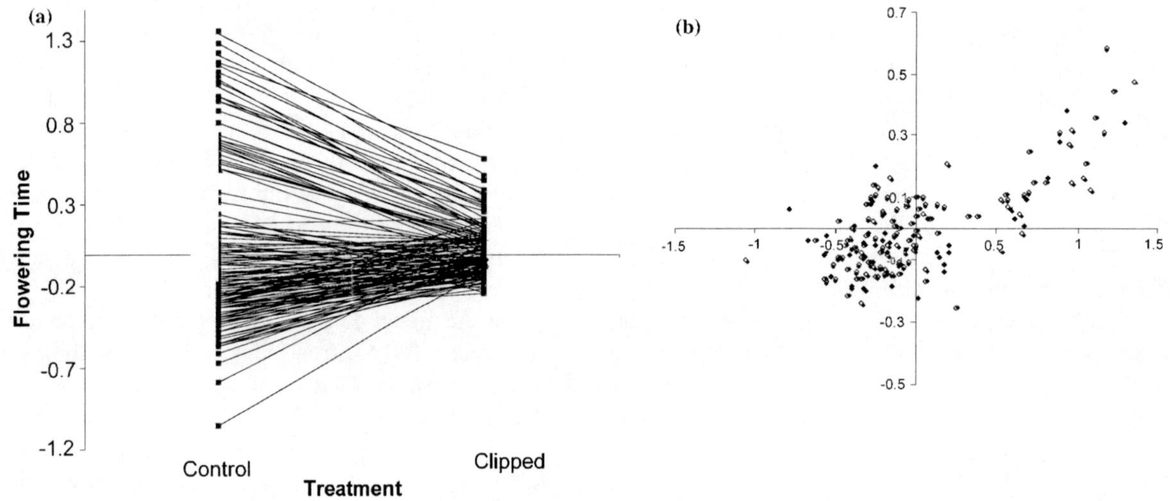

Figure 1. Reaction norm plot of RIL–treatment interaction (A) and the cross treatment genetic correlation (B) in the apical damage experiment.

QTL mapping

Tables 2 and 3 provide a listing of the QTL that significantly affected some aspect of flowering date under at least one environmental condition. Each QTL is designated as FT (Flowering Time) followed by a unique number – QTL from the L*er* x Cvi population were differentiated from those detected in the L*er* x Col by the additional identifier *cvl*. QTL presented in Tables 2 and 3 were significant at the empirically determined threshold value corresponding to $P = 0.05$ based on permutation testing. For each QTL, we present the chromosome on which it resides, the estimated cM position of the QTL, the genetic marker associated with the QTL, the additive genotypic effect (2*a*),

Table 2. Results of QTL analyses on flowering time in the L*er* x Col population using *Pseudomarker* genome scans. (A). Drought stress experiment. (B). Leaf damage experiment. Each QTL is designated as FT (flowering time) followed by a unique number

A) Drought stress experiment

Drought stress	Chromosome	Position (cM)	Marker	Additive effect 2*a* (SE)	a/σ_G	% V_G	Candidates
FT1	1	21.8	ARR7				*CRY2, FHA*
Wet				−0.78 (0.25)	0.25	7.6	
Dry				−0.83 (0.23)	0.31	8.5	
FT2	1	117.2	ve011				*EFS*
Wet				1.14 (0.25)	0.37	16.2	
Dry				0.92 (0.22)	0.34	10.6	
FT3	3	73	g2778				*VRN1*
Wet				0.91 (0.23)	0.29	10.3	
Dry*				0.73 (0.21)	0.27	6.7	
FT4	4	46.51	m226				*FCA, VRN2, FWA*
Wet				−0.93 (0.26)	0.30	10.8	
Dry				−0.74 (0.23)	0.26	6.8	

B) Leaf damage experiment

Leaf damage	Chromosome	Position (cM)	Marker	Additive effect 2*a* (SE)	a/σ_G	% V_G	Candidates
FT5	1	0	ve001				*SIN1*
Control				−0.97 (0.30)	0.32	11.8	
Damage				−0.86 (0.29)	0.29	9.2	
FT6	1	100.78	g4552				*FT*
Control				0.67 (0.36)	0.22	5.6	
Damaged				0.59 (0.35)	0.20	4.4	
FT2	1	117.2	ve011				*EFS*
Control*				0.62 (0.35)	0.20	4.8	
Damaged				0.84 (0.34)	0.28	8.8	
FT7	5	41	aw22				*ART-Sy0, FPF1*
Control				0.93 (0.30)	0.31	10.8	
Damaged*				0.93 (0.29)	0.31	10.8	

*Indicates a QTL that was not initially detected in *Pseudomarker* scans within a particular environment but was nonetheless significant in subsequent single marker analyses. Candidate gene information was obtained primarily from the website maintained by the D. Weigel lab (http://www.salk.edu/LABS/pbio-w/flower_web.html).

the standardized additive effect (a/σ_G), the proportion of the total genetic variance explained (%V_G), and identify candidate loci. Several representative posterior probability plots of QTL locations are presented in Figure 3. Localization plots for the remaining QTL are available from the authors upon request.

Altogether, fourteen significant QTL were detected (L*er* x Col, 7; L*er* x Cvi, 7). At least one QTL was detected on each linkage group with over half of the QTL located on either Chromosome I or V. The two mapping populations shared several QTL locations, suggesting that some loci may affect flowering time in both. Additive genotypic

Table 3. Results of QTL analyses on flowering time in the L*er* x Cvi population using *Pseudomarker* genome scans. Each QTL is designated as FT (flowering time) followed by a unique number and an additional identifier (*cvl*) for the L*er* x Cvi population

Apical damage	Chromosome	Position (cM)	Marker	Additive effect 2*a* (SE)	a/σ_G	% V_G	Candidates
FT1*cvl*	1	0	M1 (PVV4)				*SINI*
Control				−0.24 (0.75)	0.02	< 1	
Clipped*				−0.10 (0.23)	0.00	< 1	
FT2*cvl*	1	9.0	M2 (AXR-1)				*CRY2*, *FHA*, **EDI**
Control				−2.30 (0.75)	0.22	3.1	
Clipped				−0.40 (0.22)	0.15	3.8	
FT3*cvl*	2	47.6	M40 (Erecta)				*EAF20*, *ELF3*
Control*				−0.27 (0.52)	0.03	< 1	
Clipped				−0.93 (0.16)	0.35	14.7	
FT4*cvl*	3	0	M46 (DF.77C)				*HST*
Control				0.66 (0.54)	0.06	1.26	
Clipped*				0.28 (0.16)	0.11	< 1	
FT5*cvl*	5	18.	M91 (BH.180C)				*COL1*, *TFL2*, **FLF**
Control				5.08 (0.59)	0.49	17.4	
Clipped				0.83 (0.18)	0.31	6.2	
FT6*cvl*	5	39	M94 (GH.121L-C)				*ART1*, *FPF1*, **FLG**
Control				4.12 (0.58)	0.40	15.8	
Clipped				0.69 (0.17)	0.26	6.8	
FT7*cvl*	5	107	M110 (DF.119L)				*TOC1*, **FLH**
Control				1.21 (0.53)	0.12	2	
Clipped*				0.26 (0.16)	0.10	< 1	

*Indicates a QTL that was not initially detected in *Pseudomarker* scans within a particular environment. QTL overlapping with those detected in Alonso-Blanco et al. (1998) are indicated under the candidate column in bold using their nomenclature. Candidate gene information was obtained primarily from the website maintained by the D. Weigel lab (http://www.salk.edu/LABS/pbio-w/flower_web.html).

effects (2*a*) ranged from a low of ~0.50 to a high of 5.08 days (average, 1.10 days). The proportion of the total genetic variation explained by additive QTL effects (%V_G) ranged from 4.4–16.2% (average, 8.98%) and < 1–17.4% (average, 5.4 %) in the L*er* x Col and L*er* x Cvi experiments, respectively. In general, QTL effects were larger in the L*er* x Col compared to the L*er* x Cvi population, except for segregation in the latter of two strongly interacting QTL (FT5*cvl* and FT6*cvl*) on Chromosome V. In both experiments, each parent had some QTL alleles that accelerated and some that slowed flowering. This pattern explains the observation of transgressive segregation in both populations.

We found two QTL–QTL interactions in the L*er* x Cvi population and no interactions in the L*er* x Col. A minor interaction was found between the top of Chromosome I (FT1*cv*l, 0 cM) and the top of Chromosome 5 (FT5*cvl*, 18 cM) and corresponded to an epistastic effect (4*i*) of 4.07 days in the control treatment. Interestingly, FT1*cvl* did not have a significant additive main effect and was only detected through its interactions with FT5*cvl* and only in the control treatment. This interaction explained ~1.5% of the total variation in that

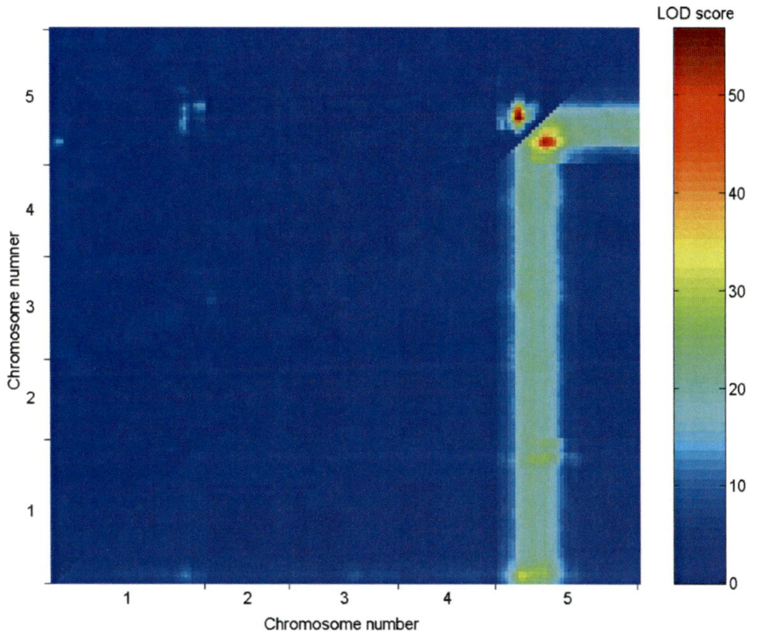

Figure 2. Results of 2-dimensional genome scan depicting epistatic loci on chromosome V in the L*er* x Cvi mapping population. The LOD score associated with a two-QTL model with interaction is plotted below the diagonal. The LOD score difference between the full two-QTL model with interaction and an additive two-QTL model is shown in the upper left above the diagonal. The values in the upper left diagonal are inflated by a factor of three to enhance visibility. For simplicity, only data from the control treatment on a 3.0 cM grid are presented.

treatment. In contrast, a very strong interaction occurred between two QTL located on Chromosome V (FT5*cvl* x FT6*cvl*) with an epistatic effect (4*i*) of 8.45 and 2.01 days in the control and clipped treatment, respectively. These loci delay flowering in the Cvi homotypic class and speed flowering in all other combinations (Figure 4). These QTL had significant additive effects and would likely have been detected even under standard QTL scans. This interaction explained ~12% of the total genetic variation in both the control and clipped treatments.

We evaluated gene–environment interaction by two methods. First, we assessed QTL-by-treatment interaction within each experiment using tests of marker-by-treatment interaction. Second, we compared the two independent mapping experiments conducted with the L*er* x Col population (drought stress and leaf damage experiments). In the absence of gene-by-environment interaction, we anticipated that we would detect similar QTL affecting flowering time in each L*er* x Col experiment (given each utilized identical genetic material and sample sizes). Detecting different flowering time QTL in these experiments would suggest QTL interactions with uncontrolled environmental variation that occurred between experiments.

By our first method, we found strong support for QTL–environment interactions in the L*er* x Cvi mapping population across apical damage treatments. Three QTL (FT2*cvl*, FT5*cvl*, FT6*cvl*) were detected in both treatments and exhibited allelic sensitivity – i.e., different magnitudes of effect between treatments without changes in the direction of effect (Tables 3 & 4). We also detected three QTL unique to the control treatment (FT1*cvl*, FT4*cvl*, FT7*cvl*) and one QTL unique to the clipping treatment (FT3*cvl*). These loci provide evidence for conditional neutrality – i.e., effects in some environments but not others. We found no evidence for antagonistic pleiotropy – i.e., opposing effects in different environments. We found that one epistatic interaction (FT1*cv*l x FT5*cv*l) was detected only in the control treatment and a second epistatic interaction (FT5*cvl* x FT6*cvl*) exhibited environmentally dependent patterns of expression ($F = 22.54$, df = 1, 304, $P < 0.0001$).

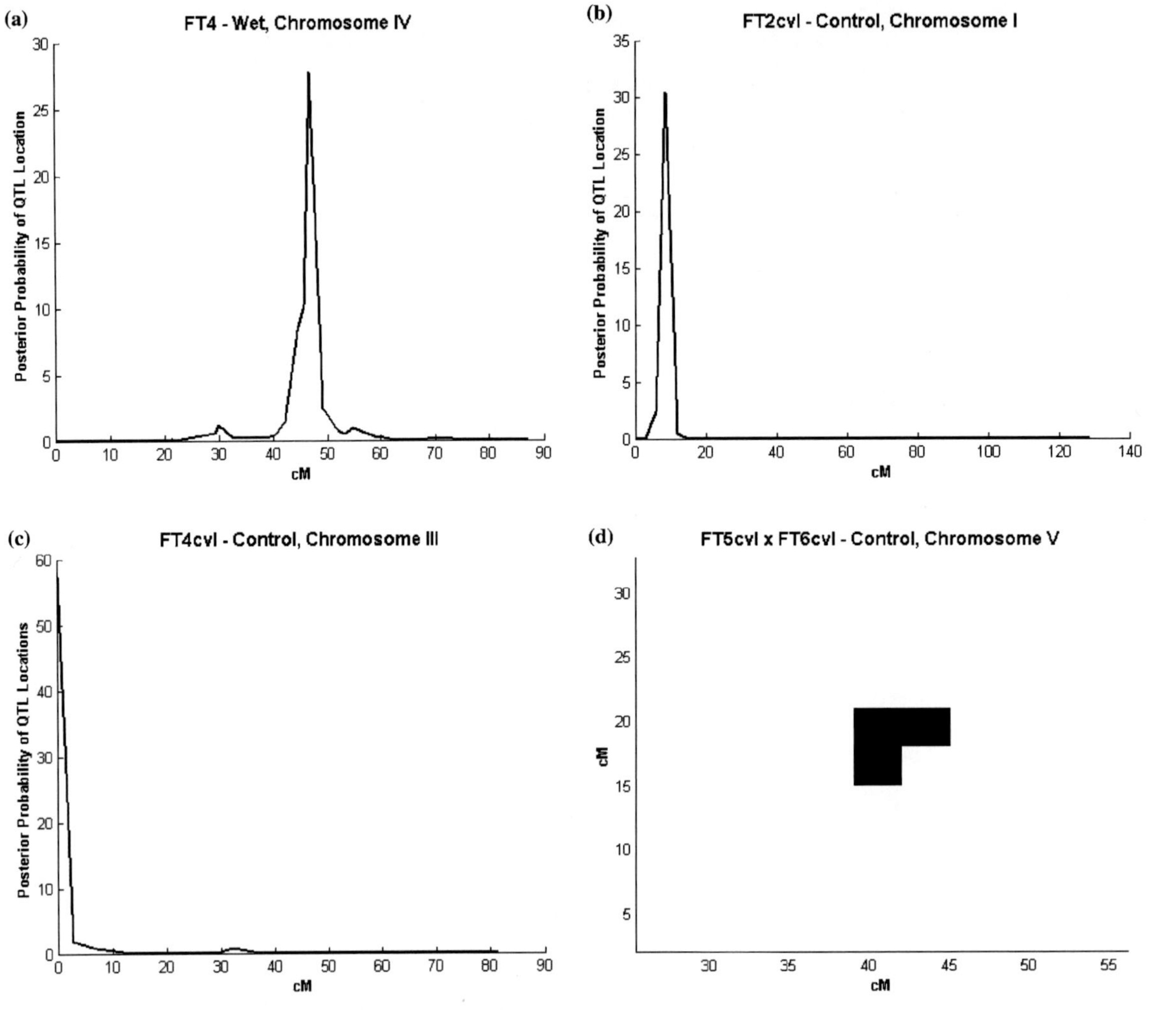

Figure 3. Localization plots for QTL in the L*er* x Col and L*er* x Cvi mapping populations. Each plot is of the posterior probability distribution of QTL locations under a given QTL model: (A) FT4 – wet; (B) FT2*cvl* – control; (C) FT4*cvl* – control; (D) FT5*cvl* x FT6*cvl* – control. Figure D represents the two-dimensional joint posterior probability of QTL locations for the interacting pair (FT5*cvl* x FT6*cvl*) on Chromosome V – the black rectangular area represents a 99% confidence interval of QTL locations.

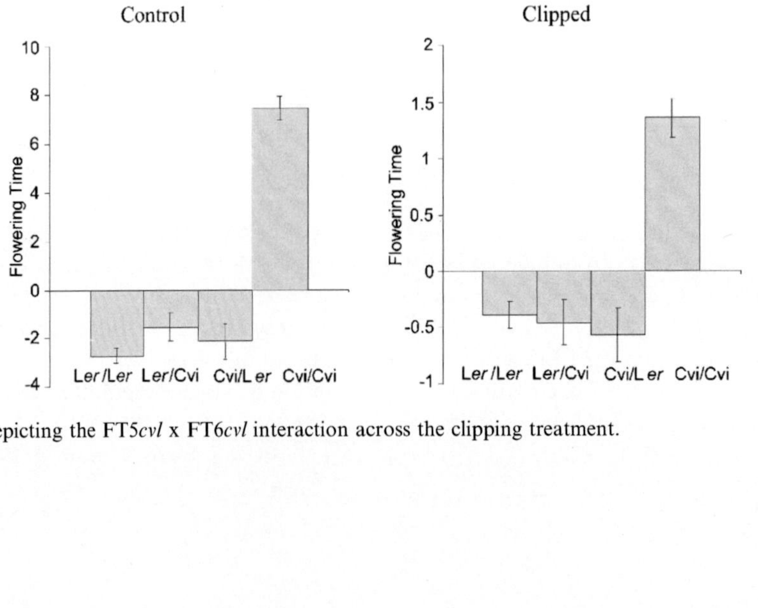

Figure 4. Bar graphs depicting the FT5*cvl* x FT6*cvl* interaction across the clipping treatment.

Table 4. Full model analysis of QTL–treatment interactions in the clipping experiment using PROC mixed in SAS. QTL are modeled as the marker or pseudomarker nearest the QTL peak

Effect	Numerator df	Denominator df	*F*-value	*P*-value
FT1*cvl*	1	304	0.30	0.5815
FT2*cvl*	1	304	24.26	<0.0001
FT3*cvl*	1	304	20.54	<0.0001
FT4*cvl*	1	304	15.02	0.0001
FT5*cvl*	1	304	143.23	<0.0001
FT6*cvl*	1	304	195.31	<0.0001
FT7*cvl*	1	304	10.74	0.0012
FT1*cvl* × FT5*cvl*	1	304	3.63	0.0576
FT5*cvl* × FT6*cvl*	1	304	106.68	<0.0001
Trt × FT1*cvl*	1	304	0.01	0.9400
Trt × FT2*cvl*	1	304	11.81	0.0007
Trt × FT3*cvl*	1	304	3.78	0.0527
Trt × FT4*cvl*	1	304	3.12	0.0785
Trt × FT5*cvl*	1	304	59.84	<0.0001
Trt × FT6*cvl*	1	304	74.15	<0.0001
Trt × FT7*cvl*	1	304	10.74	0.0012
Trt × FT1*cvl* × FT5*cvl*	1	304	1.53	0.2170
Trt × FT5*cvl* × FT6*cvl*	1	304	22.54	<0.0001

Overall, epistatic effects were considerably larger in the control treatment but the general pattern of interaction did not change with treatment (Figure 4). Taken as a whole, QTL–treatment interaction terms explained ~19% of the total genetic variation in the L*er* x Cvi experiment.

In the L*er* x Col mapping population, there was general agreement in the outcomes of mapping in each environment within each experiment. In the drought stress experiment, FT3 was not detected in the initial *Pseudomarker* search in the dry treatment, although a peak approaching significance was observed. Similarly, FT7 was not detected in the leaf damage treatment. Despite these differences, we found no evidence of QTL–treatment interactions in subsequent SAS models explicitly testing for interactions between markers and the manipulation (in all case, $P > 0.50$). Given the close correspondence of the estimated effects in contrasting treatments (Table 2), the disparities observed in the initial *Pseudomarker* searches were probably due to subtle power differences between treatments. The lack of QTL–environment interaction corresponds with the observation of no RIL x Treatment interaction at the trait level within each experiment.

Nevertheless, we found support for QTL–environment interaction in the L*er* x Col populations under our second criteria. In particular, we found a surprisingly low correspondence between QTL controlling flowering time in the drought experiment compared to the leaf damage experiment (Table 2A versus Table 2B), with only one QTL (FT2) occurring in both experiments.

Discussion

Many phenotypes of evolutionary or ecological significance exhibit continuous variation in nature and are ostensibly influenced by the segregation of many genes as well as environmental effects. For example, plant size, phenology, resistance to natural enemies, and fecundity are all traits that generally exhibit a normal distribution of values in plant populations. Quantitative traits have primarily been studied with statistical methods that ignore the underlying genetic details and instead focus on population-level patterns of genetic and phenotypic variance and covariance (Falconer & Mackay, 1996; Lynch & Walsh, 1998). Under a number of assumptions, these parameters can be used to predict short-term adaptive responses to natural selection using the familiar breeders' equation and its extensions (Lande, 1979; Lande & Arnold, 1983; Mitchell-Olds & Rutledge, 1986; Falconer & Mackay, 1998). Quantitative genetic models of adaptive evolution have been very useful heuristic tools; however, they can tell us little about the genetic details of adaptation (Barton & Turelli, 1986; Orr & Coyne, 1992; Orr 1998). For example, how many genes underlie adaptations? Do adaptations arise from the accumulation of genes of small effect or by adaptive leaps with the fixation of genes of major effect? How often does pleiotropy constrain or facilitate evolution? Can contextual genetic effects (gene–gene and gene–environment interaction) explain the maintenance of genetic diversity? We have surprisingly little data with which to evaluate these issues.

The genetic architecture of flowering time

In our studies, we detected 14 genomic locations affecting flowering time in at least one environment from a sample of only three ecotypes – this is clearly a lower limit of the actual number of loci potentially affecting flowering time. The average additive genotypic effect of these alleles was moderate, corresponding to 0.5–1.0 day and generally explained less than 10% of the total genetic variation in the RIL populations. Three QTL detected in the L*er* x Cvi population exhibited large additive genotypic effects corresponding to several days and greater than 15% of the total genetic variation within that population. Each QTL detected in our study explained a relatively small proportion of the total phenotypic variation in flowering time.

Non-additive gene interaction

A novel contribution of our study is the extensive search for non-additive gene interaction. Previous QTL studies of flowering time have included secondary tests for epistasis (Mitchell-Olds, 1996; Kuittinen et al., 1997; Ungerer et al., 2002; Alonso-Blanco et al., 1998a) and evaluated the response of flowering time to photoperiod (Jansen et al., 1995; Alonso-Blanco et al., 1998a), light intensity (Stratton, 1998), or cold vernalization treatments (Clarke et al., 1995; Jansen et al., 1995; Alonso-Blanco et al., 1998a). However, these experiments did not explicitly incorporate scans for QTL-QTL epistasis in the absence of additive QTL effects and only manipulated treatments directly linked to either the photoperiod or vernalization genetic pathways. Here, we use a novel QTL mapping method focused on detecting interacting pairs of loci with multiple-QTL models (Sen & Churchill, 2001) and evaluate flowering time in several novel and stressful environments. Taken as a whole, these experiments provide some of the best information on the quantitative genetic architecture of an ecologically important trait in plants.

QTL–QTL Interaction

Overall, we found no evidence for QTL–QTL interaction in the L*er* x Col population but two instances of QTL–QTL interaction in the L*er* x Cvi population. These interactions involved two loci on Chromosome V and a single locus on the top of Chromosome I. Previous studies have also documented the strong interaction that we observed between FT5*cv*l and FT6*cvl* (Alonso-Blanco et al., 1998a; Ungerer et al., 2002). The additive-by-additive epistatic effect (*i*) of this major QTL pair was comparable to the moderately sized additive effects (*a*) detected in this population; it explained ~12% of the total genetic variation. Ungerer et al. (2002) also detected interactions between the top of Chromosome I and a region near FT5*cvl* for several related traits (e.g., bolting time, rosette leaves at bolting) – however, our results differ in that their analysis located the interacting QTL on Chromosome I at ~7.7 cM rather than ~0 cM. Ungerer et al. (2002) also detected a significant three-way interaction between these loci. We tested for a three-way interaction between FT1*cvl*, FT5*cvl*, and FT6*cvl* but found no support for this complex pattern of epistasis. These differences may result from the fact that their search for epistasis relied on ANOVAs incorporating interactions between markers with significant additive effects. In our study, FT1*cvl* was only detected due to its interactive effect with FT5*cvl*. Additional empirical work is needed to sort out the intricate pattern of interaction variance between Chromosomes I and V in this mapping population.

QTL–environment interaction

We found considerable evidence of QTL–environment interactions. First, we found several QTL that interacted with the apical clipping treatment in the L*er* x Cvi mapping population. Here, we detected seven QTL affecting flowering time, with six exhibiting QTL–treatment interactions. Three cases corresponded to allelic sensitivity while the remaining three cases indicated conditional neutrality. Significantly, we did not observe antagonistic pleiotropy. Although QTL-apical damage interactions produced changes in the genetic variance between the treatments, we observed only minor changes in the rank of RILs between treatments. We also observed strong treatment effects on the interaction of FT5*cvl* with FT6*cvl* (Figure 4). Together, QTL–apical damage interactions explained ~19% of the total genetic variation in the L*er* x Cvi experiment. We found no QTL-treatment interactions in either the drought stress or leaf damage experiment.

Second, we observed very little overlap in the genetic architecture of flowering time in the L*er* x Col population across two independent experiments (only one QTL shared in both experiments). We feel it is unlikely that this difference arises from methodological or power considerations because the two experiments used very similar sample sizes and identical genetic material. More likely, QTL interactions with seasonal environmental differences between experiments led to shifts in the importance of different genes controlling flowering time in each experiment. Similar results have been observed by Weinig et al. (2002) for flowering time QTL in the L*er* x Col population grown in the field and growth chamber conditions. Although greenhouse conditions were very similar in each of our experiments, we did observe seasonal differences in light quality, temperature fluctuations, and photoperiod (Juenger, personal observation).

Our assessment of gene–environment interaction can be extended by comparing the numerous studies of flowering time on these mapping populations. We obtained the RIL means for flowering time (or two closely related traits; bolting time and rosette leaves at flowering) from several independent studies (Jansen et. al., 1995; Stratton, 1995; Alonso-Blanco et al., 1998; Ungerer et al., 2002) involving vernalization and light intensity manipulations. The cross-experiment genetic correlation was quite variable and averaged 0.36 (range, −0.13 to 0.88) for the L*er* x Col population (Jansen et. al., 1995; Stratton, 1995; Ungerer et al., 2002, this study) and 0.62 (range, 0.32 to 0.92) for the L*er* x Cvi population (Alonso-Blanco et al. 1998a; Ungerer et al., 2002; this study). As expected, a number of QTL were identified in each study, some unique to a particular experiment and some clearly overlapping among studies. Interestingly, the cross experiment genetic correlation was never significantly negative across a sample of 20 independent experimental conditions. This pattern suggests that genetic tradeoffs and antagonistic pleiotropy for flowering time alleles may be rare in *Arabidopsis*.

Other studies have documented relatively strong QTL-by-vernalization and QTL-by-photoperiod interactions for flowering time in *Arabidopsis* using experimental manipulations (Jansen et al., 1995; Alonso-Blanco et al., 1998a; Ungerer et al., 2002). To our knowledge, none of these studies revealed antagonistic pleiotropy and instead observed either conditional neutrality or allelic sensitivity. In conjunction with our results, these observations suggest that gene-environment interaction may influence the rate of flowering time evolution primarily through its affects on the amount of genetic variation across environments rather than through genetic trade-offs (Via, 1987). Fry et al., 1998 also reported an absence of antagonistic pleiotropy in experiments with *Drosophila* and commented on its rarity in the existing literature on QTL in many kinds of organisms.

It is clear that epistasis for flowering time can be strong, but it is uncertain how common epistatic interactions of this magnitude generally occur for ecologically important traits in plants. Our approach is a conservative evaluation of epistasis since it ignores the contribution of epistatic interactions to the additive component of genetic variance (Cheverud, 2000). Moreover, our power to detect interactions was low given the small size of our populations and our very stringent permutation-based thresholds. We believe that empirical studies of pairwise-epistasis are currently more limited by experimental population size and, therefore power, rather than analytical methods. Several groups are currently developing large *Arabidopsis* RIL and advanced intercross lines (AIL) that will significantly improve the power of 2-dimensional searches.

Candidate genes

The holy grail of QTL mapping is the isolation of the actual genetic loci controlling phenotypic variation. QTL mapping experiments are by their very nature limited to the detection of chromosomal intervals affected a phenotype and therefore cannot isolate the particular genes responsible for genetic variation. Nonetheless, many of the QTL detected in this study overlap with the positions of known flowering time mutants. For example, FT2*cvl* overlaps with the *EARLY DAY INSENSITIVE* (*EDI*) QTL detected by Alonso-Blanco et al. (1998a). Recently, El-Assal et al. (2001) cloned this QTL through the use of near isogenic lines (NILs), positional cloning, and transgenic manipulation and found it to be a novel allele of *CRYPTOCHROME2* (*CRY2*) generated from a single amino-acid substitution. *CRY2* encodes a blue-light photoreceptor that promotes flowering in long-day conditions. Similarly, deletions that disrupt the open reading frame of *FRI* and alleles of *FLC* contribute to quantitative genetic variation in flowering time (Michaels and Amasino, 1999; Johanson et al., 2000). *FRI* and *FLC* do not overlap with QTL from our study but may explain variation in crosses among other *Arabidopsis* ecotypes (Kowalski et al., 1994; Clarke et al., 1995; Kuittinen et al., 1997). These examples are some of the first cases in which QTL of moderate effect size have been cloned in plants. Tables 2 and 3 list plausible candidates for our QTL based on corresponding positions and functional information from molecular genetic studies. These hypotheses warrant further investigation through additional fine-scale mapping, the creation of NILs, association mapping, and positional cloning. Future molecular characterization of natural allelic variation at *Arabidopsis* flowering time QTL will provide much needed data on the relative role of amino-acid substitutions, coding region deletions, and transcriptional regulation in natural quantitative genetic variation of an ecologically important trait.

Evolutionary genetics of QTL

A major goal of evolutionary biology is to explain the genetic basis of adaptation. A major gap in our understanding of adaptation stems from a general paucity of data on the genetic details and adaptive function of alleles affecting quantitative traits. To date, most empirical studies of the genetics of adaptation have either analyzed genetic polymorphisms in the absence of a clear understanding of their phenotypic and fitness effects or have focused on relatively simple Mendelian traits. A critical consideration is the magnitude of a gene's affect in relation to the strength of selection on a trait. Put simply, how much do individual alleles that segregate in natural populations influence relative fitness? Unfortunately, this aspect of evolutionary quantitative genetics has been debated largely in the absence of empirical data (Barton & Turelli, 1989; Orr, 1998; Agrawal et al., 2001). Extensive quantitative genetics data on flowering time in *Arabidopsis* coupled with recent field experiments may begin to provide this much needed data. For example, Scheiner and Callahan (1999) conducted genetic (breeding value) selection experiments on bolting time in *Arabidopsis* under field conditions using families collected from natural populations. They reported standardized selection gradients on bolting time of ~0.34 (β^*, based on path analysis). This parameter describes the genetic relationship between bolting time and relative fitness in their experimental population and suggests that a shift in bolting time of one genetic standard deviation (σ_G) will result in a corresponding 0.34 σ_G shift in relative fitness. In our study, flowering time QTL had an average standardized genotypic effect (a/σ_G) of 0.24 (excluding epistatic loci). If we assumed that QTL of similar magnitude were present in the population studied by Scheiner and Callahans (1999), we could predict that substitution of an average allele at an average flowering time QTL would result in a 0.24 σ_G shift in flowering time and a corresponding ~0.08 σ_G shift in relative fitness.

This calculation ignores the evidence for strong epistatic interactions in our experiments. *Arabidopsis* has a predominantly selfing mating system and selection is therefore likely to function primarily through lineage sorting and the total genetic variation (V_G) among lines. Consequently, the interaction variance detected in our studies would primarily affect evolution through its contribution to V_G. Nonetheless, adaptive evolution may be complicated by the occurrence of strong interactions of environment. For instance, changes in the genetic variance across treatments generated by QTL–environment interactions could greatly alter the opportunity for selection across environments. Moreover, the changes in sign associated with the FT5*cvl* x FT6*cvl* interaction (Figure 4) could potentially alter the rate at which alleles at these two loci would be fixed or lost in response to selection. There are of course additional caveats to be made about such an exercise, but it points to the importance of future field studies that simultaneously incorporate selection experiments with QTL mapping analyses.

Limitations and future directions

There are numerous limitations to the study of evolutionary quantitative genetics using inbred line crosses and QTL mapping. First, these analyses evaluate only the nature of fixed genetic variation between two parental lines in the context of an artificially created experimental population. Experimental approaches will always sample a small proportion of the total naturally occurring allelic variation and will generally provide estimates of effects at potentially artificial allele frequencies (e.g., average $P = \sim 0.50$ in RIL populations). This will impart a biased view of genetic architecture, specifically with respect to the importance of gene interaction (Falconer & Mackay, 1998; Cheverud, 2000). However, these biases may act in different directions. For example, the maximum additive x additive epistatic variance is created at intermediate allele frequencies (as occurs in many experimental crosses), but pairs of loci with strong additive x additive epistasis can also nullify each other's additive effects at intermediate allele frequencies (Cheverud, 2000). Consequently, epistasis may not be detected in single locus QTL searches. Additional problems arise from experiments with few genetic lines, which have lower power and subsequently overestimate QTL effects (Beavis, 1994; Lynch & Walsh, 1998). Many of these problems will be addressed by implementing larger experiments and using outbred QTL mapping experiments, which can be analyzed in a random effects framework (Lynch & Walsh, 1998). Despite a number of drawbacks in their current implementation, however, we argue that QTL mapping experiments are an important step toward a better understanding of genetic architecture and the developmental processes linking genotype and phenotype.

The best study systems will incorporate several levels of analysis. For instance, additional QTL studies of flowering time using a variety of parental ecotypes will tell us about the generality and importance of particular QTL and their frequency among populations. Studies evaluating QTL affecting flowering time among and within local populations will provide a more detailed understanding of the degree of genetic variation available to natural selection. Coupling manipulations of putative selection agents with QTL analyses can suggest which selective forces may have produced divergent life-history strategies (e.g., spring versus winter annuals) or reproductive phenologies and detect the loci underlying adaptation. Finally, detailed linkage disequilibrium mapping and surveys within and across natural populations of candidate genes or confirmed flowering time loci will provide information on the magnitude and frequency of alleles affecting flowering time in nature. This kind of information is vital for evaluating various models of the maintenance of segregating variation and adaptation (Barton & Turelli, 1989; Orr, 1998; Agrawal et al., 2001).

Conclusions

QTL mapping studies consistently detect complex patterns of genetic architecture including numerous QTL with effects of varied magnitude (Mackay, 1995; Bradshaw et al. 1998; Schemske & Bradshaw, 1999; Westerbergh & Doebley, 2002; Ungerer et al., 2002), interactions between QTL loci (Mackay, 1995; Long et al., 1996; Routman & Cheverud, 1997; Cheverud 2000), QTL and sex (Vieira et al., 2000), and interactions of QTL with environmental heterogeneity (Jansen et al., 1995; Alonso-Blanco et al., 1998a; Fry et al., 1998; Gurganus et al., 1998; Shook & Johnson, 1999; Viera et al., 2000). The correspondence of QTL positions across experiments also provides evidence of pleiotropy (Cheverud et al., 1997; Bradshaw et al., 1998; Shook & Johnson, 1999; Kim & Rieseberg, 1999; Juenger et al., 2000). This complexity is generally unexplored in quantitative genetic analyses and ignored by Fisher's 'infinitesimal' model of adaptation (Fisher, 1930). QTL mapping is one tool that can help generate empirical data to evaluate the role of complex genetic architecture in adaptive evolution.

Acknowledgements

This work was supported by a Miller postdoctoral fellowship to T.J., a NRC Ford Foundation postdoctoral fellowship to K.A.S, and NSF DEB 981550 to E.L.S. We thank the Oxford Greenhouse staff at UC Berkeley for their excellent help with our large experiments. D. Zhao, J. Povich, S. Schultz, G. Ladd and other members of the Simms lab provided excellent help with our experiments and data collection. We thank C. Dean, M. Koornneef, D. Stratton, and M. Ungerer for sharing their data on flowering time in *Arabidopsis*. L. Taylor provided useful comments and improvements to earlier drafts of the manuscript.

References

Alonso-Blanco, C., S.E.-D. El-Assal, G. Coupland & M. Koornneef, 1998a. Analysis of natural allelic variation at flowering time loci in the Landsberg *erecta* and Cape Verde Islands ecotypes of *Arabidopsis thaliana*. Genetics 149: 749–764.

Alonso-Blanco, C., A.J.M. Peeters, M. Koornneef, C. Lister, C. Dean, N. van den Bosch, J. Pot & M.T.R. Kuiper, 1998b. Development of an AFLP based linkage map of L*er*, Col, and Cvi *Arabidopsis thaliana* ecotypes and construction of a L*er*/Cvi recombinant inbred line population. Plant J. 14: 259–271.

Agrawal, A.F., E.D. Brodie & L.H. Riesberg, 2001. Possible consequences of genes of major effect: transient changes in the G-matrix. Genetica 112–113: 33–43.

Barton, N.H & M. Turelli, 1989. Evolutionary quantitative genetics: how little do we know? Ann. Rev. Genet. 23: 337–370.

Beavis, W.D., 1994. The power and deceit of QTL experiments: lessons from comparative QTL studies, pp. 255–266 in Proc. Corn and Sorghum Industry Research Conference. American Seed Trade Association, Washington DC, USA.

Bradshaw, H.D., K.G. Otto, B.E. Frewen, J.K. Mackay & D.W. Schemske, 1998. Quantitative trait loci affecting differences in floral morphology between two species of monkeyflower (Mimulus). Genetics 149: 367–382.

Callahan, H.S. & M. Pigliucci, 2002. Shade induced plasticity and its ecological significance in wild populations of *Arabidopsis thaliana*. Ecology 83: 1965–1980.

Cheverud, J.M. & E.J. Routman, 1995. Epistasis and its contribution to genetic variance components. Genetics 139: 1455–1461.

Cheverud, J.M., E.J. Routman & D.J. Irschick, 1997. Pleiotropic effects of individual gene loci on mandibular morphology. Evolution 51: 2006–2016.

Cheverud, J.M., 2000. Detecting epistasis among quantitative trait loci, pp. 58–81 in Epistasis and the Evolutionary Process, edited by Wolf, J.B., E.D. Brodie III and M.J. Wade. Oxford University Press, New York.

Clarke, J.H., R. Mithen, J.K.M. Brown & C. Dean, 1995. QTL analysis of flowering time in *Arabidopsis thaliana*. Mol. Gen. Genet 248: 278–286.

Donohue, K., 2002. Germination timing influences natural selection on life-history characters in *Arabidopsis thaliana*. Ecology 83: 1006–1016.

Dorn, L., E.H. Pyle & J. Schmitt, 2000. Plasticity to light cues and resources in *Arabidopsis thaliana*: testing for adaptive value and costs. Evolution 54: 1982–1994.

El-Assal, S.E-D, C. Alonso-Blanco, A.J.M. Peeters, V. Raz & M. Koornneef, 2001. A QTL for flowering time in *Arabidopsis* reveals a novel allele of *CRY2*. Nature Genetics 29: 435–440.

Falconer, D.S. & T.F.C. Mackay, 1996. Introduction to Quantitative Genetics. (4th Edition). Addison Wesley Longman, Harlow, Essex.

Fenster, C.B., F.F. Galloway & L. Chao, 1997. Epistasis and its consequences for the evolution of natural populations. Trends Ecol. Evol. 12: 282–286.

Fisher, R.A., 1930. The Genetical Theory of Natural Selection. Oxford University Press, Oxford.

Fry, J.D., S.V. Nuzhdin, E.G. Pasyukova & T.F.C. Mackay, 1998. QTL mapping of genotype–environment interaction for fitness in *Drosophila melanogaster*. Genet. Research 71: 133–141.

Gurganus, M.C., J.D. Fry, S.V. Nushdin, E.G. Pasyukova, R.F. Lyman & T.F.C. Mackay, 1998. Genotype–environment interaction at quantitative trait loci affecting sensory bristle number in *Drosophila melanogaster*. Genetics 149: 1883–1898.

Jansen, R.C., J.W. van Ooijen, P. Stam, C. Lister & C. Dean, 1995. Genotype-by-environment interaction in genetic mapping of multiple quantitative trait loci. Theor. Appl. Genet. 91: 33–37.

Johanson, U., J. West, C. Lister, S. Micheals, R. Amasino & C. Dean, 2001. Molecular analysis of *FRIGIDA* a major determinant of natural variation in *Arabidopsis* flowering time. Science 290: 344–347.

Juenger, T., M. Purugganan & T.F.C. Mackay, 2000. Quantitative trait loci for floral morphology in *Arabidopsis thaliana*. Genetics 156: 1379–1392.

Kim, S.-C. & L.H. Rieseberg, 1999. Genetic architecture of species differences in annual sunflowers: implications for adaptive trait introgression. Genetics 153: 965–977.

Koornneef, M., C. Alonso-Blanco, A.J.M. Peeters & W. Soppe, 1998. Genetic control of flowering time in *Arabidopsis*. Ann. Rev. Plant Phys. Plant Mol. Biol. 49: 345–370.

Kowalski, S.P., T.-H. Lan, K.A. Feldmann & A.H. Paterson, 1994. QTL mapping of naturally-occurring variation in flowering time of *Arabidopsis thaliana*. Mol. Gen. Genet 245: 548–555.

Kuittinen, H., M.J. Sillanpaa & O. Savolainen, 1997. Genetic basis of adaptation: flowering time in *Arabidopsis thaliana*. Theor. Appl. Genet. 95: 573–583.

Lande, R. 1979. Quantitative genetic analysis of multivariate evolution, applied to brain:body allometry. Evolution 33: 402–416.

Lande, R. & S. Arnold, 1983. The measurement of natural selection on correlated characters. Evolution 37: 1210–1226.

Lander, E.S. & D. Botstein, 1989. Mapping Mendelian factors underlying quantitative traits using RFLP linage maps. Genetics 121: 185–199.

Lander, E., S. Green, P. Abrahamson, A. Barlow, M. Daley, S. Lincoln & L. Newburg, 1987. MAPMAKER: an interactive computer package for constructing primary genetic linkage maps of experimental and natural populations. Genomics 1:174–181.

Levy, Y.Y. & C. Dean, 1998. Control of flowering time. Curr. Opin. Plant Biol. 1: 49–54.

Lister, C. & C. Dean, 1993. Recombinant inbred lines for mapping RFPL and phenotypic markers in *Arabidopsis thaliana*. Plant J. 4: 745–750.

Littel, R.C., G.A. Milliken, W.W. Stroup & R. D. Wolfinger, 1996. SAS System for Mixed Models. The SAS Institute, Cary, NC, USA.

Long, A.D., S.L. Mullaney, L.A. Reid, J.D. Fry, C.H. Langley & T.F.C. Mackay, 1995. High resolution mapping of genetic factors affecting abdominal bristle number in *Drosophila melanogaster*. Genetics 139: 1273–1291.

Lynch, M. & B. Walsh, 1998. Genetics and Analysis of Quantitative Traits. Sinauer, Sunderland, MA, USA.

Mackay, T.F.C., 1995. The genetic basis of quantitative variation: numbers of sensory bristles in *Drosophila melanogaster* as a model system. Trends Genet. 11: 464–470.

Michaels, S.D. & R.M. Amasino, 1999. *Flowering locus C* encodes a novel MADS domain protein that acts as a repressor of flowering. Plant Cell 11: 949–956.

Mitchell-Olds, T. & J.J. Rutledge, 1986. Quantitative genetics in natural plant populations: a review of the theory. Am. Nat. 127: 379–402.

Mitchell-Olds, T., 1996. Genetic constraints on life-history evolution: quantitative trait loci influencing growth and flowering in *Arabidopsis thaliana*. Evolution 50: 140–145.

Mauricio, R. & M. D. Rausher, 1997. Experimental manipulation of putative selective agents provides evidence for the role of natural enemies in the evolution of plant defense. Evolution 51: 1435–1441.

Mitchell-Olds, T., 1996. Genetic constraints on life history evolution: quantitative-trait loci influencing growth and flowering in *Arabidopsis thaliana*. Evolution 50: 140–145.

Napp-Zinn, K., 1985. *Arabidopsis thaliana*, pp. 492–503 in Handbook of Flowering, edited by H.A. Halvey. CRC Press, Boca Raton, Florida, USA.

Orr, H.A. & J.A. Coyne, 1992. The genetics of adaptation: a reassessment. Amer. Natur. 140: 725–742.

Orr, H.A., 1998. The population genetics of adaptation: the distribution of factors fixed during adaptive evolution. Evolution 52: 935–949.

Pigliucci, M. & E.T. Marlow, 2001. Differentiation for flowering time and phenotypic integration in *Arabidopsis thaliana* in response to season length and vernalization. Oecologia 127: 501–508.

Routman, E. J. & J.M. Cheverud, 1997. Gene effects on a quantitative trait: two-locus epistatic effects measured at microsatellite markers and at estimated QTL. Evolution 51: 1654–1662.

Sanda, S.L. & R.M. Amasino, 1996. Interaction of *FLC* and late-flowering mutations in *Arabidopsis thaliana*. Mol. Gen. Genet. 251: 69–74.

Schemske, D.W. & H.D. Bradshaw, Jr., 1999. Pollinator preference and the evolution of floral traits in monkeyflowers (*Mimulus*). Proc. Natl. Acad. Sci. USA 96: 11910–11915.

Scheiner, S.M. & H.S. Callahan, 1999. Measuring natural selection on phenotypic plasticity. Evolution 53: 1704–1713.

Sen, S. & G. Churchill, 2001. A statistical framework for quantitative trait mapping. Genetics 159: 371–387.

Shook, D.R. & T.E. Johnson, 1999. Quantitative trait loci affecting survival and fertility related traits in *Caenorhabditis elegans* show genotype–environment interactions, pleiotropy, and epistasis. Genetics 153: 1233–1243.

Stratton, D.A., 1998. Reaction norm functions and QTL–environment interactions for flowering time in *Arabidopsis thaliana*. Heredity 81: 144–155.

Ungerer, M.C., S.S. Halldorsdottir, J.L. Modliszewski, T.F.C. Mackay & M.D. Purugganan, 2002. Quantitative trait loci for inflorescence development in *Arabidopsis thaliana*. Genetics 160: 1133–1151.

Via, S., 1987. Genetic constraints on the evolution of phenotypic plasticity, pp. 47–71 in Genetic Constraints on Adaptive Evolution, edited by V. Loeschke. Springer-Verlag, Berlin.

Vieira, C., E.G. Pasyukova, Z.-B. Zeng, B. Hackett, R.F. Lyman & T.F.C. Mackay, 2000. Genotype–environment interaction for quantitative trait loci affecting life span in *Drosophila melanogaster*. Genetics 154: 213–227.

Wade, M., 1992. Sewall Wright: Gene interaction in the shifting balance theory, pp. 35–62 in Oxford Surveys in Evolutionary Biology, Vol. 8, edited by D. Futuyma & J. Antonovics. Oxford University Press, New York.

Wade, M.J., 2000. Epistasis as a genetic constraint within populations and an accelerant of adaptive divergence among them, pp. 213–231 in Epistasis and the Evolutionary Process, edited by J.B. Wolf, E.D. Brodie III & M.J. Wade. Oxford University Press, Oxford, UK.

Wade, M.J., 2001. Epistasis, complex traits, and mapping genes. Genetica 112–113: 59–69.

Weinig, C., M. Ungerer, L. Dorn, N. Kane, Y. Toyonaga, S.S. Halldorsdottir, T.F.C. Mackay, M.,D. Purugganan, & J. Schmitt, 2002. Novel loci control variation in reproductive timing in *Arabidopsis thaliana* in natural environments. Genetics 162: 1875–1884.

Westerbergh, A. & J. Doebley, 2002. Morphological traits defining species differences in wild relatives of maize are controlled by multiple quantitative trait loci. Evolution 56: 273–283.

Wolf, J.B., E.D. Brodie & M.J. Wade (eds.), 2000. Epistasis and the Evolutionary Process. Oxford University Press, Oxford, UK.

Zeng, Z.-B., C.-H. Kao & C. Basten, 1999. Estimating the genetic architecture of quantitative traits. Genet. Research 74: 279–289.

Genetica (2005) 123: 107–124 © Springer 2005

Evolution in heterogeneous environments and the potential of maintenance of genetic variation in traits of adaptive significance

Diane L. Byers
Behavior, Ecology, Evolution, and Systematics Section, Department of Biological Sciences, Illinois State University, Campus Box 4120, Normal, IL 61790-4120, USA (Phone: +1-309-438-8167; Fax: +1-309-438-3722; E-mail: dlbyer2@ilstu.edu)

Received 22 October 2002 Accepted 7 July 2003

Key words: exploratory path analysis, genetic correlations, genetic constraints, genotype by environment interactions, heterogeneous environments, maintenance of genetic variation

Abstract

The maintenance of genetic variation in traits of adaptive significance has been a major dilemma of evolutionary biology. Considering the pattern of increased genetic variation associated with environmental clines and heterogeneous environments, selection in heterogeneous environments has been proposed to facilitate the maintenance of genetic variation. Some models examining whether genetic variation can be maintained, in heterogeneous environments are reviewed. Genetic mechanisms that constrain evolution in quantitative genetic traits indicate that genetic variation can be maintained but when is not clear. Furthermore, no comprehensive models have been developed, likely due to the genetic and environmental complexity of this issue. Therefore, I have suggested two empirical approaches to provide insight for future theoretical and empirical research. Traditional path analysis has been a very powerful approach for understanding phenotypic selection. However, it requires substantial information on the biology of the study system to construct a causal model and alternatives. Exploratory path analysis is a data driven approach that uses the statistical relationships in the data to construct a set of models. For example, it can be used for understanding phenotypic selection in different environments, where there is no prior information to develop path models in the different environments. Data from *Brassica rapa* grown in different nutrients indicated that selection changed in the different environments. Experimental evolutionary studies will provide direct tests as to when genetic variation is maintained.

Introduction

Ultimately, the extent of genetic variation in traits influencing fitness of an organism will determine the rate of evolution in these traits and the rate of fitness increase for the species (Falconer & Mackay, 1996). This is commonly referred to as Fisher's fundamental theorem, where 'the rate of increase in fitness of any organism at any time is equal to its genetic variance in fitness at that time' (Fisher, 1999). This statement underscores the importance of genetic variation in traits of adaptive significance since a lack of variation will limit their response to selection. The loss of genetic variation due to selection would be balanced by new genetic variation via mutations. Traits closely associated with fitness due to the greater intensity of selection are expected to exhibit a lower level of genetic variation than traits less associated with fitness (Mousseau & Roff, 1987).

However, genetic variation in traits associated with adaptations and fitness in wild populations has usually been found to be greater than expected considering the estimates of spontaneous mutation rates (Mousseau & Roff, 1987; Bulmer, 1989). Given the observed levels of genetic diversity of species that occur across ecological clines, for example, selection in heterogeneous environments has been one of the mechanisms proposed to maintain genetic variation. Both population and quantitative genetic models have examined the potential of selection in heterogeneous and variable environments to maintain genetic variation, with mixed conclusions (i.e., Levene, 1953, Maynard Smith & Hoekstra, 1980; Gillespie & Turelli, 1989; Prout & Savolainen, 1996; Sasaki & de Jong, 1999). Furthermore, studies have examined the expression of genetic variation given contrasting selection histories in natural environments such as clines to determine if there is support for the models (i.e., Harris & Jones, 1995; Li et al., 1998). More recently, a few studies have taken an experimental evolution approach, where environmental variation is the source of selection (i.e., Mackay, 1981, Rose et al., 1996, Bell, 1997a, Elena & Lenski, 1997). Some of the studies experimentally address the genetic mechanisms and selection dynamics underlying maintenance or loss of genetic variation.

Here I will review some of the theory as to the potential that environmental heterogeneity maintains genetic variation, as well as mechanisms. I will discuss some of the empirical evidence and illustrate an approach for examining phenotypic selection in different environmental conditions. I will conclude by reviewing the particular insights from an experimental evolutionary approach and future directions for addressing the potential role of heterogeneous environments for maintenance of genetic variation.

The dilemma of the maintenance of genetic variation in adaptive traits

Under directional selection, as may be expected of many adaptive traits, selection is expected to eliminate genetic variation (Bulmer, 1985, 1989; Falconer & Mackay, 1996). Assuming no migration (of contrasting genotypes) and no differential selection due to a variable environment, new mutations will be the main source and the ultimate source of new genetic variation. Selection acts to decrease genetic variation (permanent effect) and also to increase linkage disequilibrium (transient effect-as long as selection is occurring). Linkage disequilibrium among alleles that are not favorable can also increase the loss of genetic variation through selection. Considering the transitory nature of linkage, its dynamics can be ignored in examination of the balance between selection and mutation for genetic variation (Bulmer, 1989).

Balance between selection and mutation to produce heritability of 0.5 for a trait would require very weak selection, a very high rate of mutation per locus, or a very large number of loci affecting the trait, which are unlikely given current estimates of mutation rates (Bulmer, 1989). In addition, the models require that the existing variance be attributed to rare alleles, which does not follow empirical data from selection experiments and allozyme variation. Therefore, it has generally been concluded that the dynamics between mutation and selection cannot explain the observed genetic variation in natural populations (Bulmer, 1989). In general, models disagree if the dynamics between mutation and selection can account for the extent of genetic variation (Roff, 1997). Hence, the maintenance of quantitative genetic variation in traits under selection is still considered to be a major dilemma and an important question in evolutionary biology (Hedrick, 1986; Bulmer, 1989; Curtsinger et al., 1994; Prout & Savolainen, 1996; Roff, 1997).

Adaptive traits, in addition to having substantial genetic variation, are highly variable in their level of genetic variation (Mousseau & Roff, 1987; Houle, 1992). Spontaneous mutations in reproductive traits in *Arabidopsis thaliana* were found to have bidirectional effects (i.e., increases and decreases in seed and fruit production) which would be supportive of a diversity of mutational effects (Shaw et al., 2000). Under laboratory conditions, *Daphnia pulex* was found to accumulate mildly deleterious mutations, which if reoccurring could explain much of the standing variation in life-history traits (Lynch et al., 1998). It was further suggested that these mutations contributing to variation are likely conditionally deleterious, such that their effect on fitness traits depends on the rest of the genes and/or environment of the individual (Lynch et al., 1998). These results and conclusions suggest a role for heterogeneous environments. Traits that are consequences of complex genetic correlations and those expressed later in the life cycle are predicted to have higher mutational variances as found in these studies (Houle, 1998). While the relative importance of environmental heterogeneity is not addressed in these studies (although it is discussed in Lynch et al., 1998), the finding of mutations whose effects are conditionally dependent is consistent with models suggesting that selection in heterogeneous environments facilitates maintenance of genetic variation. A recent study by Chang and Shaw (2003) using mutation accumulation lines of *A. thaliana* directly examined if the effects of mutations were dependent on the soil nutrient environment. This study revealed no genotype by environment interaction, however; as suggested by the authors this may be due to limited statistical power. In the ecological and developmental context of the expression of fitness related traits, the variance in mutational effects may significantly contribute to the standing genetic variation. The genetic correlations (between traits and environments) will also contribute to diffuse selection, genetic constraints, and therefore the potential to maintain genetic variation.

A recent review of the magnitude and type of phenotypic selection in wild populations suggests selection is fairly weak but highly variable (Kingsolver et al., 2001). Considering the diversity of mutational effects and the weak phenotypic selection, populations may never reach the equilibrium when the balance between selection and mutation determines the extent of genetic variation.

An additional problem for the genetics of adaptive traits is that the distribution and extent of the variation is less well-known compared to genetic markers (Lynch, 1996). Several recent studies have shown that there is very little or no relationship between estimates of genetic diversity of life-history and other fitness related traits and genetic markers (Lynch et al., 1999; Reed & Frankham, 2001). This is not surprising since the expression of genetic variation in quantitative traits is often environmentally dependent, unlike genetic markers. Thus not only do I conclude that we need more experimental work on understanding the dynamics of environmental variation with maintenance of genetic variation, but that we also need to further quantify the magnitude and distribution of genetic variation in adaptive traits in natural populations.

Potential of heterogeneous and variable environments to maintain genetic variation

Patterns of genetic variation

Due to the expectation of differential selection of geographic variation beyond the isolation by distance, the effect of geographic clines has been examined for selection and local adaptation (Endler, 1986; Mousseau et al., 2000). Gene flow along environmental clines is in dynamic with selection for local adaptation which can establish a genetically structured population (i.e., Stanton et al., 1997). Furthermore, contrasting local environments along a cline can lead to disruptive selection and potentially the maintenance of genetic variation among the populations along a cline (i.e., Antonovics & Bradshaw, 1970; Kalisz & Wardle, 1994). This pattern has been observed in many species although the cause of the phenotypic variation has been quantified in a more limited number of species.

Across a variety of species (plants and animal) the percentage of polymorphic loci (allozyme variation) increases with inferred increased environmental heterogeneity (Mitton, 1997). The pattern of expression of genetic by environmental interaction and increasing negative genetic correlations in morphological and fitness traits has also suggested the importance of environmental variation for maintaining genetic diversity (i.e., Bell, 1992; Cheetham et al., 1995). While there is evidence in natural populations of environmental variation associated with different genotypes (Bossart & Scriber, 1995; Galloway, 1995; Harris & Jones, 1995; Mitton, 1997; Richard et al., 2000) it is not clear to what extent this is a cause and effect relationship; therefore an experimental approach is essential to address this question (Mackay, 1981; Rose et al., 1996; Bell, 1997a,b; Kassen, 2002).

Given these patterns of genetic variation and environmental variation, many models have been developed to determine under what conditions and by what mechanisms genetic variation may be maintained (Felsenstein, 1976; Hedrick, 1986; Bell, 1997a; Roff, 1997). Here a few of the more general models will be discussed.

Models and mechanisms

This overview of some of the models and mechanisms that facilitate the maintenance of genetic variation in heterogeneous and variable environments will focus on quantitative genetic approaches. While population genetic models (typically one to two loci) are often simpler and more accessible than quantitative genetic models, and many of the dynamics in heterogeneous environments have not been modeled for quantitative traits, it is not clear if the conclusions will apply. Most of the traits that are critical for adaptation are determined by multiple genes and interactions among genes; the population genetics models are likely missing important characteristics of these traits for maintenance of genetic variation. The characteristics would include: genetic correlations between traits as expressed in different environments; genetic correlations of a trait as expressed in different environments; and expression of genetic variation in different environments. Furthermore, genetic variation as determined by genetic markers is not a good predictor of genetic variation of life-history traits (Lynch et al., 1999; Reed & Frankham, 2001). Hence it is likely that models focusing on one gene determining a trait are also not good predictors of the dynamics, which will potentially maintain genetic variation for many adaptive traits.

Many of the quantitative genetics models focus on the role of the variation in mutational effects in contributing to maintenance of genetic variation (i.e., Houle, 1998; Charlesworth & Hughes, 1999). While this is a very important dynamic that will determine standing genetic variation, here I will mostly focus on the role of environmental variation and the variation in selection at the population level. As suggested by Charlesworth and Hughes (1999), genetic variation that is not due to mutations can be attributed to directional selection at the level of the individual, given the context of their environment and genetic background. Hence in heterogeneous environments, genetic variation at the population level will be influenced by directional selection in all of the local environments and genetic interactions.

Due to the multivariate nature of quantitative traits many of the models and discussions of mechanisms that can maintain genetic variation have focused on evolutionary constraints (e.g., Arnold, 1992). The constraints on phenotypic evolution can be due to genetic constraints, selective constraints and developmental constraints (Arnold, 1992). These evolutionary constraints are likely to play a major role in the maintenance of genetic variation. Development of multivariate statistical approaches of the breeders equation and comparisons of the **G**-matrices (genetic variance and covariances of and between different traits), for example, has allowed for examination of evolutionary constraints and mechanisms for maintaining genetic variation (Lande & Arnold, 1983; Arnold, 1992). These methods are increasingly being used for analysis of selection under field conditions as well as selection experiments, as it is critical for assessment of these potential constraints in the context of the environment of the species (Arnold, 1992; Kingsolver et al., 2001).

Many of the models examining evolution of quantitative traits in variable and heterogeneous environments have focused on the evolution of phenotypic plasticity and reaction norms (i.e., Zhivotovsky et al., 1996a,b; Sasaki & de Jong, 1999). In particular they have sought to determine if a single reaction norm can obtain the optimal phenotype across the environments. Alternatively, if no single reaction norm will be obtained then polymorphism of reaction norms and hence genetic variation will be maintained. For example, a model found that when environments changed unpredictably between development and selection, if there was density dependent selection after selection in response to the environment (soft selection) then a polymorphism of reaction norms would be maintained (Sasaki & de Jong, 1999). The unpredictability of environmental changes would limit the possibility of selection for just one reaction norm.

Comprehensive quantitative genetics models that examine the many genetic aspects of these traits that could facilitate maintenance of genetic variation in heterogeneous environments have not been done, to my knowledge. Therefore, I will discuss these characteristics of quantitative traits separately in the following sections. For each characteristic, I will discuss how genetic variation can be maintained (models and mechanisms) and present a few examples.

Environmentally dependent expression of genetic variation

While response to selection depends on the presence of genetic variation, the expression of genetic variation is often environmentally dependent (Falconer & Mackay, 1996; Roff, 1997). The lack of expression of genetic variation in one or some of the environments will prevent response to selection in the trait in that environment. The variation in expression of genetic variation across environments may be reflected in the genotype by environment interaction, which is typically thought of as a change in the relative ranking of the genotypes with environment. However, a genotype by environment interaction may result from a change in relative expression of genetic variation across environments, or alternatively stated, may reflect a change in the scale of the variation among genotypes across environment (Lynch & Walsh, 1998).

Genotype by environmental interactions are suggested to maintain genetic polymorphism in a heterogeneous environment through balancing selection (Gillespie & Turelli, 1989). The authors also suggested experimental approaches with a wide range of environments since the results of selection may depend on the environments assayed. However, there is some disagreement concerning some aspects of their model and in reanalysis it was found that without some linkage disequilibrium even a small amount of genetic variation cannot be maintained (Gimelfarb, 1990).

Environmentally dependent expression of genetic variation can lead to environmentally dependent selection. In heterogeneous environments balancing selection may potentially lead to maintenance of genetic variation as found along ecological clines. There are many examples in the literature of balancing selection associated with environmental heterogeneity that are supportive of the maintenance of genetic variation (e.g., Vavrek et al., 1996; Borash et al., 1998; Schmidt & Rand, 2001; van Kleunen & Fischer, 2001; Cheplick, 2003).

A limited numbers of studies have examined variation in gene movement in a heterogeneous environment as well as studying variation in adaptive traits. Bossart and Scriber (1995) studied *Papilio glaucus* (eastern tiger swallowtail butterfly) to determine if environmental variation (host plants–18 species) selected for the maintenance of genetic variation for important life-history traits. In addition, they used genetic markers to determine the gene flow among several populations and different hosts. The difference between the genetic markers and the quantitative genetics in oviposition preference and larval performance on the different plants was attributed to local selection (Bossart & Scriber, 1995). Differential selection due to environmental variation (different host plants) on a local scale (among trees of subpopulations for leafminers) found local adaptation for the particular tree in spite of substantial migration among trees (Mopper et al., 2000). Similarly, local selection maintained genetic variation in shell traits of a clam (*Macoma balthica*) in face of substantial dispersal as determined by genetic markers (Luttikhuizen et al., 2003).

On a larger geographic scale many studies have found patterns of selection that would favor the maintenance of genetic variation. Both phenotypic correlations (positive and negative correlations) between developmental switch for diapause with reproductive success and geographic variation (reflecting the differences in the environment) in developmental cues have been suggested to maintain genetic variation in western *Chrysoperla carnea*, lacewings (Tauber & Tauber, 1992). On a large scale, the relative growth of *A. thaliana* was found to be correlated with latitude (Li et al., 1998). This pattern of clinal variation was interpreted in response to the environmental gradient.

More detailed analysis of the genetic basis of genotype by environment interactions is now possible, particularly with model systems. Through the use of quantitative trait loci (QTL) the number of genes and distribution of effects on quantitative traits can be estimated (Falconer & Mackay, 1996; Lynch & Walsh, 1998; Mackay, 2001). If assays of QTL are conducted in multiple environments, genotype by environmental interactions in environmental dependent expression of QTL and thus the environmental dependent expression of genetic variation can be determined (Vieira et al., 2000; Mackay, 2001). For example in *Drosophila melanogaster*, variation in temperature and food showed genotype by environment and/or genotype by environment by sex interaction in 17 QTL detected for life span (Vieira et al., 2000). In addition, 10 of the QTL showed either antagonistic sexually (expressed in only one sex) or pleiotropic expression in the different environments (expressed in only one environment) which may lead to a maintenance of genetic variation in life span of adult flies.

Genetic correlations between traits

Selection on traits associated with adaptations is typically a multivariate process since often the response to selection on one trait is not independent of another trait (Lande & Arnold, 1983; Arnold, 1992). The lack of independence of two traits may be due to a gene influencing both traits (pleiotropy) or two linked genes whose alleles are in gametic phase disequilibrium. The extent that two traits are genetically associated with each other can be determined by their genetic correlation or the correlation of their breeding values (Falconer & Mackay, 1996; Lynch & Walsh, 1998). Evolutionary or genetic constraints can arise due to the lack of independence between traits due to pleiotropy or linkage and whether the response to selection of either of the traits is due to the type and direction of selection on the other trait. For example, if two traits are negatively genetically correlated and they are selected to increase in relative value it would not be possible to select for the best in both traits; this is referred to as antagonistic pleiotropy or selective constraints. Alternatively it is possible to find a faster-than-expected response to selection if the genetic correlation is in the same direction as selection (Falconer & Mackay, 1996). Antagonistic pleiotropy alone has been shown only to be able to maintain genetic variation under fairly restrictive conditions; however, it may still play a role in the maintenance of genetic variation. Phenotypic trade-offs are considered important in many species, although the underlying bases of these trade-offs are not always genetic (Curtsinger et al., 1994).

Often genetic correlation between traits is considered separately from environmental heterogeneity as a mechanism that may maintain genetic variation. However, since genetic correlation between traits can change with environmental changes, as shown in empirical studies (i.e., Donohue & Schmitt, 1999, Kause et al., 2001), I will include a general discussion of how genetic correlation between traits can maintain diversity. There has been limited theoretical work in this area that I am aware of, but there is some relevant discussion involving the evolutionary constraints associated with the **G**-matrix (Arnold, 1992). There are some empirical examples, which I will briefly discuss below, of changes of expression of genetic correlations between traits when examined in different environments.

In wild populations of the side-blotched lizard there was a negative genetic correlation between clutch size (distributive selection) and egg mass (stabilizing selection); this would be expected to maintain genetic variation given the direction of selection (Sinervo, 2000). Populations of *Cakile edentula* var. *lacustris* in drier habitats are under strong selection to decrease the number of leaves and increase water use efficiency; however since these traits are positively correlated within populations this correlation will be expected to constrain evolution, thus maintaining genetic variation (Dudley, 1996).

Genetic correlations and direction of selection between some floral traits in *Ipomopsis aggregata* resulted in antagonistic pleiotropy and thus is a potential reason for maintenance of genetic variation in some traits (Campbell, 1996). However, in other floral traits there was not apparent antagonistic pleiotropy to explain the maintenance of genetic variation found in the traits. All of these floral traits are closely tied to fitness so they would be expected to be under strong selection.

A review of the basis of phenotypic variation in quantitative traits in *Drosophila* concluded that there was evidence for negative pleiotropy maintaining genetic variation in traits influenced by selection (Roff & Mousseau, 1987). Thus in many studies there is evidence for antagonistic pleiotropy in traits where there is significant heritable variation. However, there are also examples of no apparent genetic correlations that would maintain genetic variation.

Genetic correlations between traits as expressed in different environments

If genetic correlation changes across environments it would indicate differing selection in different environments. The first example examines several species of sawflies and their expression of genetic correlations in relation to the variation in the quality of their environments. For folivorous insects in a seasonal environment the quality of their food (i.e., leaves of plants) is typically not stable throughout the growing season due to maturation of their host leaves. A recent study of specialist insects (sawflies) that feed on mountain birch (*Betula pubescens*) found the genetic correlation between larval development and larval mass changed over the season (Kause et al., 2001). Early season species growing in an environment of declining quality had a negative correlation between these traits and lower genetic variation expressed in these traits, while the mid-season species, which were in a stable environment, had a positive genetic correlation between the traits and a greater expression of genetic variation. The differences between the groups illustrate the changing pattern of selection (strongly directional in the early season species) and the environmental dependence of the genetic correlations. These seasonal changes in the genetic correlations and expression of genetic variation were attributed to changes in natural selection and environmentally-induced plasticity in the genetic architecture (Kause et al., 2001). This example illustrates the environmental dependence of the genetic architecture, which could facilitate the maintenance of genetic variation.

In the second example, for *Impatiens capensis* grown in different density environments, traits associated with the plants' response to the changes in light conditions (i.e., # of internodes and length of internode) were strongly correlated and expressed the same correlations in the different environments (Donohue & Schmitt, 1999). However, genetic correlations between traits associated with growth pattern or leaf traits sometimes had a different pattern of correlations in the different densities (Donohue & Schmitt, 1999). In this plant, traits associated with response to changes in light quality need to respond as a group and interdependently for a functional response. Therefore, it is likely that selection works on the light response traits as a group, which may not be true of other traits. Given their results, the trait influencing the growth pattern or leaf traits would be expected to have maintained more genetic variation.

Similar to the genetic correlation of two traits in one environment is the concept of across-environment genetic correlation of a trait as it is expressed in two environments. As first developed by Falconer (1952), a trait measured in two environments could be considered a character in two states or across-environmental genetic correlation. The across-environmental genetic correlation indicates the extent to which the response of the genotype is proportional or not in the two environments (Lynch & Walsh, 1998). If this across-environment genetic correlation is equal to one, then the genotype response is proportional in the two environments. The deviation of this genetic correlation from one indicates a different pattern of selection in different environments (Bell, 1997a). In heterogeneous environments the genetic correlation (across the environments) in the traits of interest can determine if the pattern of selection is similar or different (Bell, 1997a). For maintenance of genetic variation in heterogeneous environments, a genetic correlation significantly different from one is of particular interest. There are several different methods to determine across-environment genetic correlations (Windig, 1997).

Evolution in heterogeneous environments leads logically to a discussion of phenotypic plasticity (change in expression of a genotype in different environment) and what type of environmental variation may favor selection of genotypes with greater expression of phenotypic plasticity. The evolutionary potential of phenotypic plasticity can be determined from the across-environment genetic correlation (Via & Lande, 1985; Via, 1987). Selection of phenotypically plastic genotypes may be determined by the unpredictability of environmental variation, the environmental grain from the organisms' perspective (in sense of Levins, 1968), and the quality of the environments (Scheiner, 1993; Via et al., 1995; Bell, 1997a; Sasaki & de Jong, 1999). Further discussion of this is somewhat beyond the scope of this chapter but the reader should be aware of this parallel and overlapping literature and consideration of multiple effects of environmental variation.

If selection in heterogeneous environments maintains genetic variation by changes in the genetic architecture then one would predict that as environments diverge the genetic correlations across the environments would decline from one and there would be an increased finding of antagonistic pleiotropy. This pattern of across environmental genetic correlations decreasing from one as the environment diverges has been shown (Bell, 1992; Karan et al., 2000; Kassen & Bell, 2000). The decrease in the across environment genetic correlation indicates the independence of selection on the traits in the contrasting environments (Bell, 1997a).

For example, in a study of *Chlamydomonas* negative genetic correlations relative to the direction of selection may have been environmentally dependent. As found in a study of 15 *Chlamydomonas* species the across environment genetic correlations became more negative as the environments became more divergent (Kassen & Bell, 2000). In a longer-term selection experiment (20,000 generations), *Escherechia coli* was selected under stable conditions (37°C). Then these lines were grown in a wider range of temperatures to examine the extent of specialization to the one environment (37°C). The authors found that the greater difference between the new environment and the selection environment (37°C), the lower the growth rate of the *E. coli* populations. This decline in fitness as the environment diverged was attributed to antagonistic pleiotropy (Cooper et al., 2001).

In a study of *D. melanogaster* genetic architecture was expressed in different temperatures and sexes (Karan et al., 2000). As the temperatures diverged, the across environment genetic correlations were found to decrease from one. Furthermore, across environment genetic correlations and the shape of the reaction norm differed between the sexes across the range of temperatures. For example the reaction norm across the environments for thorax length was linear in shape for females but quadratic for males (Karan et al., 2000). This change in the reaction norm illustrates another aspect of response to variable environments that has the potential to contribute to genetic variation. Similarly, a population of *D. melanogaster* from an area with greater genetic variation was shown to have genotypes with a greater expression of genotype by environmental interactions than northern populations from a less diverse area when grown under a range of laboratory environments (Takano et al., 1987). Given this increased genotype by environment interaction within increased genetic variation, they concluded that the higher level of genetic variation was maintained by diversifying selection. This result is as one would predict from selection under heterogeneous environments.

Traits (signaling behavior) under sexual selection (female choice) in waxmoths, *Achroia grisella*, were found to be strongly influenced by genotype by environment interaction, such that the across-environment genetic correlation was less than one. The authors established a range of environmental treatments, which simulated natural variation found in their environment, including: variation in food quality, temperature, and photoperiod. In their range of experimental environments no genotype could obtain maximum fitness, and therefore the authors proposed that genetic variation would be maintained in this and other life-historytraits in heterogeneous environments (Jia et al., 2000).

Some species, due to their complex life cycles, typically inhabit contrasting environmental conditions. For example the aphid, *Pemphigus betae*, alternates between cottonwood trees and roots of herbaceous plants, although some clones will spend less time on the trees (Moran, 1991). A study of clones in the different environments found a negative cross-environment genetic correlation for performance, and therefore it was suggested that their life-cycle variation would maintain genetic variation for some traits (Moran, 1991).

A field experiment examined the fitness consequences and potential for evolution of a plant, *Nemophila menziesii*, given different competitive treatments (Shaw et al., 1995). For some of the competitive treatments there was a genetically based trade-off between relative successes in the contrasting environments. Thus given the range of environments in this plant, the aphids' natural community and the observed trade-offs there is the potential to evolve specialized genotypes and to maintain genetic variation.

Limitations for the maintenance of genetic variation

Many of the models that suggest that genetic variation may be maintained by evolution in heterogeneous environments have been criticized as the models require fairly strict conditions to maintain diversity (i.e. Prout, 1968; Christiansen, 1974). For example, Maynard Smith and Hoekstra (1980) pointed out that for a stable polymorphism to exist within a population in models such as Levene's (1953), the effects of the contrasting alleles favored in the different environment need to be fairly strong. An early model by Via and Lande (1985) found fairly restrictive conditions (such as no additive genetic variation in one of the environments) to maintain multiple reaction norms, but a more recent model by Sasaki and de Jong (1999) did so with less restrictive conditions. Considering the evolutionary complexity of quantitative genetic traits in heterogeneous environments, it might not be feasible to model all aspects and still have results that can be interpreted. Perhaps empirical studies can illustrate which aspects of the genetic bases of adaptive traits have the greatest potential to maintain genetic variation in heterogeneous environments.

Empirical approaches for addressing the maintenance of genetic variation

In the next two sections of this paper I will present empirical approaches which can provide insight to the complex issue of whether environmental heterogeneity leads to the maintenance of genetic variation. While the maintenance of genetic variation is possible as suggested by some models, and often supported by patterns that are observed in natural populations, the cause and effect relationship is not always clear. The approaches presented in the following sections will assist in gaining insight as to when and how diversity is maintained. For example, what are the particular environmental conditions and genetic architectures that are most likely to maintain diversity?

First, I will illustrate the use of exploratory path analysis to develop models of phenotypic selection in different environments. This statistical approach allows for the development of hypotheses when there are many variables and with limited information on the causal relationships among them. Secondly, I review a few examples of how an experimental evolutionary approach can be used with species with a quick life cycle. This approach allows for direct experimental tests of when the environment maintains genetic variation.

Quantification of differential phenotypic selection-use of exploratory path analysis

Environmental heterogeneity can be selected for different phenotypes, which gives the expression of the genetic architectures across the environments (genetic correlations and expression of heritable variation), may facilitate the maintenance of genetic variation as illustrated in the previous studies. I will present a different approach to understand phenotypic selection in heterogeneous environments. This is not to replace quantification of genetic correlation, heritability, or selection gradients or traditional path analysis but to complement and to be combined with these. This approach will allow for the development of a model of selection in a particular environment given limited causal information among the different measured variables. A question that can be addressed with this approach is if, how and the degree to which selection differs in different environments.

Traditional path analysis where different models are proposed is a very powerful method for establishing causal hypotheses to explain selection in different environments (Kingsolver & Schemske, 1991; Rausher, 1992; Mitchell, 1993; Conner et al., 1996; Scheiner & Callahan, 1999; Scheiner et al., 2000). Traditional path analysis approach has some advantage over just estimating selection gradients (multiple regression) in that the results are not biased by missing correlated traits and thus can provide a model of causal relationships. One of the disadvantages of traditional path analysis is that you need to know what set of potential models to test. If many traits are measured, the number of potential models is very large (Shipley, 1997, 2000). For example if a data set has just 4 variables there are 4096 possible path models (Shipley, 2000).

Exploratory path analysis is useful when insufficient information may be known about a system to construct a path model (Shipley, 1997, 2000). For example in the study presented below, knowledge of the relationship among traits in one environment would not necessarily apply to another environment. Given many variables there are just too many alternative path models to choose one or some to test. An exploratory path analysis approach provides a nonrandom method of developing a set of potential models. The algorithms used to determine the best or set of best relationships among the variables gives the data set. This is a formal statistical process. This analysis uses a data driven model approach so it is useful to formulate initial hypotheses and testing of these hypotheses should be done (further experiments) before biological conclusions can be drawn for the species.

The exploratory path analysis approach I used was developed for the smaller sample sizes found in most evolutionary and ecology studies (Shipley, 1997). Shipley's exploratory path analysis was a modification based on the SGS algorithm (Spirtes et al., 1993). The SGS algorithm determines potential path models (directed graphs) in a two step process. First, using the data, it examines the relationship between all pairs of variables to determine the topology of the path model. Any pair of variables that has a correlation that is not statistically different from zero will not have a path and the others will have a path (undirected edge) between them. This step continues by examining the pairs of variables (with paths) for nonsignificant first-order partial correlations and removing paths if there is no significant relationship. The algorithm continues using higher order partial correlation for examining if there are statistical relationships between pairs of variables until no more paths between variables can be removed.

The second step determines the orientation of at least some of the paths (undirected edges become directed edges). The relationship between a pair of variables as to which is dependent and independent can be determined for some sets of variables. Determination of these relationships depends on finding triplets of variables where only some of the pairs have significant relationships. The final result will be a set of path diagrams with partial directed paths that are consistent with the correlation structure of the data set (thus data directed).

SGS algorithm requires sample sizes of 1000. The modification by Shipley (1997) uses a bootstrap resampling of the data set and application of the SGS algorithm to each sample. A set of programs for this analysis are available from Shipley (1997) and were used for the following analysis.

Illustration of exploratory path analysis with rapid-cycling Brassica rapa

To examine the effect of phenotypic selection in different environments and the genetic constraints that could maintain genetic variation, I have been working with the rapid-cycling lines of *B. rapa* L. These are fast growing lines that were selected for no seed dormancy and a shortened life-cycle, resulting in a decrease in the number of days from germination to mature seed (Williams & Hill, 1986). However, there is still substantial genetic diversity in many life-history and size traits as well as allozymes (Evans, 1989, 1991). This is an outcrossing species and can grow from seed to production of mature seed in less than 2 months. The breeding system is very common in plants (Richards, 1986) and allows for more general application of the results than strictly selfing species. The short generation time allows for more experimental evolutionary approaches which have some advantages and will be discussed in the next section. Therefore, this species is a good model system for experimental studies in understanding the effects of environmental variation on phenotypic selection and on the maintenance of genetic variation.

The wild populations of this species occur in North America as a weedy species in disturbed areas and along agricultural fields. Therefore, this species occurs in a wide range of soil nutrient environments from degraded soils to run-off from farm fields. The range of nutrient treatments used in this experiment would be within the range for the species.

Experimental design

Variation in nutrient level as an environmental treatment was chosen since nutrients are frequently heterogeneous on a very local scale in natural populations and have consequences for growth and reproductive success of plants (Grime, 1994; Stratton, 1995; Pigliucci & Schlichting, 1998; Richard et al., 2000). The plants were grown in six levels of soil nutrients ranging from 4.7 to 150 ppm of nitrogen. The range of nutrients resulted in plants that were stressed from receiving too little nutrients (limited growth and reproduction) and too much (aborting seeds), but no plants died from this range of treatments.

The traits that were assessed included: rate of germination, rate of development of the leaf, size of an early leaf, number of days until first flower opens, largest leaf size at first flower, height of plant at first flower, size of the flowers, number of days of flowering, number of buds produced, number of flowers produced, number of seeds per early produced fruit, and total number of fruits produced. These traits were chosen to represent estimates of selection throughout the life-history of the plant. The total number of fruits produced is used as an estimate of fitness.

The offspring from a nested design (each of the 24 sires was crossed with 3 dams) were grown in the above nutrient environments. This mating design was chosen since it allows for estimates of heritable variation that are not confounded by maternal effects or dominance (Falconer & Mackay, 1996). This analysis will not be done using the genetic design but because of this design the same set of genotypes were used in each of the environments. In each of the nutrient environments 192 plants were grown which is an adequate sample for the number of measured traits.

As an illustration of this approach examining the variation in phenotypic selection in response to growth in different nutrient environments, I will present part of the analysis here with half of the data set (three of the six nutrient treatments) and focus on the exploratory path analysis.

Analysis

In order to address if growth in the contrasting nutrient environments results in a unique set of relationships among the measured traits and fitness the following analyses were conducted. The overall approach first established a path model for each nutrient treatment. Next the data from each of the other nutrient treatments were tested against each model to determine if the results from selection in one environment could be explained by the other models. All of the traits to be regressed on the fitness estimate (# of fruits produced) were transformed into standardized deviates with a mean of 0 and a standard deviation of 1 (Sokal & Rohlf, 1995) to allow comparison of traits on different scales. The estimate of fitness, number of fruits produced, was converted to the relative number (relative fitness) within each treatment by dividing by the mean for each treatment.

To establish a path model for each nutrient environment a multiple regression within each environmental treatment determined if any of the traits were associated with the estimate of fitness (total number of fruits produced). The traits that had a direct selection estimate (the slope of the multiple regression for a trait) whose probability value was 0.2 or less were included in the path analysis. This was done to remove traits that had no significant relationship with number of fruits (Conner et al., 1996; Conner & Rush, 1997). The data within each nutrient environment was also tested for multicollinearity which was not found. These analyses were conducted using SAS software and the PROC REG procedure (SAS 2001).

Since I did not have a particular set of hypotheses at the start of this experiment to construct a path model and I was only interested if the environments differed I chose to use an exploratory path analysis approach (Shipley, 1997, 2000). This is an appropriate approach for estimating potential path models of the plants in the different environments.

To obtain a model for each of the nutrient environments I used the EPA program available from B. Shipley (2000). The one constraint that I imposed on the path models is that traits determined earlier in the life-cycle cannot be influenced by traits determined later. The exploratory path programs cannot include this limitation; hence I modified the simplest significant model from the exploratory program by removing paths from later to earlier traits. The modified path models were tested using the PROC CALIS procedure of SAS (2001). The χ^2 statistic was used to determine if the model fit the data, and if so the probability value for this test would be greater than 0.05 (Hatcher, 1994).

While the path models do not appear to be the same, the following analyses were done to compare the models to the data from the other nutrient environments. To determine if the models derived from the data in one nutrient environment would also fit the data from another environment, I used PROC CALIS and included the variables, the paths, and the direction of the paths for the particular nutrient environment. Each data set was tested with each of the models. In many ways this is a fairly conservative test since I did not constrain the model to have particular values for any path coefficients. I was primarily interested in determining if the overall model would fit the data from a different environment. A nonsignificant χ^2 probability value ($p > 0.05$) indicates the model fits the data.

Results and conclusions

The exploratory models for three of the nutrient treatments show very different patterns (Figure 1). This would indicate that selection in these contrasting nutrient environments is different. In the lowest nutrient treatment, there are more paths associated with the earlier traits than in the other nutrient environments. The moderate nutrient environment has fewer paths among the traits perhaps indicating that there is less integration among the traits. In order for contrasting environments to facilitate the maintenance of genetic variation the environments need to have different patterns of phenotypic selection. This potential is nicely illustrated here with the different patterns of selection associated with the different path diagrams.

Testing each of the models using the data from the other environments revealed that the two extreme environments had unique models that did

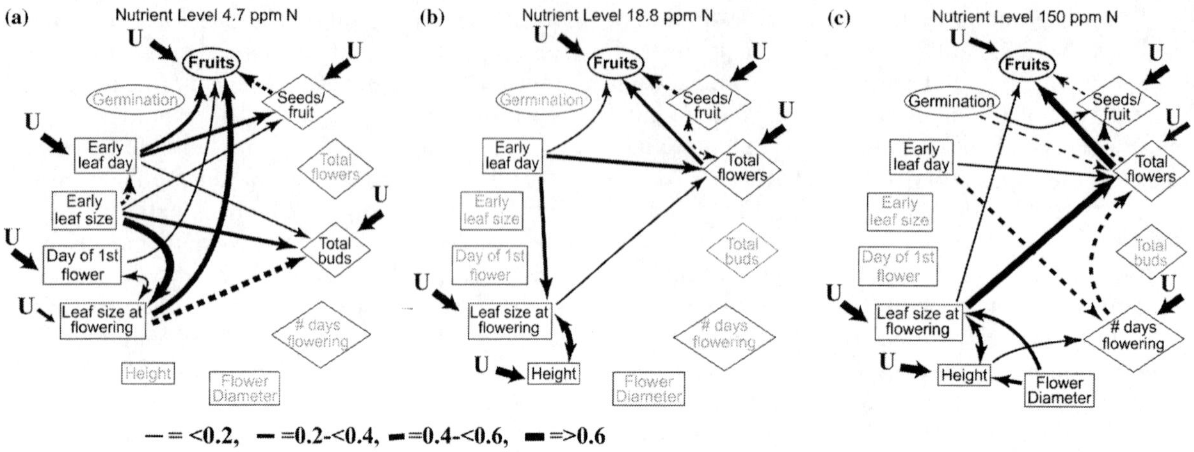

Figure 1. Exploratory path diagrams for three different nutrient levels. The faded boxes are traits that were not included in the model ($p > 0.2$ for the selection gradient associated with that trait). For clarity, only the significant paths are shown. Dashed lines indicate negative coefficients. Solid lines are positive coefficients. The thickness of the line indicates the size of the effect (absolute value). U = unexplained variation.

Table 1. Tests the fit of the best exploratory path model and of other nutrients by other chosen models for each nutrient environment

Source of Data (Nppm)	Test Model		
	4.7 Nppm χ^2, AIC	18.8 Nppm χ^2, AIC	150 Nppm χ^2, AIC
4.7	7.69, −2.31 (0.1741)	8.77, 2.77 (0.0325)	189.30, 153.30 (0.0001)
18.8	28.48, 18.48 (0.0001)	1.94, −4.06 (0.5852)	77.66, 41.66 (0.0001)
150	17.43, 7.43 (0.0037)	6.83, 0.83 (0.0775)	22.48, −13.52 (0.2115)

For each test the χ^2 value (significance) is reported and Akaike's Information Criterion (AIC) where the smaller values indicate a better fit of the model to the data.

not fit the others' data (Table 1). Akaike's Information Criterion indicates that the other models were a poor fit to data. The moderate nutrient environment model could not be rejected for the higher nutrient data. This pattern of results suggests the more extreme environments have a greater difference in pattern of selection than the more moderate environments. It is suggested by this result that more strongly contrasting environments would be more likely to maintain greater genetic diversity than more similar environments. This result is consistent with others' findings of decreasing values of the across environment genetic correlations as the environments diverge (i.e., Kassen & Bell, 2000).

Since this is using an exploratory approach this should be seen as development of hypotheses that need further testing. For example, to test if selection in low nutrient environments is stronger on early traits, while selection in high nutrient environments is more on later traits, one could grow a new set of genetic lines in a couple of low and high nutrient environments. Measuring the same set of traits as in this study would allow for traditional path models to be constructed and tested for these patterns of selection.

Another hypothesis resulting from this analysis is that more strongly contrasting environments result in a greater contrast in phenotypic selection, and therefore a greater potential to maintain genetic variation. In order to test this hypothesis I would suggest an experimental evolutionary approach of growing and selecting *B. rapa* (or any other quick life-cycle species) in heterogeneous environments that are very similar or very different. After a number of generations of selection the extent of genetic variation in the different selection treatments would be determined. Experimental evolutionary approaches have the advantage of more clearly determining the cause and effect relationship than other approaches.

Experimental evolutionary studies

A particularly powerful method for examining and determining if heterogeneous environments maintain genetic variation is an experimental evolutionary approach (for a review see Kassen, 2002). For example, starting with a set of genotypes that are then exposed to uniform or heterogeneous environments and then assaying for genetic variation is a much more direct test. For most of the above studies it is not possible to determine if environmental variation is the cause of the maintenance of the genetic variation. In part, what is missing is the history of selection pressure and genetic responses through time that has produced the observed phenotypic and genetic patterns. Natural environments are very heterogeneous, and consequently selection histories are very complex. Therefore, to determine the importance of environmental heterogeneity and genetic architecture on maintenance of genetic variation I propose that it is essential to take an experimental evolutionary approach. Only by using an experimental approach can the selection history and changes in the genetic architecture be known, thus allowing knowledge of some of the genetic detail as it affects the phenotype (Rose et al., 1996; Bell, 1997a). An experimental approach will provide simplified experimental conditions resulting in our ability to directly test evolutionary predictions and make conclusions. Since this general approach requires an organism with a very short generation time, it has only been used for a limited number of model systems.

There are two general types of experimental studies of evolution: (1) shorter-term experiments where examination of selection, response, and maintenance of genetic variation in different environmental treatments will mostly be due to initial genetic variation; and (2) longer-term experiments where the results will be influenced by initial genetic variation but also genetic variation as a result of mutation. Many of the same mechanisms maintaining genetic variation discussed above have been examined in experimental evolution studies.

The expression of negative genetic correlations and genotype by environment interactions have been found in several experimental evolution studies to be environmentally dependent, and unusual artifacts (loss of a trade-off) may arise due to selection in laboratory conditions (Leroi et al., 1994a,b). The environmental conditions for selection can also influence the outcome of maintenance of genetic variation and/or selection for phenotypic plasticity. Scheiner and Yampolsky (1998) used experimental populations of *D. pulex* in temporally varying environments, which had a limited effect on the maintenance of genetic diversity. However, in this experiment there was apparently low heritable genetic variation for the traits of interest in the particular environments. Therefore, both the environmental conditions and the genetic lines need to be carefully chosen to allow for tests of the theories or mechanisms. Here I will briefly discuss some examples.

A population of *E. coli*, initially derived from a single individual was grown for 2000 generations at 37°C (Bennett et al., 1992). Then this line was subjected to further selection in one of three constant temperatures (32, 37, or 42°C) or a daily alternating temperature (32 or 42°C) to determine if specialists for the new environmental conditions would arise. The newly selected lines were then grown in competition with the initial selected lines (37°C) in the three constant temperatures to determine if the new lines were specialists for their temperature. The *E. coli* outcompeted the ancestor line (37°C) in the temperature for which they were selected which would support specialization. However, the expected negative correlations or trade-offs with their relative success in other environments was not found (Bennett et al., 1992). A later assay of the 37°C selected lines (20,000 generations), when grown at more extreme temperatures (20 or 41°C) showed evidence of antagonistic pleiotropy (Cooper et al., 2001). In addition the expression of genetic variation was greatest in the extreme environments. This series of studies supports the findings of others that in more extreme environments there was an increase in the expression of antagonistic pleiotropy.

Bell and colleagues have undertaken an extensive series of experimental evolution studies in *Chlamydomonas*. In part, they found that heterogeneous environments maintain a greater genetic diversity through frequency dependent selection and negative cross environment genetic correlations (e.g., Bell, 1991; Bell & Reboud, 1997; Kassen & Bell, 1998; Kassen & Bell, 2000). In one study, he examined the effect of selection in uniform vs. heterogeneous nutrient environments and the relative effect on the growth rate of *Chlamydomonas* populations (Bell, 1997b). There was a loss of genetic variation in the more uniform environments and the genetic correlations across the environments were negative suggesting specialization would result and genetic variation would be maintained in the heterogeneous environments. Furthermore, he suggested that with environmental heterogeneity, the theoretical end points (equilibrium) are just not obtained, and therefore genetic diversity may be easier to maintain than predicted by theory.

Concluding remarks and future directions

The dilemma of the maintenance of genetic diversity has for some time been a major focus of evolutionary biologists and perhaps with some of the newer tools such as QTL we can further resolve this issue. The two suggested approaches presented in this paper, exploratory path analysis and experimental evolution, could be used for gaining insights as to when genetic variation is maintained. Currently, we have substantial evidence for genetic constraints through the genetic architecture (genetic correlations) but it is unclear how extensive these constraints are across all species. Here I will just list areas that I believe are in particular need of further research.

(1) What is the distribution of mutational effects in wild populations? The distribution of mutational effects is mostly being quantified in model systems in uniform environments. For understanding the maintenance of diversity in naturally variable populations the variation of mutational effects would likely impact the genetic diversity maintained by selection balance. Currently the extent of variation in mutational effects in variable natural habitats is not known.

(2) Do wild populations ever reach the theoretical equilibrium where the standing genetic variation will be determined by the balance between mutation and selection? Perhaps the genetic variation present in many populations is primarily due to not reaching the theoretical equilibrium. Certainly many of the mechanisms discussed here would constrain reaching the equilibrium. A recent review on phenotypic selection in the wild found that selection was mostly fairly weak but also found it was highly variable across studies (Kingsolver et al., 2001). Likely much of this variation in estimates of selection gradients is due to variation in the low statistical power of many of the studies; it also may reflect that selection is highly variable. Stronger field estimates of the dynamics of selection and constraints due to the genetic architecture in the context of the natural environment are needed.

(3) Only a limited number of species and types of environmental variation have been explored through the use of the experimental evolution method. This approach has great potential since the initial genotypes and the experimental environments can be more carefully controlled. This approach is particularly strong for testing predictions of models.

(4) Although the experimental evolutionary approach is useful it should not replace field experiments since the lab cannot mimic the complexity of natural conditions. As pointed out by Gillespie and Turelli (1989), the experimental detection of genotype by environment interactions and the preservation of genetic variation depends on the type and range of environmental variation. Therefore experiments, whether in the lab or field, when possible should reflect the relevant range of environments for the species (e.g., Shaw et al., 1995; Jia et al., 2000).

(5) Further work on the expression of QTL in heterogeneous environments is needed to give further insight to the genetic basis of the genotype by environment interaction. The limited work in this area indicates that the expression of QTL is very influenced by the environment (e.g., Vieira et al., 2000).

(6) Further information on both the distribution of genetic variation and genetic correlations

of traits important for adaptation and the extent that their expression is environmentally dependent is needed.

(7) There is a need for a set of more comprehensive models on the evolution of adaptive traits. Models need to include the complex genetic base of quantitative genetic traits and in that context address how and when heterogeneous environments will maintain genetic variation.

Acknowledgements

A Faculty Research Support Grant from Illinois State University supported this work. I am grateful to Charles Bristow, Steven Juliano, Costas Mannouris and the anonymous reviewers for commenting and improving earlier versions of this paper. Thanks to Aaron Warsaw for patiently crossing many of the *Brassica* plants and counting numerous fruits, and to Don Schmidt for the excellent care of the plants in the greenhouse.

References

Antonovics, J. & A.D. Bradshaw, 1970. Evolution in closely adjacent plant populations VIII. Clinal patterns at a mine boundary. Heredity 25: 349–362.

Arnold, S.J., 1992. Constraints on phenotypic evolution. Am. Nat. 140: S85–S107.

Bell, G., 1991. The ecology and genetics of fitness in *Chlamydomonas* IV. The properties of mixtures of genotypes of the same species. Evolution 45: 1036–1046.

Bell, G., 1992. The ecology and genetics of fitness in *Chlamydomonas* V. The relationship between genetic correlation and environmental variance. Evolution 46: 561–566

Bell, G., 1997a. Selection: the Mechanism of Evolution. Chapman and Hall, New York.

Bell, G.A.C., 1997b. Experimental evolution in *Chlamydomonas* I. Short-term selection in uniform and diverse environments. Heredity 78: 490–497.

Bell, G. & X. Reboud, 1997. Experimental evolution in *Chlamydomonas* II. Genetic variation in strongly contrasted environments. Heredity 78: 498–506.

Bennett, A.F., R.E. Lenski & J.E. Mittler, 1992. Evolutionary adaptation to temperature I. Fitness responses of *Escherichia coli* to changes in its thermal environment. Evolution 46: 16–30.

Borash, D.J., A.G. Gibbs, A. Joshi & L.D. Mueller, 1998. A genetic polymorphism maintained by natural selection in a temporally varying environment. Am. Nat. 151: 148–156.

Bossart, J.L. & J.M. Scriber, 1995. Maintenance of ecologically significant genetic variation in the tiger swallowtail butterfly through differential selection and gene flow. Evolution 49: 1163–1171.

Bulmer, M.G., 1985. The Mathematical Theory of Quantitative Genetics. Oxford Univ. Press., New York.

Bulmer, M.G., 1989. Maintenance of genetic variability by mutation-selection balance: a child's guide through the jungle. Genome 31: 761–767.

Campbell, D.R., 1996. Evolution of floral traits in a hermaphroditic plant: field measurements of heritabilities and genetic correlations. Evolution 54: 1442–1453.

Chang, S.-M. & R.G. Shaw, 2003. The contribution of spontaneous mutation to variation in environmental response in *Arabidopsis thaliana:* response to nutrients. Evolution 57: 984–994.

Charlesworth, B. & K.A. Hughes, 1999. The maintenance of genetic variation in life-history traits, pp. 369–391 in Evolutionary Genetics: From Molecules to Morphology, Vol. 1., edited by R.S. Singh & C.B. Krimbas. Cambridge University Press.

Cheetham, A.H., J.B.C. Jackson & L.A.C. Hayek, 1995. Quantitative genetics of bryozoan phenotypic evolution 3. Phenotypic plasticity and the maintenance of genetic variation. Evolution 49: 290–296.

Cheplick, G.P., 2003. Evolutionary significance of genotypic variation in developmental reaction norms for a perennial grass under competitive stress. Evol. Ecol. 17: 175–196.

Christiansen, F.B., 1974. Sufficient conditions for protected polymorphism in a subdivided population. Amer. Nat. 108: 157–166.

Conner, J.K., 1996. Understanding natural selection: an approach integrating selection gradients, multiplicative fitness components, and path analysis. Ethol. Ecol. Evol. 8: 387–397.

Conner, J.K. & S. Rush, 1997. Measurements of selection on floral traits in black mustard, *Brassica nigra*. J. Evol. Biol. 10: 327–335.

Conner, J.K., S. Rush & P. Jennetten, 1996. Measurements of natural selection on floral traits in wild radish (*Raphanus raphanistum*) I. Selection through lifetime female fitness. Evolution 50: 1127–1136.

Cooper, V.S., A.F. Bennett & R.E. Lenski, 2001. Evolution of thermal dependence of growth rate of *Escherichia coli* populations during 20,000 generations in a constant environment. Evolution 55: 889–896.

Curtsinger, J.W., P.M. Service & T. Prout, 1994. Antagonistic pleiotropy, reversal of dominance, and genetic polymorphism. Am. Nat. 144: 210–228.

Donohue, K. & J. Schmitt, 1999. The genetic architecture of plasticity to density in *Impatiens capensis*. Evolution 53: 1377–1386.

Dudley, S.A., 1996. The response to differing selection on plant physiological traits: evidence for local adaptation. Evolution 50: 103–110.

Elena, S.F. & R.E. Lenski, 1997. Long-term experimental evolution in *Escherichia coli* VII. Mechanisms maintaining genetic variability within populations. Evolution 51: 1058–1067.

Endler, J.A., 1986. Natural Selection in the Wild. Princeton University Press, Princeton, NJ.

Evans, A.S., 1989. Phenotypic Plasticity of Physiological and Life History Characteristics: Evolutionary Implications in Rapid-cycling *Brassica campestris* L. Ph.D. Thesis. The University of Chicago, Chicago, IL.

Evans, A.S., 1991. Leaf physiological aspects of nitrogen-use efficiency in *Brassica campestris* L.: quantitative genetic variation across nutrient treatments. Theor. Appl. Genet. 81: 64–70.

Falconer, D.S., 1952. The problem of environment and selection. Am. Nat. 86: 293–298.

Falconer, D.S. & T.F.C. Mackay, 1996. Introduction to Quantitative Genetics, 4th edn., Longman Group, Ltd., Essex, England.

Felsenstein, J., 1976. The theoretical population genetics of variable selection and migration. Annu. Revi. Genet. 10: 253–280.

Fisher, R.A., 1999. The Genetical Theory of Natural Selection. Complete Variorum edn. Oxford University Press, Oxford.

Galloway, L.F., 1995. Response to natural environmental heterogeneity: maternal effects and selection on life-history characters and plasticities in *Mimulus guttatus*. Evolution 49: 1095–1107.

Gillespie, J.H. & M. Turelli, 1989. Genotype-environment interactions and the maintenance of polygenic variation. Genetics 121: 129–138.

Gimelfarb, A., 1990. How much genetic variation can be maintained by genotype-environment interactions? Genetics 124: 443–445.

Grime, J.P., 1994. The role of plasticity in exploiting environmental heterogeneity, in Exploitation of Environmental Heterogeneity by Plants, edited by M.M. Caldwell & R.W. Pearcy. Academic Press, San Diego.

Harris, D.J. & J.S. Jones, 1995. Genotype-specific habitat selection and thermal ecology in *Nucella lapillus* (L.) (the dogwhelk). Heredity 74: 311–314.

Hatcher, L., 1994. A Step-by-Step Approach to Using the SAS System for Factor Analysis and Structural Equation Modeling. SAS Institute Inc. Pub. Cary, NC.

Hedrick, P.W., 1986. Genetic polymorphism in heterogeneous environments: a decade later. Ann. Rev. Ecol. Syst. 17: 535–566.

Houle, D., 1992. Comparing evolvability and variability of quantitative traits. Genetics 130: 195–204.

Houle, D., 1998. How should we explain variation in the genetic variance of traits? Genetica 102/103: 241–251.

Jia, F.-Y., M.D. Greenfield & R.D. Collins, 2000. Genetic variance of sexually selected traits in waxmoths: maintenance by genotype by environment interaction. Evolution 54: 953–967.

Kalisz, S. & G.M. Wardle, 1994. Life history variation in *Campanula Americana* (Campanulaceae): population differentiation. Am. J. Bot. 81: 521–527.

Karan, D., J.-P. Morin, P. Gibert, B. Moreteau, S.M. Scheiner & J.R. David, 2000. The genetics of phenotypic plasticity. IX. Genetic architecture, temperature, and sex differences in *Drosophila melanogaster*. Evolution 54: 1035–1040.

Kassen, R., 2002. The experimental evolution of specialists, generalists, and the maintenance of diversity. J. Evol. Biol. 15: 173–190.

Kassen, R. & G. Bell, 1998. Experimental evolution in *Chlamydomonas*. IV. Selection in environments that vary through time at different scales. Heredity 80: 732–741.

Kassen, R. & G. Bell, 2000. The ecology and genetics of fitness in *Chlamydomonas* X. The relationship between genetic correlation and genetic distance. Evolution 54: 425–432.

Kause, A., I. Saloniemi, J.-P. Morin, E. Haukioja, S. Hanhimäki & K. Ruohomäki, 2001. Seasonally varying diet quality and the quantitative genetics of development time and body size in birch feeding insects. Evolution 55: 1992–2001.

Kingsolver, J.G. & D.W. Schemske, 1991. Path analyses of selection. Trends Ecol. Evol. 6: 276–280.

Kingsolver, J.G., H.E. Hoekstra, J.M. Hoekstra, D. Berrigan, S.N. Vignieri, C.E. Hill, A. Hoang, P. Gilbert & P. Beerli, 2001. The strength of phenotypic selection in natural populations. Am. Nat. 157: 245–261.

Lande, R. & S.J. Arnold, 1983. The measurement of selection on correlated characters. Evolution 37: 1210–1226.

Leroi, A.M., A.K. Chippindale & M.R. Rose, 1994a. Long-term laboratory evolution of a genetic life-history trade-off in *Drosophila melanogaster*. 1. The role of genotype-by-environment interaction. Evolution 48: 1244–1257.

Leroi, A.M., W.R. Chen & M.R. Rose, 1994b. Long-term laboratory evolution of a genetic life-history trade-off in *Drosophila melanogaster*. 2. Stability of genetic correlations. Evolution 48: 1258–1268.

Levene, H., 1953. Genetic equilibrium when more than one ecological niche is available. Am. Nat. 87: 331–333.

Levins, R., 1968. Evolution in Changing Environments. Monographs in Population Biology. Vol. 2. Princeton Univ. Press, Princeton, NJ.

Li, B., J.-I. Suzuki & T. Hara, 1998. Latitudinal variation in plant size and relative growth rate in *Arabidopsis thaliana*. Oecologia 115: 293–301.

Luttikhuizen, P.C., J. Drent, W. van Delden & T. Piersma, 2003. Spatially structure genetic variation in a broadcast spawning bivalve: quantitative vs. molecular traits. J. Evol. Biol.16: 260–272.

Lynch, M., 1996. A quantitative-genetic perspective on conservation issues, pp. 471–501 in Conservation Genetics: Case Histories From Nature, edited by J.C. Avise & J.L. Hamrick. Chapman and Hall, New York.

Lynch, M. & B. Walsh, 1998. Genetics and Analysis of Quantitative Traits. Sinauer Associates, Sunderland, MA.

Lynch, M., L. Latta, J. Hicks & M. Giorgianni, 1998. Mutation, selection, and the maintenance of life-history variation in a natural population. Evolution 52: 727–733.

Lynch, M., M. Pfrender, K. Spitze, N. Lehman, J. Hick, D. Allen, L. Latta, M. Ottene, F. Bogue & J. Colbourne, 1999. The quantitative and molecular genetic architecture of a subdivided species. Evolution 53: 100–110.

Mackay, T.F.C., 1981. Genetic variation in varying environments. Gen. Res. 37: 79–93.

Mackay, T.F.C., 2001. The genetic architecture of quantitative traits. Ann. Rev. Gen. 35: 303–339.

Maynard Smith, J. & R. Hoekstra, 1980. Polymorphism in a varied environment: how robust are the models? Gen. Res. 35: 45–57.

Mitchell, R.J., 1993. Path analysis: pollination, pp. 211–231 in Design and Analysis of Ecological Experiments, edited by

S.M. Scheiner & J. Gurevitch. Chapman and Hall, New York.
Mitton, J.B., 1997. Selection in Natural Populations. Oxford University Press, New York.
Mopper, S., K. Landau & P. Van Zandt, 2000. Adaptive evolution and neutral variation in a wild leafminer metapopulation, pp. 116–138 in Adaptive Genetic Variation in the Wild, edited by T.A. Mousseau, B. Sinervo & J. Endler. Oxford University Press, Oxford.
Moran, N.A., 1991. Phenotype fixation and genotypic diversity in the complex life cycle of the aphid *Pemphigus betae*. Evolution 45: 957–970.
Mousseau, T.A. & D.A. Roff, 1987. Natural selection and the heritability of fitness components. Heredity 59: 181–197.
Mousseau, T.A., B. Sinervo & J. Endler, 2000. Adaptive Genetic Variation in the Wild. Oxford University Press, New York.
Pigliucci, M. & C.D. Schlichting, 1998. Reaction norms of *Arabidopsis*. V. Flowering time controls phenotypic architecture in response to nutrient stress. J. Evol. Biol. 11: 285–301
Prout, T., 1968. Sufficient conditions for multiple niche polymorphism. Am. Nat. 102: 493–496.
Prout, T. & O. Savolainen, 1996. Genotype-by-environment interaction is not sufficient to maintain variation: levene and the leafhopper. Am. Nat. 148: 930–936.
Rausher, M.D., 1992. The measurement of selection on quantitative traits: biases due to environmental covariance between traits and fitness. Evolution 46: 616–626.
Reed, D. H. & R. Frankham, 2001. How closely correlated are molecular and quantitative measures of genetic variation? A meta-analysis. Evolution 55: 1095–1103.
Richard, M., T. Bernhardt & G. Bell, 2000. Environmental heterogeneity and the spatial structure of fern species diversity in one hectare of old-growth forest. Ecography 23: 231–245.
Richards, A.D., 1986. Plant Breeding Systems. Allen and Unwin, London.
Roff, D.A., 1997. Evolutionary Quantitative Genetics. Chapman and Hall, New York.
Roff, D.A. & T.A. Mousseau, 1987. Quantitative genetics and fitness: lessons from *Drosophila*. Heredity 58: 103–118.
Rose, M.R., T.J. Nusbaum & A.K. Chippindale, 1996. Laboratory evolution: the experimental wonderland and the Cheshire cat syndrome, pp. 221–241 in Adaptation, edited by M.R. Rose & G.V. Lauder. Academic Press, Inc., San Diego, CA.
SAS, 2001. Statistical Analysis System, SAS Institute Inc., SAS/STAT software version 8.2 Cary, NC.
Sasaki, A. & G. de Jong, 1999. Density dependence and unpredictable selection in a heterogeneous environment: compromise and polymorphism in the ESS reaction norm. Evolution 53: 1329–1342.
Scheiner, S.M., 1993. Genetics and evolution of phenotypic plasticity. Ann. Rev. Ecol. Syst. 34: 35–68.
Scheiner, S.M. & H.S. Callahan, 1999. Measuring natural selection on phenotypic plasticity. Evolution 53: 1704–1713.
Scheiner, S.M., R.J. Mitchell & H.S. Callahan, 2000. Using path analysis to measure natural selection. J. Evol. Biol. 13: 423–433.
Scheiner, S.M. & L.Y. Yampolsky, 1998. The evolution of *Daphnia pulex* in a temporally varying environment. Genet. Res. 72: 25–37.
Schmidt, P.S. & D.M. Rand, 2001. Adaptive maintenance of genetic polymorphism in an intertidal barnacle: habitat- and life-stage-specific survivorship of *Mpi* genotypes. Evolution 55: 1336–1344.
Shaw, R.G., G.A.J. Platenkamp, F.H. Shaw & R.H. Podolsky, 1995. Quantitative genetics of response to competitors in *Nemophila menziesii*: a field experiment. Genetics 139: 397–406.
Shaw, R.G., D.L. Byers & E. Darmo, 2000. Spontaneous mutational effects on reproductive traits of *Arabidopsis thaliana*. Genetics 155: 369–378.
Shipley, B., 1997. Exploratory path analysis with applications in ecology and evolution. Am. Nat. 149: 1113–1138.
Shipley, B., 2000. Cause and Correlation In Biology, a User's Guide To Path Analysis, Structural Equations and Causal Inference. Cambridge U. Press, Cambridge.
Sinervo, B., 2000. Adaptation, natural selection, and optimal life-history allocation in the face of genetically based trade-offs, pp. 41–64 in Adaptive Genetic Variation in the Wild, edited by T.A. Mousseau, B. Sinervo & J. Endler. Oxford University Press, Oxford.
Sokal, R.R. & F.J. Rohlf, 1995. Biometry, 3rd edn., W.H. Freeman and Co., New York.
Spirtes, P., C. Glymour & R. Scheines, 1993. Causation, prediction and search. Springer-Verlag, New York.
Stanton, M.L., C. Galen & J. Shore, 1997. Population structure along a steep environmental gradient: consequences of flowering time and habitat variation in the snow buttercup, *Ranunculus adoneus*. Evolution 51: 79–94.
Stratton, D.A., 1995. Spatial scale of variation in fitness of *Erigeron annuus*. Am. Nat. 146: 608–624.
Takano, T., S. Kusakabe & T. Mukai, 1987. The genetic structure of natural populations of *Drosophila melanogaster*. XX. Comparison of genotype-environment interaction in viability between a northern and a southern population. Genetics 117: 245–254.
Tauber, T.A. & M.J. Tauber, 1992. Phenotypic plasticity in *Chrysoperla*: genetic variation in the sensory mechanism and in correlated reproductive traits. Evolution 46: 1754–1773.
Vavrek, M.C, J.B. Mcgraw & H.S. Yang 1996. Within-population variation in demography of *Taraxacum officinal* – maintenance of genetic diversity. Ecology 77: 2098–2107.
Van Kleunen, M. & M. Fischer, 2001. Adaptive evolution of plastic foraging responses in a clonal plant. Ecology 82: 3309–3319.
Vieira, C., E.G. Pasyukova, Z.-B. Zeng, J.B. Hackett, R.F. Lyman & T.F.C. Mackay, 2000. Genotype-environment interaction for quantitative trait loci affecting life span in *Drosophila melanogaster*. Genetics 154: 213–227.
Via, S., 1987. Genetic constraints on the evolution of phenotypic plasticity, pp 47–71 in Genetic Constraints on Adaptive Evolution, edited by V. Loeschcke. Springer-Verlag, Berlin.
Via, S., R. Gomulkiewicz, G. de Jong, S.M. Scheiner, C.D. Schlichting & P.H. van Tienderen, 1995. Adaptive phenotypic plasticity: consensus and controversy. Trends Ecol. Evol. 10: 212–217.

Via, S. & R. Lande, 1985. Genotype-environment interaction and the evolution of phenotypic plasticity. Evolution 39: 505–522.

Williams, P.H. & C.B. Hill, 1986. Rapid-cycling populations of Brassicas. Science 232: 1385–1389.

Windig, J.J., 1997. The calculation and significance testing of genetic correlations across environments. J. Evol. Bio. 10: 853–874.

Zhivotovsky, L.A., M.W. Feldman & A. Bergman, 1996a. Fitness patterns and phenotypic plasticity in a spatially heterogeneous environment. Gen. Res. 68: 241–248.

Zhivotovsky, L.A., M.W. Feldman & A. Bergman, 1996b. On the evolution of phenotypic plasticity in a spatially heterogeneous environment. Evolution 50: 547–558.

Genetica (2005) 123: 125–136 © Springer 2005

The genetic basis of adaptation: lessons from concealing coloration in pocket mice

Michael W. Nachman
Department of Ecology and Evolutionary Biology, University of Arizona, Tucson, AZ 85721,USA
(Phone: +1-520-626-4595; Fax: +1-520-621-9190; E-mail: nachman@u.arizona.edu)

Received 18 August 2003 Accepted 1 January, 2004

Key words: Chaetopidus, melanocortin receptor, natural selection, pigmentation

Abstract

Recent studies on the genetics of adaptive coat-color variation in pocket mice (*Chaetodipus intermedius*) are reviewed in the context of several on-going debates about the genetics of adaptation. Association mapping with candidate genes was used to identify mutations responsible for melanism in four different populations of *C. intermedius*. Here, I review four main results (i) a single gene, the melanocortin-1-receptor (*Mc1r*), appears to be responsible for most of the phenotypic variation in color in one population, the Pinacate site; (ii) four or fewer nucleotide changes at *Mc1r* appear to be responsible for the difference in receptor function; (iii) studies of migration-selection balance suggest that the selection coefficient associated with the dark *Mc1r* allele at the Pinacate site is large; and (iv) different (unknown) genes underlie the evolution of melanism on three other lava flows. These findings are discussed in light of the evolution of convergent phenotypes, the average size of phenotypic effects underlying adaptation, the evolution of dominance, and the distinction between adaptations caused by changes in gene dosage versus gene structure.

Introduction

More than a century after the publication of 'The Origin of Species' many questions about the genetics of adaptation remain unanswered. Darwin (1859) provided a mechanism for evolution, but he was unaware of Mendel, and thus early evolutionary theory was developed without an accurate understanding of the nature of inheritance. The integration of Mendelian inheritance with evolutionary theory was provided by the work of Haldane, Fisher, and Wright, who, among many other things, developed the first models of the dynamics of allele frequency change under various forms of selection (Fisher, 1930; Wright, 1931; Haldane, 1932). In these models, fitness is typically summarized by a single parameter, the selection coefficient, which is usually associated with a particular allele at a single locus. Early empirical studies of adaptation proceeded somewhat independently of the theoretical studies of Fisher, Wright and Haldane. Empiricists such as Dobzhansky (1937, 1970), Dice (1940), Mayr (1942, 1963), Lack (1947), Stebbins (1950) and others began to describe geographic and temporal patterns of phenotypic variation, and many of these patterns provided convincing, though indirect, evidence for selection.

Natural selection acts on the phenotype, but it is the genotype that is passed from one generation to the next. Nonetheless, even today, relatively few studies have been able to make links between genotype and phenotype for traits under selection. To a considerable extent, theoretical studies (often dealing mostly with genotypes) and empirical studies (often dealing mainly with phenotypes) have remained divorced from each other. In principle, finding the genes underlying adaptation might allow us to bring these two approaches together; that is, to study the ecology of adaptation in the context of explicit population genetic models.

Some of the best examples of the genetic basis of phenotypic responses to selection involve anthropogenic influences, either intentionally through artificial selection, or accidentally through human-induced changes to the environment. It is well known that the first chapter of The Origin of Species (Darwin, 1859) describes extensive changes in phenotype caused by selective breeding. There is now an enormous literature on both plant and animal breeding, and in some cases, the specific genes underlying response to artificial selection have been identified (e.g., Doebley, Stec & Hubbard, 1997; Wang et al., 1999; Newton et al., 2000). Examples of responses to human disturbance include insecticide, herbicide, and drug resistance (Palumbi, 2001; Reznick & Ghalambor, 2001), and in many cases, the genes underlying theses traits have also been identified (e.g., Fidock et al., 2000; Raymond et al., 2001; Walsh, 2000; Cowen, Anderson & Kohn, 2002; Daborn et al., 2002; Wootton et al., 2002; Hughes, 2003). One potential limitation of both kinds of studies for developing a more general understanding of the genetic basis of adaptation is that selection caused by anthropogenic influence is likely to be unusually strong (Darwin, 1859; Reznick & Ghalambor, 2001). Ideally we would like to be able to make links between genotype and phenotype for fitness-related traits in a more natural setting.

Many general questions about the genetics of adaptation remain, and in principle, might be answered by identifying the genes underlying adaptive phenotypes. For example, do adaptations result from the fixation of many mutations individually of small effect (Fisher, 1932), or do they involve single mutations of large effect, as documented for insecticide resistance (e.g. Daborn et al., 2002)? Are most adaptive mutants dominant as suggested by Haldane (1924), and do they correspond to gain-of-function mutations at the molecular level (Wright, 1934)? What kinds of molecular changes result in adaptation; are most adaptations the result of changes in protein structure or changes in gene regulation (Britten & Davidson, 1969)? How common are pleiotropy and epistasis? Do epistatic interactions typically involve other mutations in the same gene or mutations in different genes (Kondrashov, Sunyaev & Kondrashov, 2002)? With the ultimate goal of addressing these and related questions, we have taken a candidate-gene approach to understand the genetic basis of adaptive melanism in the rock pocket mouse, *Chaetodipus intermedius*. While some of these questions can be addressed without identifying the specific mutations underlying a trait, others cannot. Using a candidate-gene approach also has some serious limitations, as discussed below. First, I describe the relevant natural history of pocket mice, including variation in pigmentation. Second, I describe the genetics and biochemistry of mammalian pigmentation and the power and limitations of a candidate-gene approach in this system. Finally, I describe some of our chief findings and their implications for addressing the questions above.

Pigmentation variation in rock pocket mice

The rock pocket mouse, *Chaetodipus intermedius*, is a small rodent that inhabits rocky areas and desert scrub at low elevations principally in the Sonoran and Chihuahuan deserts. Its range includes southern Arizona, southern New Mexico, western Texas, and adjacent areas in northern Mexico. Pocket mice are in the family Heteromyidae, a New World family of rodents that includes six genera (*Chaetodipus*, *Perognathus*, *Dipodomys*, *Microdipodops*, *Liomys*, and *Heteromys*) and has its center of diversification in xeric habitats of Central and North America. Heteromyid rodents are distantly related to murid rodents, such as laboratory mice (*Mus domesticus*). Like many species of heteromyids, rock pocket mice are well adapted for deserts: they are strictly nocturnal and remain in underground burrows during the heat of the day. Pocket mice are so named because of external cheek pouches which are used to carry seeds during bouts of foraging. Pocket mice can subsist entirely on a dry diet and do not require free water. *C. intermedius* is restricted to rocky habitats, and is broadly sympatric with *C. penicillatus*, its sister species, which is found in more sandy habitats.

In most parts of its range, *C. intermedius* has a light, sandy-colored dorsal pelage and lives on light-colored rocks. In several different regions throughout its range, however, *C. intermedius* is found on lava flows which are typically dark in color. The mice on these lava flows typically have a melanic dorsal pelage. Examples of typical habitat are shown in Figure 1, and variation in coat color is shown in Figure 2. The lava flows on which the mice are found tend to be geographically isolated from one another and vary in size from a few km^2 to over 1500 km^2, and they vary in age from less than 1000 years old to nearly 2 million years old (Hoekstra & Nachman, 2003). Lava flows are typically separated from one another by intervening habitat consisting either of light-colored rocks, which is suitable habitat for *C. intermedius*, or sand, which is unsuitable habitat for *C. intermedius*. This system was first described in detail in the 1930's by Benson (1933) and Dice and Blossom (1937) who documented a strong positive association between the color of the mice and the color of the substrate on which the mice live. Dice and Blossom noted that owls are major predators of these mice, and suggested that the variation in mouse coat color served as concealing coloration from predators.

Figure 1. Typical habitats for *C. intermedius* showing light rocks (a) and dark lava (b).

While the phenotypic variation in color would seem to be a good example of crypsis to avoid predation, an obvious question, given that pocket mice are nocturnal, is whether owls discriminate between light and dark mice (on either light or dark backgrounds) while foraging at night. Dice (1947) conducted such experiments with two species of owls (Barn owl and Long-eared owl) in enclosures using varying degrees of illumination. Dice showed that owls capture approximately twice as many conspicuously colored mice as concealingly colored mice, even in near total darkness. Interestingly, this difference was seen only in enclosures containing a complex substrate with places for the mice to hide. When the experiment was done in an enclosure with a bare substrate, owls did not discriminate between conspicuously colored and concealingly colored mice. Moreover, on bare substrate, owls captured equal numbers of mice in low-light and in total darkness, suggesting that in this simplified situation owls hunt effectively using only hearing (Dice, 1947). These experiments were conducted using dark-colored and light-colored deer mice (*Peromyscus maniculatus*), rather than pocket mice, and comparable experiments have not been conducted with rock pocket mice. Nonetheless, the difference between light and dark *C. intermedius* is greater than the difference between light and dark *P. maniculatus*, so it seems likely that similar or more extreme results would be obtained with pocket mice. The close match between mouse color and substrate color across a wide range of populations (Dice & Blossom, 1937), the fact that owls are known to be major predators of pocket mice, and the fact that owls can effectively discriminate between light and dark mice even in low light conditions all suggest that the variation in coat color of *C. intermedius* is an adaptation to avoid predation. It is unlikely that variation in coat-color plays a significant role in thermoregulation since these mice are nocturnal and typically do not emerge from their burrows until ambient temperatures are below body temperature.

Candidate genes: the pigmentation process in mammals

This system is amenable to genetic analysis because of the wealth of information on the

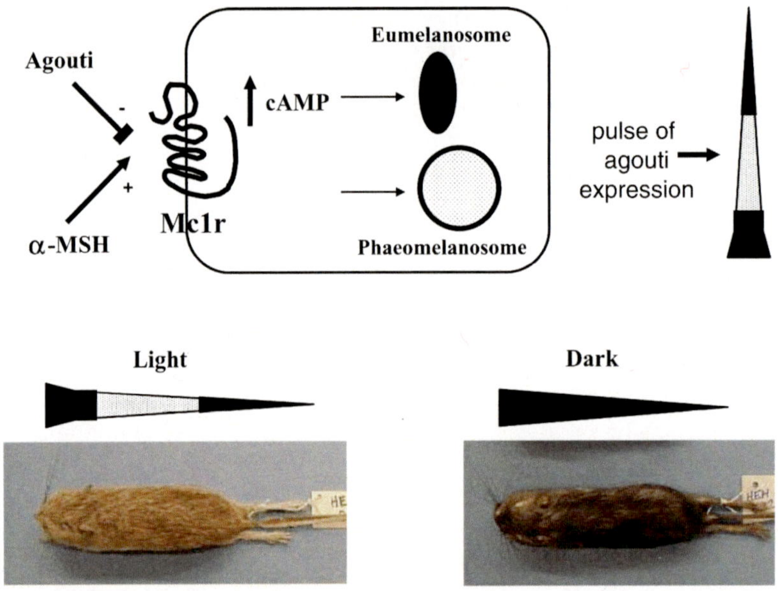

Figure 2. Regulatory control of melanogenesis (top) and typical light and dark *C. intermedius* (bottom). Alpha-MSH signals MC1R, resulting in higher levels of cAMP and production of eumelanin. Agouti is an antagonist that increases production of phaeomelanin. Agouti expression during the haircycle results in a banding pattern on individual hairs, a phenotype known as the 'agouti' hair (shown at right). Light *C. intermedius*, typically found on light-colored rocks, have agouti hairs on their dorsum, while dark *C. intermedius*, typically found on lava, have unbanded, uniformly melanic hairs on their dorsum. See text for further details.

genetics, development, and biochemistry of pigmentation, largely from studies on laboratory mice (reviewed in Silvers, 1979; Jackson, 1994, 1997; Barsh, 1996).

The deposition of pigment in hair and skin is the end-point of a process that involves the coordinated action of many genes and cell types. Melanocytes, the pigment-producing cells, originate in the neural crest and migrate during development throughout the dermis. The melanoblast cell lineage that gives rise to melanocytes is committed early in development and subsequent expression of many gene products is regulated in a cell-specific manner (Steel et al., 1992; Erickson, 1993; Bronner-Fraser, 1995). Within melanocytes are specialized organelles known as melanosomes (reviewed in Prota, 1992); they are the site of melanogenesis. There are two primary types of melanosomes and they differ both structurally and biochemically: eumelanosomes are ellipsoidal and are the site of synthesis of black or brown eumelanin whereas phaeomelanosomes are spherical and are the site of synthesis of yellow or red phaeomelanin (Figure 2). Once full of melanin, melanosomes are secreted from the melanocyte as pigment granules. Several lines of evidence suggest a close relationship between melanosomes and lysosomes and it is possible that melanosomes are modified lysosomes (Jackson, 1994, 1997). For example, many mouse mutations which affect melanosome function also disrupt lysosome function (e.g. Barbosa et al., 1996; Feng et al., 1997), raising the possibility that evolution of some pigmentation genes will be constrained by pleiotropic effects. Finally, synthesis of melanin within melanosomes involves the interactions of many loci, and some aspects of melanogenesis are under hormonal regulation.

Mouse pigmentation mutations have been identified in all steps of this process (Prota, 1992; Jackson,1994). For example, there are mutant phenotypes such as *piebald*, *steel*, and *white spotting* that result from improper development or migration of melanocytes, leaving portions of the body without pigment-producing cells. Other mutations, such as *beige* and *pale ear*, interfere with the proper structure and function of melanosomes. Some mutations, such as *albino*, *brown*, or *slaty*, interfere directly with proteins involved in synthesis of melanin. Finally, mutations at the *agouti*, *extension*, and *mahogany* loci disrupt the control and regulation of melanogenesis.

Approximately 80 genes have been identified that affect coat-color in the mouse (Jackson, 1997), and a large and growing number of these have now been characterized at the molecular level.

When employing a candidate-gene approach to finding the genes underlying a particular trait, it is typical to look for laboratory mutants that mimic naturally occurring variation (Palopoli & Patel, 1996; Haag & True, 2001). In this regard, there are several mouse coat-color mutants that suggest themselves as particularly relevant for understanding coat-color variation in *Chaetodipus*. In mammals, there are two basic kinds of melanin: eumelanin, which produces a dark brown or black color, and phaeomelanin, which produces a cream, yellow, or red color. The switch between production of eumelanin and phaeomelanin is controlled largely by the interaction of two key proteins, the melanocortin-1 receptor (MC1R) and the agouti signaling protein (Figure 2). MC1R is a transmembrane G-protein-coupled receptor that is highly expressed in melanocytes. Alpha-melanocyte-stimulating-hormone (α-MSH) activates MC1R, resulting in elevated levels of cAMP and increased production of eumelanin. The agouti protein is an antagonist of MC1R; local expression of agouti results in suppression of synthesis of eumelanin and increased production of phaeomelanin. Many dominant agouti mutations result in increased agouti expression and largely yellow phenotypes. In contrast, recessive, loss-of-function agouti mutations result in nonagouti, all black phenotypes. Dominance relationships among *Mc1r* alleles are opposite in order to those at agouti: recessive, loss-of-function Mc1r mutations typically result in yellow phenotypes (although slightly different phenotypically from the dominant yellow of agouti).

Wild mice have light bellies as a result of constitutive ventral agouti expression and associated production of phaeomelanin. In contrast, hairs on the dorsum of wild mice have a banded pattern, with a black tip, a middle yellow band, and a black base (the agouti hair). This banding is due to a pulse of agouti expression during the mid-phase of the hair cycle, resulting in deposition of phaeomelanin during the middle of hair growth and deposition of eumelanin at the beginning and end of hair growth (Figure 2). Mutations at both *agouti* (Vrieling et al., 1994; Bultman et al., 1994) and at *Mc1r* (Robbins et al., 1993) have been identified that produce black, unbanded dorsal hairs in the laboratory mouse but light hairs on the belly. Importantly, we observed a very similar phenotype in *C. intermedius* from lava flows; we found unbanded, uniformly melanic hairs in all dark *C. intermedius*, and banded dorsal hairs in all light *C. intermedius* (Figure 2), suggesting a possible role for either *agouti* or *Mc1r*.

A candidate-gene approach has both advantages and limitations. One clear advantage is that it may be possible to find the genes underlying a trait rather easily. Moreover, studies on laboratory mutants can provide important clues to the development, biochemistry, or cell biology that will help explain the mechanism by which a given genetic change produces a particular phenotype in nature. An obvious but important limitation of this approach is that, by itself, it will only lead to genes for which candidates are available. In the absence of a comprehensive mapping study, it is difficult to know how many undiscovered loci may contribute to the phenotypic variation of interest. Another limitation of a candidate-gene approach is that most laboratory mutants are changes of relatively large effect. If most of adaptive evolution typically occurs through many changes of small effect, we might expect that in most circumstances developmental mutants from the laboratory will not be useful mimics of naturally occurring variation (Haag & True, 2001). This is a question open to validation empirically by studies such as those described here. Perhaps the most powerful approach to study the genetic architecture of phenotypic variation in nature is to use a combination of mapping and candidate genes.

The genetic basis of adaptive melanism in pocket mice

We have sequenced portions of several genes known to produce coat-color mutants in the laboratory mouse and conducted association studies between polymorphisms in these genes and phenotypic variation in natural populations of *C. intermedius* (Nachman, Hoekstra & D'Agostino, 2003; Hoekstra & Nachman, 2003; Hoekstra, Drumm & Nachman, 2004). The general strategy has been to compare melanic mice collected on lava flows with light-colored mice collected on adjacent light-colored rocks (usually within a few kilometers of the lava). We have explored genetic and phenotypic variation in this way at four paired sites, representing four different lava flows in Arizona and New Mexico (Figure 3). Several key results have emerged: (i) a single gene, *Mc1r*, appears to be responsible for most of the phenotypic variation in color in one population, the Pinacate site; (ii) four or fewer nucleotide changes at *Mc1r* appear to be responsible for the difference in receptor function; (iii) studies of migration-selection balance suggest that the selection coefficient associated with the dark *Mc1r* allele at the Pinacate site is large; and (iv) different (unknown) genes underlie the evolution of melanism on three other lava flows. These are briefly described below.

Several lines of evidence implicate *Mc1r* in coat-color variation at the Pinacate site (Nachman, Hoekstra & D'Agostino, 2003). First, there is a perfect association between *Mc1r* genotype and coat-color phenotype among all mice in this population. The *Mc1r* D allele is distinguished from the *Mc1r* d allele by four amino acid substitutions and one synonymous substitution, and mice with DD or Dd genotypes have melanic, unbanded dorsal hairs while mice with dd genotypes are light-colored, with agouti hairs on their dorsum. Second, the darkening *Mc1r* D allele is dominant over the *Mc1r* d allele, consistent with dominance relationships seen among *Mc1r* alleles in the laboratory mouse. Third, all four amino acid substitutions that distinguish the D and d alleles are charge-changing substitutions and are found in regions of the receptor that may be important for

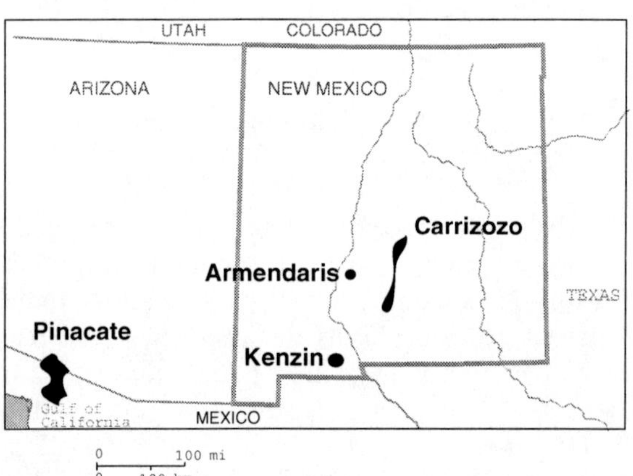

Figure 3. Four lava flows on which *C. intermedius* were studied. In each case, mice were collected on lava and on nearby light-colored rocks.

ligand binding or for interactions with other proteins. Fourth, the four amino acid sites at which substitutions distinguish *Mc1r* D and *Mc1r* d alleles are otherwise invariant across all other species of pocket mice (unpublished results), suggesting that these sites are functionally important. Fifth, the pattern of nucleotide variation seen at *Mc1r* is consistent with the recent action of natural selection; *Mc1r* D chromosomes have approximately one tenth as much variation as *Mc1r* d chromosomes. Sixth, genotype–phenotype associations decay immediately upstream and downstream of *Mc1r*, indicating that the observed association between *Mc1r* alleles and coat-color is not a consequence of linkage to some other, nearby locus. Finally, cAMP assays of receptor function *in vitro* show that the *Mc1r* D allele encodes a hyperactive receptor relative to the *Mc1r* d allele (Nachman, Hoekstra & D'Agostino, 2003). All of these observations strongly support the involvement of *Mc1r* in coat-color variation at the Pinacate site.

It is noteworthy that the differences in coat color are associated with a relatively small number of amino acid changes. At present, it is unknown whether each of the four *Mc1r* amino acid substitutions contributes to the difference in phenotype, or whether a subset of these four mutations is responsible for the difference in coat color. It does seem likely, however, that most of the coat-color variation can be explained by *Mc1r* genotype without a significant contribution from other genes. Most of the phenotypic variance correlates with *Mc1r* genotypic differences; there is little variation in coat-color within each of the three *Mc1r* genotypic classes (DD, Dd, dd). In principle, a gene linked to *Mc1r* could also contribute to the variation in phenotype, but this seems unlikely because of the rapid decay of linkage disequilibrium immediately upstream and downstream of *Mc1r*.

To estimate the strength of selection on *Mc1r* D and d alleles, we conducted a transect across the Pinacate site, collecting animals on light-colored rock as well on the lava flow (Hoekstra, Drumm & Nachman, 2004). At this site, the light rocks are separated from the lava by ~5 km of sand, which is not suitable habitat for *C. intermedius*. In general, most of the mice trapped on the lava were dark, and most of the mice trapped on the light-colored rocks were light. However, a small number of mis-matched mice were found, both on the lava and on the light rocks, suggesting that migration between the two substrates occurs. We estimated migration rates from the degree of mitochondrial DNA differentiation between mice on light rocks and on lava. We assumed that the frequencies of mis-matched *Mc1r* alleles (D on light rock, and d on lava) were determined by the balance between the input of new alleles due to migration and their elimination by selection (migration-selection balance). Selection coefficients estimated this way were large (~2–40%) for light alleles (*Mc1r* d) on dark rock, but were considerably smaller (< 1%) for dark alleles (*Mc1r* D) on light rock.

To study the genetic basis of melanism in different geographic regions, we captured *C. intermedius* on four different lava flows (Figure 3) and found dark mice on all of them (Hoekstra & Nachman, 2003). The Pinacate site is in Arizona and is separated from the three lava flows in New Mexico by over 700 km. We sequenced *Mc1r* in dark mice from each lava flow and in light mice from light-colored rocks adjacent to each lava flow; we found that *Mc1r* does not seem to be involved in pigmentation variation at any of the three New Mexico sites. The four amino acid substitutions that define the *Mc1r* D allele were not observed in any dark mice from New Mexico. Moreover, no other associations between *Mc1r* polymorphisms and color variation were observed. Dark mice from all four lava flows are similar phenotypically in having unbanded, entirely melanic hairs on the dorsum, but they differ somewhat in the amount of reflectance off the dorsum as measured with a spectrophotometer: in general, melanic mice from the New Mexico sites are darker than melanic mice from the Pinacate site.

Implications for the understanding the genetics of adaptation

These results help us understand the genetic details of adaptive melanism in mice and provide a good example of evolution by natural selection. Beyond serving as an example, can these findings shed light more generally on the evolutionary process? Below I discuss several evolutionary principles in the context of these observations. In some cases, knowing the specific genetic changes underlying a trait of interest allows us to address issues that would be otherwise intractable; in other cases, a candidate-gene approach is one of several methods that can be used to address a particular problem.

Constraint and convergence

A key issue in evolution is the extent to which adaptive change is constrained by developmental pathways. If there are many ways to arrive at a given phenotype we might expect convergent evolution to be common. If, on the other hand, pathways are highly constrained, we might expect a similar "genetic solution" in different instances of the same 'evolutionary problem'. The observation that *Mc1r* is responsible for dark color in *C. intermedius* on one lava flow but not in three others has two immediate implications. First, it shows conclusively that dark color has evolved multiple times in this species. The alternative hypothesis, that dark color evolved once and spread through long-distance migration among lava flows, is clearly ruled out. Second, it provides evidence for convergence: nearly identical phenotypes have evolved through changes in different genes. We still have not identified the genes responsible for dark color in *C. intermedius* from the three New Mexico sites, but the candidate-gene approach may continue to prove useful in finding them.

In some respects, we knew *a priori*, that different genes might underlie similar color variation in different populations. In the laboratory mouse, mutations at different pigmentation genes can produce similar phenotypes. For example, some gain-of-function *Mc1r* mutations resemble, at least superficially, some loss-of-function *agouti* mutations. But laboratory studies are typically unable to reveal small or even modest fitness differences, and consequently the full range of pleiotropic effects is difficult to assess in the laboratory. If different mutants produce similar coat-color but affect fitness in other ways, their probability of fixation in natural populations may be dramatically different. Our data show that in rock pocket mice, not only are there different genes that may contribute to dark color, but there are different solutions that are evolutionarily viable.

Fisher's microscope

A long-standing debate in evolution concerns the average amount of phenotypic change caused by adaptive mutations. Darwin (1859) argued that most adaptations result from numerous small changes. This view was given theoretical support from Fisher (1930) who showed that mutations of large effect had a higher probability of being deleterious than mutations of small effect, and that mutations of very small effect had an equal chance of being advantageous or deleterious. To illustrate this point, Fisher used the analogy of a microscope that is slightly out of focus: a large change will almost certainly make the situation worse, but a small change may improve the focus. Fisher's model contains many simplifying assumptions; for example, it considers a phenotype consisting of n orthogonal characters, whereas real characters are often correlated. It also assumes that organisms are evolving in an adaptive landscape that contains a single, fixed optimum. Importantly, Fisher only considered the probability that an individual mutation will be advantageous or deleterious, and as Kimura (1983) pointed out, this is different from the *rate* of adaptive substitution, which includes both the number of mutations and their probabilities of fixation. Kimura (1983) showed that while mutations of large effect have a lower probability of being beneficial, they have a higher probability of being fixed than mutations of small effect. Assuming that the 'size' of a mutation (i.e. the magnitude of its phenotypic effect) is proportional to its effect on fitness (s), Kimura (1983, p. 155) derived the distribution of substitution rates for mutations of different sizes and argued that adaptation might consist mainly of mutations of intermediate effect. This literature has been nicely summarized by Orr (1998) who expanded on the results of Fisher and Kimura to show that the distribution of mutational effects fixed during an 'adaptive walk' is typically exponential and can include one or more mutations of fairly large effect.

How do empirical observations conform with theory? Orr and Coyne (1992, p. 725) summarized the data available 10 years ago and argued that while 'some adaptations are apparently based on many genes of small effect, others clearly involve major genes'. QTL studies, especially in plants (Mauricio, 2001), often find a mixture of minor and major genes contributing to phenotypic variation, but it is not uncommon to find a few genes that account for a substantial amount of the phenotypic variation. Other evidence comes from organisms in disturbed environments, where single mutations of large effect seem to be the rule for explaining traits such as industrial melanism, insecticide resistance, and antibiotic resistance (e.g. Fidock et al., 2000; Walsh, 2000; Raymond et al., 2001; Cowen, Anderson & Kohn, 2002; Daborn et al., 2002; Wootton et al., 2002; Hughes, 2003) Clearly in this situation, selection is very strong, so that negative pleiotropic effects, like the physiological cost of resistance, may be easily outweighed by the benefits of resistance. The extent to which mutations of large effect are also seen in more natural situations is still unclear (Orr & Coyne, 1992; Charlesworth, 1994; Orr, 1999).

Pocket mice provide several important lessons here. First, the phenotypic difference between light and dark mice is striking and large, and the fit of mice to their environment seems to be quite good. Spectrophotometry measurements of reflectance from mice and from the rocks on which they are found show a strong positive correlation (Dice & Blossom, 1937; Hoekstra & Nachman, 2003). In the Pinacate site, this close fit seems to be due almost entirely to a single locus, *Mc1r*; the presence or absence of banded 'agouti' hairs on the dorsum appears to be a discrete rather than a quantitative trait, and is perfectly associated with *Mc1r* genotype. The situation is slightly more complicated than this, however, since, mice with different *Mc1r* genotypes (DD, Dd, dd) also differ in total reflectance, and Dd mice are roughly intermediate in reflectance between DD and dd mice. Thus, there appears to be some quantitative variation in reflectance among mice with uniformly melanic, unbanded hairs. Nonetheless, the amount of this variation is much greater between *Mc1r* genotypic classes than within genotypic classes, again suggesting a major role for *Mc1r*. The difficulty of breeding pocket mice has precluded a mapping study to identify QTL, and thus we do not know how many other loci (of presumably minor effect) may be contributing to the observed variation. Nonetheless, it is clear that *Mc1r* is a major gene, and therefore that major genes are not restricted to phenotypes associated with artificial selection or human disturbance (see also Haag & True, 2001).

The second lesson is that while *Mc1r* is a major gene, the dark allele (D) differs from the light allele (d) by four amino acid substitutions and one silent substitution. We do not know the relative contributions of each of these mutations (the synonymous substitution may, of course, have no effect). At one extreme, a single mutation may be responsible for the phenotypic variation, and at the other extreme, each of four mutations may contribute to the phenotypic variation, and they may be either additive or epistatic. This distinction is instructive: conventional mapping studies typically identify chromosomal regions of importance but do not identify the number of mutations within those regions that contribute to the phenotype of interest. Thus the support for genes of major effect from QTL studies must be tempered with the caveat that these genes may, in fact, contain multiple mutations of smaller effect. We hope to disentangle the relative contribution of each mutation in *Mc1r* using site-directed mutagenesis and an *in vitro* cAMP assay for receptor function. These studies should also enable us to ask whether these mutations act together in an additive or epistatic manner. In this regard, knowing the identity of the gene enables us to address questions that would be impossible otherwise.

Haldane's sieve

Haldane (1924) showed that selection on rare, autosomal recessive mutations is ineffective because they are most often found in heterozygotes where they are hidden from selection. This stands in contrast to autosomal dominant mutations, which, when present in heterozygotes, are visible to selection. From this result, Haldane argued 'it seems therefore very doubtful whether natural selection in random mating organisms can cause the spread of autosomal recessive characters unless they are extraordinarily valuable to their possessors' (Haldane 1924, p. 38). This notion, later termed Haldane's sieve by Turner (1981), was supported by the observation that many known adaptations resulted from dominant mutations, despite the fact that many laboratory mutants were recessive (Haldane, 1924). Haldane also pointed out that the situation is quite different for sex-linked genes and for high levels of selfing, where recessive mutations may spread under selection, and both of these ideas have been explored in greater detail by Charlesworth, Coyne and Barton (1987) and Charlesworth (1992). Much was written on the evolution of dominance during the first 50 years of population genetics (reviewed in Merrell, 1969) but the following observation now seems well supported: many mutations in the laboratory with large phenotypic effects are recessive while many adaptations in animal populations that result from genes of major effect are usually dominant or semi-dominant. This result appears consistent with the preferential fixation of beneficial dominant mutations. An alternative possibility, however, is that most favorable mutations are dominant rather than recessive, and thus the large number of dominant mutations underlying adaptation would simply reflect their greater occurrence rather than their higher probability of fixation. Beneficial mutations may often result from gain-of-function, and dominance may simply correspond to gain of function at the biochemical level (Wright, 1929, 1934). Finally, Orr and Betancourt (2001) have recently shown that the situation is quite different if one considers adaptive fixations resulting from standing variation rather than from new mutations; when positive selection favors a previously deleterious allele at mutation-selection balance, the probability of fixation is largely independent of the degree of dominance.

How do our observations in pocket mice fit with these theoretical considerations? It is worth pointing out that *Mc1r* is autosomal rather than X linked in all mammals where it has been mapped, so it seems likely that it is autosomal in pocket mice as well; thus, the special considerations for dominance in sex-linked genes do not need to be considered. First, adaptive melanism at the Pinacate site appears to be caused by a dominant or semi-dominant allele at a single major gene. This observation is entirely consistent with the observation of dominance for genes underlying adaptations to human disturbance (e.g. Haldane, 1924; Jasieniuk, Brule-Babel & Morrison, 1996). The studies on pocket mice also underscore the difficulty of correctly ascertaining the degree of dominance. The presence or absence of a sub-terminal band of phaeomelanin on individual hairs is a Mendelian trait, with the melanic hair (*Mc1r* D) fully dominant over the agouti hair (*Mc1r* d). To the human eye, this difference appears to be the most significant aspect of color variation in these mice; all observers easily group mice into 'light' and 'dark' categories based on the presence or absence of agouti hairs on the dorsum (Figure 2). However, spectrophotometry measurements indicate that *Mc1r* Dd mice are intermediate in total reflectance between *Mc1r* DD and *Mc1r* dd mice, an attribute that is not easily detected by the human eye (Hoekstra & Nachman 2003, Figure 2C). It remains unclear whether *Mc1r* DD and *Mc1r* Dd genotypes have the same fitness. Knowing the gene underlying adaptive melanism also makes it possible to relate dominance to biochemical function. Our studies measuring *Mc1r* function *in vitro* show that the *Mc1r* D allele encodes a hyperactive receptor relative to the *Mc1r* d allele, and thus dominance in this case corresponds to the gain of biochemical function (Wright, 1934). However, as described above, darkening alleles are known from both dominant, gain-of-function mutations at *Mc1r* and recessive, loss-of-function mutations at *agouti* in the laboratory mouse. In principle, we might expect that either could serve as a substrate for adaptive evolution in natural populations, and thus there is *no a priori* reason for thinking that most adaptive pigmentation mutations arise from gain-of-function mutants. So far, however, we have only been able to identify gain-of-function (dominant) mutants in the wild; it will be interesting to see whether recessive alleles are responsible for melanic phenotypes in other populations. Finally, can we say anything about the likelihood that melanic mice arise from new mutations rather than from standing variation? In several species of mammals, occasional melanic individuals are observed, raising the possibility that melanic forms are present at low frequency in mutation-selection balance. Although we have never observed melanic *C. intermedius* at sites that are far from dark rocks (based on approximately 1000 mice), the possibility that selection acted on pre-existing variation cannot be excluded.

Gene regulation and gene structure

A question of considerable recent interest concerns the degree to which adaptive evolution derives from changes in gene dosage versus changes in gene product. Britten and Davidson (1969) argued that much of evolution may be caused by modifications to regulatory networks, and current microarray technology has allowed investigators to explore large-scale changes in gene expression between closely related species (e.g. Enard et al., 2002). Knowing the identify of the gene underlying a trait allows us to address this question directly. Adaptive melanism in the Pinacate mice is caused by changes in the amino acid sequence of *Mc1r*, and these changes alone produce a receptor that functions differently. Importantly, however, these changes have many downstream effects. In mice with *Mc1r* DD genotypes, there appears to be no production of phaeomelanin in dorsal melanocytes. Thus while changes at *Mc1r* are clearly structural, they cause changes in the expression pattern of many downstream genes. This highlights a potential difficulty with using differences in expression to identify causative mutations.

Linking phenotype to genotype

The candidate-gene approach has been useful here for making several connections between genotype and phenotype. In addition to the description of phenotypic differences associated with different *Mc1r* genotypes, we have made some preliminary estimates of the strength of selection on *Mc1r* D and *Mc1r* d alleles. In principle, this should allow us to compare both the magnitude of phenotypic effect and the value of s for different alleles. However, because the *Mc1r* D and d alleles differ by four amino acid substitutions and each of these may have been a separate step in the 'adaptive walk', we may not be able to link the effect size with s for individual mutations. Nonetheless, the approach used here has allowed us to shed light on the biochemistry, population genetics, and ecological genetics associated with the evolution of melanism and it serves as an example of the utility (and limitations) of this method. This approach will clearly not work in all situations; when adaptive differences are quantitative and caused by many genes of small effect, a mapping study may prove more useful. But for traits where good candidate genes are available and phenotypic differences are relatively simple, studies of candidate genes may be quite useful for understanding the evolutionary process.

Acknowledgements

I am indebted to H.E. Hoekstra for many stimulating discussions on the genetic basis of coat-color. She also kindly provided all of the figures. This work was supported by the National Science Foundation.

References

Barbosa, M.D.F.S., Q.A. Nguyen, V.T. Tchernev, J.A. Ashley, J.C. Detter, S.M. Blaydes, S.J. Brandt, D. Chotai, C. Hodgman, R.C.E. Solari, M. Lovett, & S.F. Kingsmore, 1996. Identification of the homologous beige and Chediak-Higashi-Syndrome genes. Nature 382: 262–265.

Barsh, G.S., 1996. The genetics of pigmentation: from fancy genes to complex traits. Trend. Genet. 12: 299–305.

Benson, S.B., 1933. Concealing coloration among some desert rodents of the southwestern United States. Univ. Calif. Pub. Zool. 40: 1–70.

Britten, R.J. & E.H. Davidson, 1969. Gene regulation for higher cells: a theory. Science 165: 349–357.

Bronner-Fraser, M., 1995. Origins and developmental potential of the neural crest. Exp. Cell Res. 218: 405–417.

Bultman, S.J., M.L. Klebig, E.J. Michaud, H.O. Sweet, M.T. Davisson, & R.P. Woychik, 1994 Molecular analysis of reverse mutations from nonagouti (a) to black-and-tan (a^t) and white-bellied agouti (A^w) reveals alternative forms of agouti transcripts. Genes Develop. 8: 481–490.

Charlesworth, B., 1992. Evolutionary rates in partially self-fertilizing species. Am. Nat. 140: 126–148.

Charlesworth, B., 1994. The genetics of adaptation: lessons from mimicry. Am. Nat. 144: 839–847.

Charlesworth, B., J.A. Coyne & N.H. Barton, 1987. The relative rates of evolution of sex chromosomes and autosomes. Am. Nat. 130: 113–146.

Cowen, L.E., J.B. Anderson & L.M. Kohn, 2002. Evolution of drug resistance in *Candida albicans*. Ann. Rev. Microbiol. 56: 139–165.

Daborn, P.J., J.L. Yen, M.R. Bogwitz, G. LeGoff, E. Feil, S. Jeffers, N. Tijet, T. Perry, D. Heckel, P. Batterham et al., 2002. A single P450 allele associated with insecticide resistance in *Drosophila*. Science 297: 2253–2256.

Darwin, C., 1859. The Origin of Species by Mean of Natural Selection. John Murray, London.

Dice, L.R., 1940. Ecologic and genetic variability within species of Peromyscus. Am. Nat. 74: 212–221.

Dice, L.R., 1947. Effectiveness of selection by owls of deer-mice (*Peromyscus maniculatus*) which contrast in color with their background. Contr. Lab. Vert. Biol., Univ. Michigan 34: 1–20.

Dice, L.R. & P.M. Blossom, 1937. Studies of mammalian ecology in southwestern North America with special attention to the colors of desert mammals. Publ. Carnegie Inst. Washington 485: 1–129.

Doebley, J., A. Stec & L. Hubbard, 1997. The evolution of apical dominance in maize. Nature 386: 485–488.

Dobzhansky, T., 1937. Genetics and the Origin of Species. Columbia University Press, New York.

Dobzhansky, T., 1970. Genetics of the Evolutionary Process. Columbia University Press, New York.

Enard W., P. Khaitovich, J. Klose, S. Zollner, F. Heissig, P. Giavalisco, K. Nieselt-Struwe, E. Muchmore, A. Varki, R. Ravid, G.M. Doxiadis, R.E. Bontrop & S. Paabo, 2002. Intra- and interspecific variation in primate gene expression patterns. Science 296: 340–343.

Erickson, C.A., 1993. From the crest to the pheriphery: control of pigment-cell migration and lineage segregation. Pigment Cell Res. 6: 336–347.

Feng, G.H., T. Bailin, J. Oh & R.A. Spritz, 1997. Mouse pale ear (ep) is homologous to human Hermasky-Pudlak syndrome and contains a rare AT-AC intron. Hum. Mol. Genet. 6: 793–797.

Fidock, D.A., T. Nomura, A.K. Talley, R.A. Cooper, S.M. Dzekunov, M.T. Ferdig, L.M.B. Ursos, A.B.S. Sidhu, B. Naude, K.W. Deitsch, X.Z. Su, J.C. Wootton, P.D. Roepe & T.E. Wellems, 2000. Mutations in the P-falciparum digestive vacuole transmembrane protein PfCRT and evidence for their role in chloroquine resistance. Mol. Cell. 6: 861–871.

Fisher, R.A., 1930. The Genetical Theory of Natural Selection. Oxford University Press, Oxford.

Haag, E.S. & J.R. True, 2001. Perspective: from mutants to mechanisms? Assessing the candidate gene paradigm in evolutionary biology. Evolution 55: 1077–1084.

Haldane, J.B.S., 1924. A mathematical theory of natural and artificial selection, Part 1. Trans. Camb. Philos. Soc. 23: 19–41.

Haldane, J.B.S., 1932. The Causes of Evolution. Longmans, Green & Co, Ltd., New York.

Hoekstra, H.E., K.E. Drumm & M.W. Nachman, 2004. Ecological genetics of adaptive color polymorphism in pocket mice: geographic variation in selected and neutral genes. Evolution 58: 1329–1341.

Hoekstra, H.E. & M.W. Nachman, 2003. Different genes underlie adaptive melanism in different populations of rock pocket mice. Mol. Ecol. 12: 1185–1194.

Hughes, D., 2003. Exploiting genomics, genetics and chemistry to combat antibiotic resistance. Nat. Rev. Genet. 4: 432–441.

Jackson, I.J., 1994. Molecular and developmental genetics of mouse coat color. Ann. Rev. Genet. 28: 189–217.

Jackson, I.J., 1997. Homologous pigmentation mutations in human, mouse and other model organisms. Human Mol. Genet. 6: 1613–1624.

Jasieniuk, M., A.L. Brule-Babel & I.N. Morrison, 1996. The evolution and genetics of herbicide resistance in weeds. Weed Sci. 44: 176–193.

Kimura, M., 1983. The Neutral Theory of Molecular Evolution. Cambridge University Press, Cambridge.

Kondrashov, A.S., S. Sunyaev & F.A. Kondrashov, 2002. Dobzhansky-Muller incompatibilities in protein evolution. Proc. Nat. Acad. Sci. 99: 14878–14883.

Lack, D.L., 1947. Darwin's Finches. University Press, Cambridge.

Mauricio, R., 2001. Mapping quantitative trait loci in plants: uses and caveats for evolutionary biology. Nat. Rev. Genet. 2: 370–381.

Mayr, E., 1942. Systematics and the Origin of Species. Columbia University Press, New York.

Mayr, E., 1963. Animal Species and Evolution. Harvard University Press, Cambridge, Massachusetts.

Merrell, D.J., 1969. The evolutionary role of dominant genes. Genet. Lect. (Oregon State University, Corvalis) 1: 167–194.

Nachman, M.W., H.E. Hoekstra & S.L. D'Agostino, 2003. The genetic basis of adaptive melanism in pocket mice. Proc. Nat. Acad. Sci. USA 100: 5268–5273.

Newton, J.M., A.L. Wilkie, L. He, S.A. Jordan, D.L. Metallinos, N.G. Holmes, I.J. Jackson & G.S. Barsh, 2000. Melanocortin 1 receptor variation in the domestic dog. Mamm. Genome 11: 24–30.

Orr, H.A., 1998. The population genetics of adaptation: the distribution of factors fixed during adaptive evolution. Evolution 52: 935–949.

Orr, H.A. & J.A. Coyne, 1992. The genetics of adaptation: a reassessment. Am. Nat. 140: 725–742.

Orr, H.A., 1999. The evolutionary genetics of adaptation: a simulation study. Genet. Res. Camb. 74: 207–214.

Orr, H.A. & A.J. Betancourt, 2001. Haldane's sieve and adaptation from the standing genetic variation. Genetics 157: 875–884.

Palopoli, M.F. & N.H. Patel, 1996. Neo-Darwinian developmental evolution: can we bridge the gap between pattern and process? Curr. Op. Genet. Dev. 6: 502–508.

Palumbi, S.R., 2001. Humans as the world's greatest evolutionary force. Science 293: 1786–1790.

Prota, G., 1992. Melanins and Melanogenesis. Academic Press Inc., San Diego, CA.

Raymond, M., C. Berticat, M. Weill, N. Pasteur & C. Chevillon, 2001. Insecticide resistance in the mosquito *Culex pipiens*: what have we learned about adaptation? Genetica 112–113: 287–296.

Reznick, D.N. & C.K. Ghalambor, 2001. The population ecology of contemporary adaptations: what empirical studies reveal about the conditions that promote adaptive evolution. Genetics 112–113: 183–198.

Robbins, L.S., J.H. Nadeau, K.R. Johnson, M.A. Kelly, L. Rosellirehfuss, E. Baack, K.G. Mountjoy & R.D. Cone, 1993. Pigmentation phenotypes of variant extension locus alleles result from point mutations that alter receptor function. Cell 72: 827–834.

Silvers, W.K., 1979. The Coat Colors of Mice: A Model for Mammalian Gene Action and Interaction. Springer-Verlag, New York.

Stebbins, G.L., Jr., 1950. Variation and Evolution in Plants. Columbia University press, New York.

Steel, K.P., D.R. Davidson & I.J. Jackson, 1992. TRP-2/DT, a new early melnoblast marker shows that the steel growth factor (c-kit ligand) is a survival factor. Development 115: 1111–1119.

Turner, J.R.G., 1981. Adaptation and evolution in *Heliconius*: a defense of NeoDarwinism. Ann. Rev. Ecol. Syst. 12: 99–121.

Vrieling, H., D.M.J. Duhl, S.E. Millar, K.A. Miller & G.S. Barsh, 1994. Differences in dorsal and ventral pigmentation result from regional expression of the mouse agouti gene. Proc. Natl. Acad. Sci. 91: 5667–5671.

Walsh, C., 2000. Molecular mechanisms that confer antibacterial drug resistance. Nature 406: 775–781.

Wang, R.L., A. Stec, J. Hey, L. Lukens, & J. Doebley, 1999. The limits of selection during maize domestication. Nature 398: 236–239.

Wootton, J.C., X. Feng, M.T. Ferdig, R.A. Cooper, J. Mu, D.I. Baruch, A.J. Magill & X.Z. Su, 2002. Genetic diversity and chloroquine selective sweeps in Plasmodium falciparum. Nature 418: 320–323.

Wright, S., 1929. Fisher's theory of dominance. Am. Nat. 63: 274–279.

Wright, S., 1931. Evolution in Mendelian populations. Genetics 16: 97–159.

Wright, S., 1934. Physiological and evolutionary theories of dominance. Am. Nat. 68: 25–53.

Genetica (2005) 123: 137–145 © Springer 2005

The genetics of adaptation in *Drosophila sechellia*

Corbin D. Jones
Department of Biology, Carolina Center for Genome Science, University of North Carolina, Chapel Hill, NC 27599, USA (Phone: (919) 843-5162; E-mail: cdjones@email.unc.edu)

Received 31 May 2002 Accepted 16 July 2003

Key words: adaptation, *Drosophila*, *D. melanogaster*, *D. sechellia*, *D. simulans*, *Morinda citrifolia*, oviposition, pesticide resistance, QTL

Abstract

Drosophila sechellia is an island endemic of the Seychelles. After its geographic isolation on these islands, *D. sechellia* evolved into a host specialist on the fruit of *Morinda citrifolia* – a fruit often noxious and repulsive to *Drosophila*. Specialization on *M. citrifolia* required the evolution of a suite of adaptations, including resistance to and preference for some of the toxins found in this fruit. Several of these adaptive traits have been studied genetically. Here, I summarize what is known about the genetics of these traits and briefly describe the ecological and geographical context that shaped the evolution of these characters. The data from *D. sechellia* suggest that adaptations are not as genetically complex as historically thought, although almost all of the adaptations of *D. sechellia* involve several genes.

Introduction

Renewed interest in the genetics of adaptation is improving our understanding of how individual genes affect adaptive phenotypic differences between closely related species. This work has focused on identifying the number and phenotypic effects of genes involved in adaptive differences between species (for simplicity, 'adaptive' refers to a derived condition that arose as a result of selection). In particular, many of these studies have tried to determine if adaptive evolution typically results from the action of many genes of small phenotypic effect or from a few genes of large phenotypic effect.

Historically, evolutionists and quantitative geneticists preferred a polygenic view of adaptive evolution that assumed that phenotypic change involved many factors of very small effect each. This view is being challenged by recent data from quantitative trait locus (QTL) analyses. QTL analysis allows genetic dissection of traits in species that can be crossed to form hybrids carrying random combinations of chromosomal regions from the parental species. Once these hybrids are created, the species identity of chromosomal regions is inferred from genetic markers, and then the phenotype of each genotype is scored. From these data, one can map, count, and estimate the effects of genes underlying the trait studied. Such analyses have repeatedly shown that morphological differences often involve only a handful of chromosome regions of substantial effect each. Most QTL studies, however, have focused on agriculturally and economically important organisms. Unfortunately, the genetics of agricultural traits, with their long history of strong artificial selection by humans, may not be representative of the genetics of phenotypic differences that evolved in nature. Nevertheless, there is increasing evidence suggesting that 'natural adaptations' may also involve a modest number of genes. Moreover, it appears that the distribution of gene effects underlying morphological evolution may be roughly exponential – an idea supported by evolutionary theory (Kearsey & Farquhar, 1998; Orr, 1998, 2001). In many cases, genes of small effect are clearly involved, but a few factors of large effect typically account for much of the phenotypic differences between species.

An ideal model species for studying the genetics of adaptive divergence would (1) have recently evolved adaptive traits, (2) be closely related to a genetic model system, and (3) allow the creation of transgenic animals. Remarkably, *D. sechellia* has all three of these attributes, and so provides a rare opportunity to address the genetics of adaptation. Here, I review what we have learned about the relationship of *D. sechellia* to its sister species, its natural history, and the genetic basis of its adaptations. These data highlight how useful *D. sechellia* is as a model system for studying the genetics of adaptation.

Species relationships

D. sechellia is a member of the *D. melanogaster* subgroup and is most closely related to *D. simulans* and *D. mauritiana*. Which of these two species is the closer relative is not known, although recent evidence tentatively suggests that *D. sechellia* speciated before the split between *D. simulans* and *D. mauritiana* (Kliman et al., 2000). The genetics of reproductive isolation in this group has been recently reviewed by Coyne and Orr (1998; see related Macdonald & Goldstein, 1999). Thus, I will only discuss the basic biology of interspecific hybrids relevant to conducting genetic analyses of *D. sechellia*.

Both *D. simulans* and *D. mauritiana* produce fertile females and sterile males when crossed to *D. sechellia* regardless of the direction of the cross. (*Wolbachia* bacteria, while present in some strains of all three species do not appear to greatly affect the fertility or viability of hybrids (Giordano, O'Neill & Robertson, 1995). This means that backcross hybrids can be generated between these species. This allows us to take advantage of the genetic tools available in these species including a number of genetic markers, a few chromosomal abberations, and some marker P-element insertion lines (True, Weir & Laurie, 1996; Flybase, 1999). It has also been shown that transgenic flies can be made in these species (Scavarda & Hartl, 1984; True, Weir & Laurie, 1996).

Typical for the *D. simulans* clade, *D. melanogaster* females when crossed to *D. sechellia* males produce only sterile F1 daughters, whereas *D. melanogaster* males when crossed to *D. sechellia* females produce only sterile F1 sons. A number of hybrid rescue mutations have been discovered in *D. melanogaster* and *D. simulans* (Ashburner, 1989). These mutations typically lead to the production of both sterile males and females. Some combinations of these mutations can weakly restore the fertility of hybrids (Davis et al., 1996; Barbash, & Ashburner 2003). *D. sechellia* seems to be more recalcitrant to hybrid rescue that its sister species (Barbash, Roote & Ashburner, 2000; Barbash & Ashburner, 2003). This means that only those *D. melanogaster* genetic tools that are informative in F1 hybrids (e.g., deficiencies) are useful.

Genetics in *D. sechellia*

Relative to *D. melanogaster* (or even *D. simulans*) the genetic tools available in *D. sechellia* are sparse. Several visible genetic markers are available and, recently, a number of molecular markers have been developed (Rux & Coyne, 1991; Colson, MacDonald & Goldstein 1999; Flybase, 1999). However, most mapping studies using visible markers have taken advantage of the far more plentiful tools available in *D. simulans* via interspecific hybrids. Unfortunately, these studies are still of limited resolution and power.

In principle, it is possible to use many of the chromosomal deficiencies and duplications available in *D. melanogaster* to map traits in F1 hybrids between it and *D. sechellia*. In practice, however, this mapping approach is frustrated by three facts. (1) The viability of F1 hybrids between *D. melanogaster* and *D. sechellia* is poor and gets worse in hybrids with a chromosomal aberration (Barbash, Roote & Ashburner 2000; Jones, unpublished). (2) F1 *melanogaster/sechellia* hybrids show a number of morphological abnormalities including degenerated reproductive organs, bristle loss, malformed cuticle, and other morphological defects (Takano, 1998). (3) *D. sechellia* is not completely chromosomally homosequential with *D. melanogaster*, which means a few regions cannot be adequately analyzed using deficiencies (Lemeunier & Ashburner, 1984).

Recently, Colson, MacDonald and Goldstein (1999) expanded the number of genetic tools available in *D. sechellia* by developing a set of microsatellite markers that distinguish *D. sechellia* from *D. simulans*. The future development of molecular markers like these has been greatly facilitated by the *D. melanogaster* genome project – and will soon be further simplified by the genomic sequencing of *D. simulans*.

Natural history

D. sechellia is endemic to the Seychelles archipelago, a collection of coralline and granitic islands in the Indian ocean several hundred kilometers off the east coast of Africa. These islands are home to a number of endemic plants and animals. Permanent settlement of these islands by humans began about 400 years ago, although these islands may have been visited occasionally before then. With human settlement, a number of species were introduced. DNA evidence suggests, however, that *D. sechellia* inhabited these islands well before humans arrived (Kliman et al., 2000).

As first reported by Tsacas and Bächli (1981), *D. sechellia* is typically found near the fruit of the rubiaceous shrub, *Morinda citrifolia*. This small tree is common in the Seychelles, often inhabiting shorelines but also found at higher elevations (Sauer, 1967; Robertson, 1989). It has been cataloged on many of the islands in the Seychelles archipelago (Robertson, 1989) and has also been found on Mauritius and Madagascar (Sauer, 1961; Baker, 1970). *Morinda* is also common throughout the Indian Ocean, Malaysia, and the islands of the Pacific. When *Morinda* arrived in the Seychelles is not known. It is most likely that *Morinda* fruit – which can survive salt water for more than a year – floated to the shore islands some time in the ancient past (Sauer, 1967).

When and how *D. sechellia* arrived in the Seychelles is not known either. Presumably, a *D. simulans*-like ancestor was blown from the coast of Africa (or Madagascar) and settled on an island of the Seychelles archipelago. From here, it colonized several other islands of the Seychelles. (*D. sechellia* has been collected on Praslin, Cousin, Frigate and Mahé islands.)

After arriving in the Seychelles, *D. sechellia* shifted from being a *D. simulans*-like generalist to specializing on the fruit of *M. citrifolia*. *Morinda* fruit is toxic to many insects (Legal & Plawecki, 1995). Why *D. sechellia* specialized on this normally toxic plant is not clear. *Morinda* fruit is abundant year round and maybe the only readily available host on the smaller islands of the Seychelles – although the main island, Mahé, surely provides a variety of other hosts. Alternatively, *D. sechellia* may have been driven to use *Morinda* by interspecific competition from other fruit flies such as *D. malerkotliana* or *D. simulans*, which are sympatric with *D. sechellia* (Louis & David, 1986; R'Kha et al., 1997). Another possibility is that *D. sechellia* may have moved to a toxic host to avoid predation by parasitoid wasps such as *Leptopilina* species, which are also found on the Seychelles (Louis & David, 1986). At this point, simply not enough is known to suggest which is the more plausible scenario.

To use *Morinda* fruit as its host, *D. sechellia* evolved resistance to the toxins in this fruit. R'Kha, Capy and David (1991) showed that media containing *Morinda* fruit pulp was toxic to *D. simulans, D. mauritiana, D. melanogaster, D. ananassae*, and *D. malerkotliana*, but not to *D. sechellia*. Legal, Chappe and Jallon (1994) showed that octanoic acid, which constitutes 58% of the identifiable volatile compounds in ripe *Morinda* fruit (hexanoic acid, a closely related compound, comes in a distant second at 19%), is the primary source of the toxicity of the fruit (Legal, Chappe & Jallon, 1994; Farine et al., 1996). They also showed that *D. sechellia* is highly resistant to the toxic effects of octanoic acid.

As *Morinda* fruit rots, levels of octanoic acid decline. Interestingly, *D. sechellia* shows much less resistance to the volatiles that become common in rotten fruit (Legal, Chappe & Jallon, 1994). In fact, *D. sechellia* is less tolerant of ethanol than its close relatives (Mercot et al., 1994). This result is intriguing as most *Drosophila* are saprophagous – that is, they feed on decaying, partially fermented resources. *D. sechellia*, on the other hand, appears to be better adapted at using the relatively unspoiled ripe *Morinda*.

Field and laboratory studies have shown that *D. sechellia* is strongly attracted to ripe *Morinda* and that this attraction is primarily mediated through octanoic acid, although other volatile compounds play a role as well (Louis & David, 1986; R'Kha, Capy & David, 1991; Higa & Fuyama, 1993; Amlou, Moreteau & David, 1998b; Legal, Moulin & Jallon, 1999). Relatively low concentrations of octanoic acid (0.1% by weight) have been shown to repulse *D. simulans, D. mauritiana, and D. melanogaster*, yet attract *D. sechellia* (Figure 1) (Amlou, Moreteau & David 1998b; Legal, Moulin & Jallon, 1999). Hexanoic acid also has this effect, but only when higher concentrations are used, which is surprising given greater vapor pressure of hexanoic acid (Amlou, Moreteau & David, 1998b). These data and the fact that octanoic acid is three times more abundant in *Morinda* fruit than hexanoic acid suggest that octanoic acid is the primary attractant in nature.

The host preference behavior of *D. sechellia* involves chemotaxis, oviposition site preference, and stimulation of egg production. Louis and David (1986) demonstrated that *D. sechellia* is attracted to *Morinda* fruit in the field and the lab. In a set of release and recapture experiments, R'Kha, Capy and David (1991) showed that *D. sechellia*, unlike *D. simulans*, will find and choose *Morinda* fruit over a banana bait 98% of the time, even when released 150 m away. Legal, Moulin and Jallon (1999) suggested that part of this attraction is likely due to octanoic and hexanoic acid.

R'Kha, Capy and David (1991) also showed that *D. sechellia* exhibited strong attraction oviposition site preference for media containing *Morinda* fruit. Subsequently, it has been shown that *D. sechellia's* oviposition site preference is strongly influenced by octanoic and hexanoic acid (Higa & Fuyama, 1993; Amlou, Moreteau & David, 1998b; Legal, Moulin & Jallon, 1999). *D. simulans* and *D. melanogaster* both avoid laying eggs on media containing either of these acids. Interestingly, ethyl esters of these acids, which are common components of rotting fruit, do not cause the same species specific behaviors (Legal, Moulin & Jallon, 1999).

The presence of *Morinda* also appears to stimulate egg production in *D. sechellia* (R'Kha Capy & David, 1991). In general, *D. sechellia* shows a 5–10 fold lower rate of egg production than its sister species (Coyne, Rux & David, 1991; R'Kha, et al., 1997). This effect is partially explained by the fact that *D. sechellia* has only 50–60% as many ovarioles as its sister species. Additionally, when not allowed to oviposit on *Morinda*, the number of eggs produced by each ovariole in *D. sechellia* females is about 60% that of its close relatives. When allowed to oviposit on *Morinda*, however, the rate of egg production per ovariole increases, again suggesting the *D. sechellia* prefers to use *Morinda* fruit as its host.

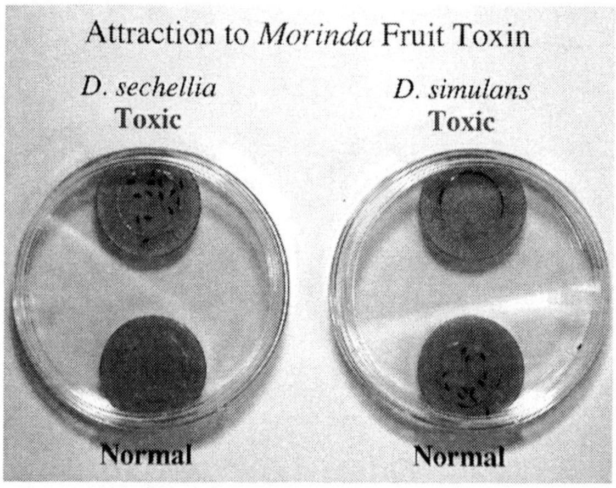

Figure 1. Comparison of the effects of media tainted with the Morinda fruit toxin, octanoic acid, on the behavior of *D. sechellia* and *D. simulans*. Here, adult *D. simulans* and *D. sechellia* are presented with a choice of media: one with octanoic acid (TOXIC) and one without octanoic acid (NORMAL). *D. simulans* avoids the media containing octanoic acid. In contrast, *D. sechellia* prefers the media containing octanoic acid. This suggests that *D. sechellia* is not simply indifferent to the presence of octanoic acid; instead, these flies actively seek the tainted media.

Genetic analyses of adaptive traits in *D. sechellia*

D. sechellia has evolved a suite of adaptations to overcome the challenge of specializing on the fruit of *M. citrifolia*. Table 1 and the following section summarizes what is known about the genetic basis of these ecologically important traits.

Adult resistance

Adult flies must survive exposure to the toxins in *Morinda* fruit in order to feed and oviposit on *Morinda* fruit. Several studies have looked at the genetic basis of this toxin resistance. R'Kha, Capy & David (1991) performed a preliminary genetic analysis of *D. sechellia* adult resistance to *Morinda* fruit. They showed that resistance to *Morinda* was dominant to susceptibility in F1 hybrids between *D. sechellia* and *D. simulans*. They estimated, using a biometric approach, the number of effective factors to be at least three, but did not employ any genetic markers to determine which chromosomes

Table 1.

Phenotype	Dominance of *D. sechellia* trait[a]	Minimum number of genes	Genes or regions of largest effect[b]
Adult resistance[c]	D	5	3R: 91A–93D
Larval resistance[c]	A–D	3	3R
Oviposition-site preference	R	2	2L
Ovariole number[d]	R–A	2	3
Egg production[d]	R	3	2 [e]
Larval morphology	R	1	1:*ovo/shaven-baby*

[a] Dominance of the *D. sechellia* phenotype in hybrid with *D. simulans*. D means that the *D. sechellia* trait is dominant, A mean the trait is additive (intermediate in dominance), and R means the trait is recessive.
[b] Chromosome or chromosome arm of region harboring factor of greatest effect, including more precise locations if known.
[c] These resistance traits may involve some of the same genes.
[d] Egg production includes factors affecting ovariole number.
[e] This phenotype maps to the centromere of this chromosome.

carried these resistance factors. Using octanoic acid and hexanoic acid instead of *Morinda* fruit, Amlou et al. (1997) repeated the analysis of R'Kha, Capy and David (1991). Again, resistance was shown to be dominant, but the resistance factors were not mapped. Interestingly, Amlou et al. (1997) also suggested that resistance must be a fairly polygenic trait, as they were not able to introgress resistance from *D. sechellia* into *D. simulans*. This result, however, could also be a byproduct of linkage between major resistance factors and hybrid infertility and inviability factors, of which there are many (Coyne, Rux & David, 1991; Joly et al., 1997). As hybrid flies were repeatedly backcrossed to *D. simulans*, selection for viability and fertility in hybrids would reduce the likelihood that chromosome regions from *D. sechellia* would be successfully introgressed.

I used 15 genetic markers to analyze the genetic basis of *D. sechellia*'s resistance to the primary toxin in *Morinda* fruit, octanoic acid (Jones, 1998). As Amlou et al. (1997) had shown, I found that resistance was dominant in F1 hybrids. Subsequently, a series of backcrosses were used to identify chromosome regions harboring factors affecting resistance. These genetic analyses suggested that at least five loci are involved in resistance. Although the Y and the dot fourth do not carry genes affecting resistance, the three major chromosomes harbor resistance factors. I also identified large chromosome regions having no effect on resistance, suggesting that *D. sechellia*'s resistance is neither very simple nor highly polygenic. Instead, resistance appears to be oligogenic.

The third chromosome has the greatest effect and carries at least two factors. One of these factors maps to a small interval between cytological bands 91A and 93D. This region represents about 2–3% of the genes in the *D. sechellia* genome, yet the resistance factor in this interval explains ~15% of the difference in resistance between *D. simulans* and *D. sechellia*. In this region, *Choline acetyltransferase* (*Cha*) stands out as a candidate resistance gene. *Cha* is essential for the production of the neurotransmitter, acetylcholine. *Cha* has been shown to be inhibited by octanoic acid *in vitro* (Ninomiya & Kayama, 1998). Currently, it is not known whether or not *Cha* from *D. sechellia* is less inhibited by octanoic acid than *Cha* from *D. simulans*. It is clear, however, that several *D. sechellia* specific amino acid changes have occurred in this gene (Jones & Begun, unpublished results). The paucity of DNA polymorphism in *D. sechellia* unfortunately makes impractical the standard population genetic tests for directional selection acting at this locus.

Larval resistance

Not surprisingly, *D. sechellia* larvae – which must grow and develop in *Morinda* – are also highly resistant to the toxins in *Morinda* fruit (R'Kha, Capy & David, 1991; Amlou, Moreteau & David, 1998a). The larvae of *D. simulans*, *D. mauritiana*, and *D. melanogaster* are not resistant. Although they did not use genetic markers, Amlou, Moreteau and David (1998a) investigated the basic genetics of egg and larval resistance to octanoic and hexanoic acids. In *D. melanogaster, D. simulans*, and *D. mauritiana*, they showed that low doses of these toxins delayed larval development and that high doses of these toxins were lethal to larvae. *D. sechellia* larvae, on the other hand, were only affected at much higher concentrations of these acids. Amlou Moreteau and David (1998a) also suggested that resistance was mostly recessive and depended somewhat on the actual concentration of the toxins. This was surprising as R'Kha, Capy and David (1991) had previously reported that embryonic resistance to *Morinda* fruit was a partially dominant trait and may have had a maternal component (although it was not possible to rule out an X chromosome effect in their study). The difference between these two results may be because Amlou, Moreteau and David (1998a) always used *D. simulans* females as mothers in F1 crosses. Thus, a maternal effect would obscure the dominance of larval resistance genes. To clarify this situation, Jones (2001) used reciprocal F1 hybrids, compound-*X* chromosomes, and reciprocal backcrosses, to show that egg-to-adult resistance to octanoic acid does indeed involve a maternal effect and exhibits intermediate dominance at toxin levels approximately equal to those found in *Morinda* fruit.

In a series of interspecific backcrosses using 11 genetic markers, I mapped factors affecting egg-to-adult ('larval') resistance in *D. sechellia* (Jones, 2001). Resistance again appears to be oligogenic. Neither the *X* chromosome, which contains 20% of *D. sechellia*'s genome, nor the fourth chromosome appear to affect resistance. The third chromosome, however, harbors at least one partially dominant resistance factor. The second chromosome carries at least two mostly dominant resistance factors but no recessive factors. These data hint that larval resistance may only involve a subset of the factors affecting adult resistance (e.g., the factors on the second and third chromosomes).

Oviposition-site preference and olfaction

As noted above, several studies have shown that *D. sechellia* is attracted to toxic volatile compounds in *Morinda* fruit. Higa and Fuyama (1993) mapped some of the factors involved in *D. sechellia's* preference for *Morinda*. Higa and Fuyama concentrated on analyzing the attraction of *D. sechellia* to hexanoic acid. To identify chromosome regions affecting this behavior, they crossed a *D. simulans* line carrying two dominant genetic markers to *D. sechellia*. The resulting F1 females were backcrossed to *D. sechellia*. Reciprocal F1 crosses were used to determine the effect of the *X* chromosome. The olfactory preference of backcross progeny was then measured in a water trap assay. From these data, Higa and Fuyama suggested that *D. sechellia's* preference is recessive to *D. simulans*' avoidance and that only the second chromosome affects preference.

Because Higa and Fuyama's analysis was low resolution and only looked at hexanoic acid, which is much less abundant in *Morinda* than octanoic acid, I investigated the genetics of oviposition site preference in *D. sechellia* (Jones, unpublished results). Earlier work indicated that *D. sechellia* showed a strong preference for *Morinda* and its toxins and that this preference was likely a recessive trait (R'Kha, Capy & David, 1991; Amlou, Moreteau & David, 1998b; Legal, Moulin & Jallon, 1999). I also showed that the preference of *D. sechellia* for toxic media is recessive to *D. simulans*' avoidance of toxic media. Using 10 genetic markers, I identified chromosome regions affecting preference. The left arm of the second chromosome harbors at least one factor strongly affecting preference. This factor may be the same factor that affected hexanoic acid in Higa and Fuyama's earlier study. (Sugaya, Higa & Fuyama, (1995), however, report in an abstract that they deficiency mapped the hexanoic factor to a region on the distal end of the right arm of second chromosome, which is far from the factor I identified). I have also shown that the right arm of the third chromosome also carries at least one factor affecting preference. The *X* chromosome, on the other hand, does not affect preference. The fact that the *X*, which comprises 20% of *D. sechellia*'s genome, has no effect on preference also suggests that the genetic basis of this host specialization is oligogenic, not polygenic.

Several authors (such as Hawthorne & Via (2001)) have conjectured that genes for host preference and those for host resistance should be genetically linked. The idea is that if there are genetically based trade-offs in performance on different hosts and genetically based preferences for different hosts, then maintaining a genetic correlation between the appropriate preference and performance factors may be advantageous (or at least, facilitate the invasion of a new host preference). One way to achieve such a genetic correlation is via genetic linkage. In *D. sechellia*, linkage between preference and performance (resistance, in this case) may occur on chromosome 3. However, the data from *D. sechellia* also suggests that resistance and preference factors need not always be linked as the *X* chromosome does not affect preference, and yet the *X* harbors resistance factors.

Ovariole number and egg production

Kambysellis and Heed (1971), in their well known paper on Hawaiian *Drosophila*, suggested that *Drosophila* with strong host preferences tend to have fewer ovarioles than their non-specialist relatives and that this difference may be an adaptation to the nutritional content of the hosts. Matching this pattern, *D. sechellia* has fewer ovarioles than its generalist sister species (although it is not known whether or not this difference is adaptive). Coyne, Rux and David (1991) genetically analyzed this trait. They showed that *D. sechellia* has about 50% as many ovarioles as *D. simulans* and that ovariole number exhibited intermediate dominance in F1 hybrids between these two species. Hodin and Riddiford (2000) showed that part of this difference was due to interspecific differences in cell number and differentiation early on in ovariole development. Coyne, Rux and David (1991), using four genetic markers, showed that at least two loci are involved, one on each autosome. The *X* chromosome and the left arm of the second chromosome have little effect on ovariole number. Their result suggests that this morphological difference between these two species is not highly polygenic.

R'Kha et al. (1997) showed that *D. sechellia* not only has fewer ovarioles, but that it produces 40% fewer eggs per ovariole, when restricted to ovipositing on standard *Drosophila* medium. When allowed to oviposit on media containing *Morinda* fruit, *D. sechellia*'s rate of egg production increases, although it remains relatively low compared to that of its sister species. Recently, I investigated the genetic basis of this difference in egg production. I have shown that all major chromosomes harbor factors affecting egg production (Jones, 2004), which suggests that interspecific difference in egg production may be more polygenic than ovariole number.

Larval morphology

Recently, Sucena and Stern (2000) discovered a conspicuous morphological difference between *D. sechellia* and its sister species. A carpet of fine hairs typically covers the posterior region of the anterior compartment of most segments of the dorsum of first-instar larvae of *D. melanogaster*, *D. simulans*, and *D. mauritiana*. Remarkably, these hairs have been lost in *D. sechellia*. The adaptive significance of the loss of these hairs is not known, but may help *D. sechellia* larvae penetrate the exterior of barely ripe *Morinda* fruit (as Legal et al., (1986) noted, the fruit is very firm when unripe). Sucena and Stern (2000), through a series of mapping experiments and complementation tests, have identified *ovo/shaven-baby* as the gene responsible for this difference. Sucena and Stern (2000) have also shown that the *D. sechellia* allele is recessive and have evidence that the *D. sechellia* phenotype is due to a change in the *cis*-regulatory regions of *ovo/shaven-baby*. The nature of this change, however, is not currently known. Nevertheless, Sucena and Stern's analysis of *ovo/shaven-baby* in *D. sechellia* is a remarkable example of how the powerful tools of *D. melanogaster* can be used to genetically dissect a striking – and likely adaptive – difference between species. Sucena and Stern's result suggests that the genetics of *natural* adaptations may be fairly simple and may involve changes in regulatory sequence rather than protein coding sequence.

A number of interesting questions about *ovo/shaven-baby* and bristle loss remain. For instance, as the locus *ovo/shaven-baby* is also known to play a role during oogenesis in females, could *ovo/shaven-baby* also be playing a role in the ovariole and egg production differences between *D. sechellia* and its sister species? Could the bristle loss be a pleiotropic effect of these other adaptations (or vice versa)? Once Sucena and Stern identify the regulatory changes responsible for the bristle loss phenotype, it should be possible to answer these questions using transgenic animals.

Conclusions

D. sechellia is quickly becoming one of the major systems for investigating the genetics of natural adaptations in animals. This is largely because *D. sechellia* is genetically tractable and has evolved several remarkable adaptations. Steady progress is being made towards understanding the genetic basis of these adaptations. With the exception of *ovo/shaven-baby*, however, the genes underlying the adaptations of *D. sechellia* are not yet known.

Despite not knowing the actual genes underlying these adaptations, several genetic patterns are becoming clear. First, these interspecific differences map to typically a few regions of large effect. This suggests, but does not prove, that these traits are *not* highly polygenic. Second, it is also clear that – with the notable exception of *ovo/shaven-baby* – more than one gene affects most of these adaptive phenotypes. Third, some genes involved in one trait clearly have pleiotropic effects on other related traits (for instance, adult and larval resistance, and ovariole number and egg production rate). While this observation is confounded by how these traits were initially defined, the fact that traits that should be related logically are related genetically suggests that the observed pleiotropy reflects an underlying genetic pattern. Finally, there is no clear trend for dominance of adaptive species differences. Based on the data in *D. sechellia*, one might speculate that *D. sechellia* traits involving a 'loss' of a feature, such as decline in rate of egg production and loss of bristles, tend to be recessive in hybrids. In contrast, those traits that are a 'gain' of a feature, such as increased resistance, tend to be more dominant in hybrids. Again, these observations are confounded by how a phenotypic trait is defined as a 'gain' or as a 'loss.' Semantic issues aside, however, it will be interesting to see if this dominance pattern holds for adaptations in other species.

Data from *D. sechellia* has contributed to progress in understanding the genetics of adaptive species differences. A number of questions, however, still need to be answered: even if only a few genes are involved, how many changes occurred in these genes? Are these changes regulatory or structural? Do alleles involved in adaptive species differences exist in the standing genetic variation of related species? Are new mutations often the source of adaptive alleles? How often are new genes involved in adaptations? Answering questions such as these requires identifying the genes underlying adaptive differences between species. This is possible in *D. sechellia* and will likely occur within the next several years.

Acknowledgements

R. Mauricio for organizing the Georgia Genetics Symposium. D. J. Begun and H. Allen Orr for discussion of the ideas in this paper. I also thank the two regular and two student reviewers of this paper for their many insightful comments. This work was supported by the National Science Foundation (C. D. Jones and D. J. Begun).

References

Amlou, M., B. Moreteau & J.R. David, 1998a. Larval tolerance in the *Drosophila melanogaster* species complex toward the two toxic acids of the *D. sechellia* host plant. Hereditas 129: 7–14.

Amlou, M., B. Moreteau & J.R. David, 1998b. Genetic analysis of *Drosophila sechellia* specialization: oviposition behavior toward the major aliphatic acids of its host plant. Behav. Genet. 28: 455–464.

Amlou, M., E. Pla, B. Moreteau & J.R. David, 1997. Genetic analysis by interspecific crosses of the tolerance of *Drosophila sechellia* to major aliphatic acids of its host plant. Genet. Sel. Evol. 29: 511–522.

Ashburner, M., 1989. *Drosophila*: a laboratory handbook. Cold Spring Harbor Laboratory Press, Cold Spring Harbor, NY.

Baker, J.G., 1970. Flora of Mauritius and the Seychelles. Lehre, Cramer, New York.

Barbash, D.A. & M. Ashburner, 2003. A novel system of fertility rescue in *Drosophila* hybrids reveals a link between hybrid lethality and female sterility. Genetics 163: 217–226.

Barbash, D.A., J. Roote & M. Ashburner, 2000. The *Drosophila melanogaster* hybrid male rescue gene causes inviability in male and female species hybrids. Genetics 154: 1747–1771.

Colson, I., S.J. MacDonald & D.B. Goldstein, 1999. Microsatellite markers for interspecific mapping of *Drosophila simulans* and *D. sechellia*. Mol. Ecol. 8: 1951–1955.

Coyne, J.A. & H.A. Orr, 1998. The evolutionary genetics of speciation. Philos. T. Roy. Soc. B 353: 287–305.

Coyne, J.A., J. Rux & J.R. David, 1991. Genetics of morphological differences and hybrid sterility between *Drosophila sechellia* and its relatives. Genet. Res. 57: 113–122.

Davis, A.W., J. Roote, T. Morley, K. Sawamura, S. Herrmann & M. Ashburner, 1996. Rescue of hybrid sterility in crosses between *D. melanogaster* and *D. simulans*. Nature 380: 157–159.

Farine, J.P., L. Legal, B. Moreteau & J.L. le Quere, 1996. Volatile components of ripe fruits of *Morinda citrifolia* and their effects on *Drosophila*. Phytochemistry 41: 433–438.

Flybase, 1999. The FlyBase database of the *Drosophila* genome projects and community literature. Nucleic Acids Res. 27: 85–88.

Giordano, R., S.L. O'Neill & H.M. Robertson, 1995. *Wolbachia* infections and the expression of cytoplasmic incompatibility in *Drosophila sechellia* and *D. mauritiana*. Genetics 140: 1307–1317.

Hawthorne, D.J. & S. Via, 2001. Genetic linkage of ecological specialization and reproductive isolation in pea aphids. Nature 412: 904–907.

Higa, I. & Y. Fuyama, 1993. Genetics of food preference in *Drosophila sechellia*. I. Responses to food attractants. Genetica 88:129–136.

Hodin, J. & L.M. Riddiford 2000. Different mechanisms underlie phenotypic plasticity and interspecific variation for a reproductive character in drosophilids (Insecta: Diptera). Evolution 54: 1638–1653.

Joly, D., C. Bazin L.W. Zeng & R.S. Singh, 1997. Genetic basis of sperm and testis length differences and epistatic effect on hybrid inviability and sperm motility between *Drosophila simulans* and *D. sechellia*. Heredity 78: 354–362.

Jones, C.D., 1998. The genetic basis of *Drosophila sechellia*'s resistance to a host plant toxin. Genetics 149: 1899–1908.

Jones, C.D., 2001. The genetic basis of larval resistance to a host plant toxin in *Drosophila sechellia*. Genet. Res. 78: 225–233.

Jones, C.D., 2004. Genetics of egg production in *Drosophila sechellia*. Heredity 92: 235–241.

Kambysellis, M.P. & W.B. Heed. 1971. Studies of oogenesis in natural populations of Drosophilidae. I. Relation of ovarian development and ecological habits of the Hawaiian species. Am. Nat. 105: 31–49.

Kearsey, M.J. & A.G. Farquhar, 1998. QTL analysis in plants; where are we now? Heredity 80: 137–142.

Kliman, R.M., P. Andolfatto, J.A. Coyne, F. Depaulis, M. Kreitman, A.J. Berry, J. McCarter, J. Wakeley, & J. Hey, 2000. The population genetics of the origin and divergence of the *Drosophila simulans* complex species. Genetics 156: 1913–1931.

Legal, L. B. Chappe, & J.M. Jallon, 1994. Molecular basis of *Morinda citrifolia*, L.: toxicity on *Drosophila*. J. Chem. Ecol. 20: 1931–1943.

Legal, L., B. Moulin, & J.M. Jallon, 1999. The relation between structures and toxicity of oxygenated aliphatic compounds homologous to the insecticide octanoic acid and the chemotaxis of two species of *Drosophila*. Pestic. Biochem. Physiol. 65: 90–101.

Legal, L. & M. Plawecki, 1995. Comparative sensitivity of various insects to toxic compounds from *Morinda citrifolia*, L. Ent. Problemy 26: 155–159.

Lemeunier, F. & M. Ashburner, 1984. Relationships within the melanogaster species subgroup of the genus Drosophila (Sophophora). Chromosoma 89: 343–351.

Louis, J. & J.R. David, 1986. Ecological specialization in the *Drosophila melanogaster* species subgroup: a case study of *D. sechellia*. Acta Oecol. 7: 215–229.

Macdonald, S.J. & D.B. Goldstein, 1999. A quantitative genetic analysis of male sexual traits distinguishing the sibling species *Drosophila simulans* and *D. sechellia*. Genetics 153: 1683–1699.

Mercot, H., D. Defaye, P. Capy, E. Pla, & J.R. David, 1994. Alcohol tolerance ADH activity, and ecological niche of *Drosophila species*. Evolution 48: 746–757.

Ninomiya, Y. & Y. Kayama, 1998. Inhibition of choline acetyltransferase activity by serum albumin modified with octanoic acid and other fatty acids. Neurochem. Res. 23: 1303–1311.

Orr, H.A., 1998. The population genetics of adaptation: the distribution of factors fixed during adaptive evolution. Evolution 52: 935–949.

Orr, H.A., 2001. The genetics of species differences. Trends Ecol. Evol. 16: 343–350.

R'Kha, S., P. Capy & J.R. David, 1991. Host-plant specialization in the *Drosophila melanogaster* species complex: a physiological, behavioral, and genetical analysis. Proc. Natl. Acad. Sci. USA 88: 1835–1839.

R'Kha, S., B. Moreteau, J.A. Coyne & J.R. David, 1997. Evolution of a lesser fitness trait: egg production in the specialist *Drosophila sechellia*. Genet. Res. Camb. 69:17–23.

Robertson, S.A. 1989. Flowering plants of Seychelles : an annotated check list, of angiosperms and gymnosperms, with line drawings. Royal Botanic Gardens, Kew, UK.

Rux, J.G. & J.A. Coyne, 1991. Recently-acquired or recently-arisen mutations in the sibling species *D. simulans, D. mauritiana and D. sechellia*. D. I. S. 70: 282.

Sauer, J.D., 1961. Coastal plant geography of Mauritius. Coastal studies series; no. 5. Louisiana State University Press, Baton Rouge.

Sauer, J.D., 1967. Plants and man on the Seychelles Coast; a study in historical biogeography. University of Wisconsin Press, Madison, WI.

Scavarda, N.J. & D.L. Hartl, 1984. Interspecific DNA transformation in *Drosophila*. Proc. Natl. Acad. Sci. USA 81: 7515–7519.

Sucena, E. & D.L. Stern, 2000. Divergence of larval morphology between *Drosophila sechellia* and its sibling species due to *cis*-regulatory evolution of ovo/shaven-baby. Proc. Natl. Acad Sci USA 97: 4530–4534.

Sugaya, S., I. Higa & Y. Fuyama, 1995. Food preference of *Drosophila sechellia* VI. Jpn J. Genet. 70: 739.

Takano, T.S., 1998. Loss of notum macrochaetae as an interspecific hybrid anomaly between *Drosophila melanogaster* and D. simulans. Genetics 149: 1435–1450.

True, J.R., B.S. Weir, & C.C. Laurie, 1996. A genome-wide survey of hybrid incompatibility factors by the introgression of marked segments of *Drosophila mauritiana* chromosomes into *Drosophila simulans*. Genetics 142: 819–837.

Tsacas, L. & G. Bächli, 1981. *Drosophila sechellia*, n.sp., huitieme espece du sous-goupe melanogaster des Iles Sechelles [Diptera, Drosophilidae]. Revue Fr. Ent. NS 3: 146–150.

Genetica (2005) 123: 147–156 © Springer 2005

Back to the future: genetic correlations, adaptation and speciation

Sara Via[1,2] & David J. Hawthorne[2]
[1]*Department of Biology, University of Maryland, College Park, MD 20742, USA (Phone: 301-405-8941; Fax: 301-314-9358; E-mail: svia@umd.edu);* [2]*Department of entomology, University of Maryland, College Park, MD 20742, USA*

Received 1 March 2003 Accepted 1 January 2004

Key words: ecological specialization, ecological speciation, genetic constraints, genotype × environment interaction, QTL mapping, sympatric speciation, tradeoffs

Abstract

Genetic correlations can affect the course of phenotypic evolution. Although genetic correlations among traits are a common feature of quantitative genetic analyses, they have played a very minor role in recent linkage-map based analyses of the genetic architecture of quantitative traits. Here, we use our work on host-associated races in pea aphids to illustrate how quantitative trait locus (QTL) mapping can be used to test specific hypotheses about how genetic correlations may facilitate ecological specialization and speciation.

Introduction

Phenotypic traits are genetically correlated if they are affected by the same genes or sets of genes through pleiotropy or linkage disequilibrium (Lande, 1979; Lynch & Walsh, 1998). Genetic correlations can have important evolutionary consequences on phenotypic evolution, because changes in allele frequencies due to selection on one trait produce correlated responses to selection in other traits influenced by the same genes or sets of genes. Correlated responses can lead to evolutionary change in neutral traits that are correlated with traits under selection, or they may constrain evolution by slowing the joint evolution of multiple characters (Lande, 1979; Via & Lande, 1985). However, if the signs of the correlations produce correlated responses in the direction of multivariate selection, genetic correlations can also facilitate adaptive evolution (Lande, 1979). In heterogeneous environments, appropriate patterns of genetic correlations among key traits expressed in different environments may speed population divergence and make speciation more likely (review in Via, 2001). This paper concerns an example in which genetic correlations among key traits may have facilitated simultaneous divergence and reproductive isolation between populations of the same species of an herbivorous insect (pea aphid) that use different host plants as a food resource (background in Via, 1991).

Early quantitative geneticists understood the effects that genetic correlations could have on the evolution of the phenotype. From the 1930s through the 1970s, quantitative genetics was largely the province of animal and plant breeders, who elaborated the theory and statistical analysis of individual quantitative traits (e.g., Falconer, 1952; Jinks, 1954; Kempthorne, 1957; Robertson, 1959a, b; Van Vleck & Henderson, 1961; Hill & Robertson, 1966; Eberhart & Russell, 1966; Hill, 1970). They also devised selection indices that exploit genetic correlations among traits in order to speed the response to artificial selection on trait groups (e.g., Kempthorne, 1957).

In the mid-1970s, the theory of quantitative genetics came back to the attention of evolutionary biologists when Lande (1975, 1976) illustrated how the 'breeder's equation' can be used to describe phenotypic evolution ($R = h^2S$, where R is the response to selection in one generation, h^2 is the proportion of phenotypic variance that is genetically based, and S is the difference between the phenotypic mean of the parents of the next generation and the population as a whole before selection). Soon, the application of quantitative genetics theory to phenotypic evolution was expanded to the multivariate case (Lande, 1979, 1980a), and the crucial roles of genetic correlations in life history evolution (Lande, 1982), sexual dimorphism (1980c), sexual selection (1980c, 1981), speciation (Lande, 1980b), and evolution in heterogeneous environments (Via & Lande, 1985) were studied.

Quantitative genetics describes phenotypic evolution in terms of parameters that can be estimated in natural populations (trait means, genetic variances and covariances, selection gradient), in contrast to the largely unmeasurable gene frequencies and selection coefficients of classical population genetics (p, q, s). This provided empiricists with new tools for the study of the genetic basis of phenotypic evolution in continuously varying traits in natural populations. By the mid-1980s a cottage industry of evolutionary biologists was estimating genetic variances and covariances in natural populations (review in Roff, 1997).

Within the past decade, the increased accessibility of DNA markers and improved analytical tools have made it possible to use linkage maps to localize loci that influence characters of importance in adaptation and speciation [so-called quantitative trait loci (QTL), see Bradshaw et al., 1995; Via & Hawthorne, 1998, Hawthorne & Via, 2001]. To date, most QTL analyses have focused on basic issues of genetic architecture: how many QTL influence particular traits, where they are located, and what is the magnitude and type of their effects on the traits of interest (Tanksley, 1993; Liu, 1997; Paterson, 1997). When different environments have been considered, interest has largely centered on the extent of variation in expression of QTL among environments, measured as QTL × environment interactions (e.g., Fry et al., 1996; Juenger et al., this volume). In contrast, the role of QTL in genetic correlations among traits has received relatively little attention.

We assert that QTL analyses may be useful in understanding the profound impact of genetic correlations on adaptation and speciation. Using an analysis of adaptation in heterogeneous environments as an example, we consider ways in which unique insights on the nature and evolutionary impact of genetic correlations among traits can be obtained from QTL mapping analyses. By focusing attention on how individual chromosomal blocks may influence multiple traits, map-based analyses may allow us to take another step toward understanding the roles of genetic correlations in phenotypic evolution.

Genetic correlations, adaptation and speciation

Evolutionary biologists considering genetic correlations usually stress their constraining influence on phenotypic evolution (e.g., Lande, 1982; Via & Lande, 1985). However, phenotypic evolution can be greatly facilitated when selection favors trait combinations that happen to be most likely, given the pattern of genetic correlations among traits. For example, if selection on two traits is in the same direction (i.e., favoring large or small values of both traits, (Figure 1(A)), then a positive genetic correlation will facilitate response to selection, while a negative one will constrain it. The opposite is true if selection on the two traits is in opposite directions (Figure 1(B)). In this paper, we discuss how genetic correlations among demographic and behavioral traits in a heterogeneous environment may act to speed population divergence and facilitate speciation.

Genetic correlations in heterogeneous environments

The genetics of traits expressed in different environments can be quantified in two ways. First, if alleles affecting a particular character vary in their phenotypic effects in different environments, or if different alleles are expressed in different environments, a genotype × environment interaction will result for that trait (Falconer & Mackay, 1996, p. 132). Alternatively, a character expressed in two environments may be considered to be two genetically correlated character states (Falconer, 1952; Via & Lande, 1985; Falconer & Mackay, 1996, p. 321). A lack of perfect correlation (i.e., $r < +1$) between character states in different environments indicates that alleles affecting the trait differ in their effects in different environments.

The relationship between g × e and the genetic correlation across environments is relatively straightforward (Falconer, 1952; Via, 1987). If most

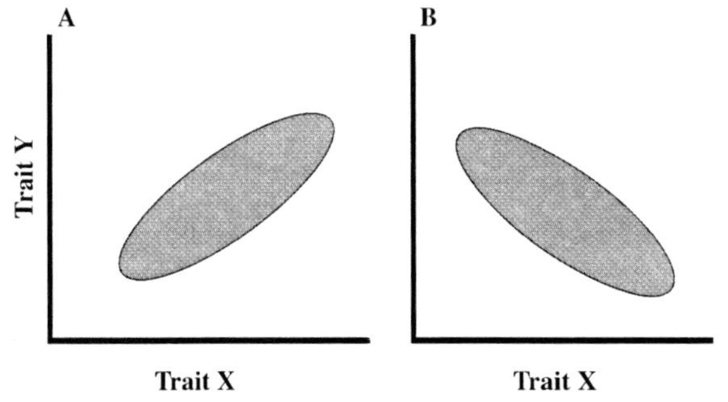

Figure 1. When traits are genetically correlated, there is more genetic variation for some trait combinations than for others. The sign of the genetic correlation and the direction of selection determines whether multivariate evolution will be facilitated or constrained. (A) Evolution is facilitated in when selection favors increases or decreases in *both* X and Y, because this is the axis of most genetic variation. Evolution of larger or smaller values of only one of the traits is constrained, because there is relatively lower genetic variation for that trait combination. (B) Under negative genetic correlation evolution is facilitated when selection acts to change the traits in opposite directions, while evolution of joint increases or decreases of the two traits is constrained.

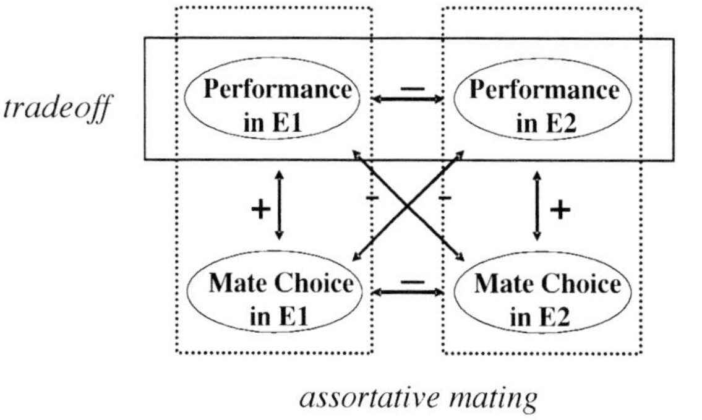

Figure 2. Hypothesized network of genetic correlations among traits expressed in two environments (E1 and E2) that would speed population divergence and facilitate speciation. The negative genetic correlation between performance in the two environments in the solid box quantifies a genetic tradeoff in specialization. The positive genetic correlations between performance and mate choice in each environment lead to assortative mating. Note, however, that correlated responses to selection occur through every link in the network, not just through the three correlations marked by the boxes.

alleles have the same effect on the expression of the character in each environment, then the genetic correlation between the character states will be high and positive, and there will be little or no genotype × environment interaction, indicating little potential for independent evolutionary change of the phenotype in each environment. In contrast, if different loci influence a trait in different environments or if the expression of pleiotropic alleles is environment-dependent, the genetic correlation between character states in different environments will be $< +1$, and a significant genotype × environment interaction will be seen (Via, 1987). In this case, partial genetic independence of the trait expressed in different environments provides the possibility for evolution of a different mean phenotype in each environment.

Considering genetic correlations among character states in different environments adds a very useful dimension to the study of how populations in a heterogeneous environment diverge. Even though individuals may experience only a single environment, they carry alleles that could be expressed differently in other environments [just as males and females each carry some alleles that are be expressed differently in the other gender (Lande, 1982b)]. Thus, selection not only affects traits in the environment in which it occurs, it also produces correlated responses in phenotypes that would be expressed in other environments. Given gene flow between environments, the correlated responses to selection in one environment may either constrain or facilitate evolution in the other environments (Via & Lande, 1985).

If we know the genetic correlations among traits within and between environments, we can predict how genetic correlations between character states in different environments may affect the trajectory of evolution in populations experiencing a spatial patchwork (e.g., Via & Lande, 1985). Therefore, even though estimating a genotype × environment interaction is an effective test for a difference in gene (or QTL) expression across environments, we contend that g × e is not as useful a metric as a genetic correlation for predicting the course of phenotypic evolution in a heterogeneous environment, because its effect cannot be quantified as easily in a genetic model (Via, 1987).

What causes genetic correlations?

Genetic correlations can be caused either by pleiotropic effects of individual alleles on several characters, or by linkage disequilibrium between alleles at a set of loci that affect a pair of traits. Although not required, close physical linkage greatly facilitates the retention of linkage disequilibrium (Lynch & Walsh, 1998). For example, if alleles at two loci that are physically close come into a favorable combination by mutation or a chance chromosome rearrangement, subsequent selection favoring individuals with that gene combination could cause the linkage disequilibrium between them to increase before it is eroded by recombination. Given the reduction in recombination among linked loci, a genetic correlation between traits caused by a favorable allelic combination at linked loci could potentially arise under selection that is too weak to counter free recombination.

The role of genetic correlations in ecological specialization

It has long been thought that the evolution of high performance in one environment may come at the cost of adaptation to other environments, causing an ecological tradeoff (e.g., Futuyma & Moreno, 1988). Taking tradeoff thinking to the genetic level has lead to the assumption that the cause of genetically-based ecological tradeoffs is the antagonistic effects of alleles on performance in different environments. In this view, it isn't possible to have high fitness in all environments, because alleles that increase performance in one environment result in decreased performance in other environments. Antagonism of allelic effects can't be tested using standard quantitative genetics, because measurement of tradeoffs as negative genetic correlations across environments reflects the composite effects of genes across the entire genome. In contrast, linkage mapping and QTL analyses permit evaluation of the extent to which particular chromosomal blocks may have antagonistic fitness effects in different environments.

One of the conundrums of the empirical study of ecological specialization has been that empirical evidence for tradeoffs has been elusive – few negative genetic correlations in performance in different environments have been found (Rausher, 1988; Fry, 1996; Agrawal, 2000). Perhaps this means that antagonistic effects of alleles in different environments are few, that they can't be detected with typical experiments, or that antagonistic effects at some loci are cancelled out by overriding positive effects in both environments at other loci. Because pea aphid host races are one of the few examples in which genetically based tradeoffs are probable (Via, 1991 and below), this is a good system within which to explore the genetic causes of performance tradeoffs in different environments in more detail.

When does specialization lead to speciation?

Sometimes, characters involved in ecological specialization also affect patterns of mating. Choosing a mate on the basis of traits that confer specialized resource use (such as body size and shape in stickleback fishes in postglacial lakes, Schluter, 1998, 2001), leads to assortative mating among individuals specialized to the same environment. Such traits can carry ecological specialization into speciation because as populations diverge under selection, mating becomes increasingly assortative, leading to a progressive decline in gene flow between the increasingly specialized taxa (Schluter, 1998, 2001). Extending this idea, any positive genetic correlation ($0 < r_G < 1$) between two traits affecting resource use and mate choice should speed population divergence and speciation to an extent proportional to the value of the correlation (Hawthorne & Via, 2001). This is a variant of an argument first proposed by Rice (1987).

To more fully understand the evolutionary implications of genetic correlations, it is useful to combine the cross-environment genetic correlations in resource use (e.g., Agrawal, 2000) and the genetic correlations between use of a given resource and mate choice (e.g., Diehl & Bush, 1989; Schluter, 2001) into a single network (Figure 2, modified from Hawthorne & Via, 2001). This approach suggests that even modest pairwise genetic correlations among the traits in the network could lead to multiple complementary correlated responses to selection that would promote divergence and reproductive isolation.

Despite the conceptual simplicity of this approach, estimation of such networks of genetic correlations would fill a large gap in our understanding of the role of local adaptation and ecological specialization in the process of speciation. How prevalent are such networks of complementary genetic correlations that may facilitate adaptation and speciation? Is divergence and the evolution of reproductive isolation between sympatric populations more likely when this type of genetic architecture is present?

We estimated the correlational network for resource use and habitat acceptance (which determines mate choice) in a segregating F_2 generation of a cross between two clones representing the specialized host races of pea aphids on alfalfa and red clover. We then used QTL mapping to test the hypothesis that pleiotropy or close linkage can be distinguished from linkage disequilibrium of unlinked loci as a cause of genetic correlations among key ecologically important characters in pea aphids (see also Via & Hawthorne, 1998, 2002; Hawthorne & Via, 2001). The following specific hypotheses were tested by examining the degree to which QTL for performance and behavioral acceptance of each host co-localized on the pea aphid linkage map:

(1) Are genetic tradeoffs in host use by specialized pea aphids based in antagonistic effects of alleles or linked sets of alleles? To test for genetic tradeoffs in performance of pea aphids on the two focal host plants, we asked if QTL influencing performance on alfalfa and red clover map to the same location but have opposite effects. Though we cannot distinguish pleiotropy from close linkage, we can conclude that if QTL for performance in the different environments map to only unlinked genomic locations (with other likelihood ratios far below the significance threshold), an observed negative genetic correlation across environments cannot be explained by antagonistic pleiotropy or close linkage.

(2) Correlations between resource use and mate choice: If QTL influencing performance in one of the environments map to the same chromosomal blocks as habitat acceptance (which determines mate choice, e.g., Caillaud and Via, 2000) for that environment and have the same directionality of effects, they could contribute to a positive genetic correlation through pleiotropy or close linkage. In contrast, if QTL for performance and habitat choice map to unlinked chromosomal blocks, then any genetic correlation observed between them must be due to linkage disequilibrium maintained by strong selection.

Materials and methods

The system

Pea aphids [*Acyrthosiphon pisum pisum* Harris (Homoptera: Aphididae)] are cyclically parthenogenetic insects that feed on the phloem of legumes (Eastop, 1973). The main hosts for these insects in our New York and Iowa study areas are commercially farmed alfalfa and red clover. Reciprocal transplant experiments of pea aphids between these hosts show clear ecological specialization: clones have much higher performance on the natal plant than the alternate plant, and clones from a given plant do better on that plant than do clones transferred from the alternate host plant (Via, 1991, 1999). Moreover, within regional populations, there is a strong negative genetic correlation in fecundity across environments: genotypes that do well on alfalfa tend to do poorly on clover (Figure 3(A, B), modified from Via, 1991).

However, because much of the negative genetic correlation across environment is between-populations (Figure 3(B)), its mechanism is unclear. Is this apparent tradeoff due to divergence along lines established by genetic correlations within populations due to pleiotropy, or has it resulted from LD accumulated during the partially independent evolution of specialization in the races within each environment? In the latter case, a negative between populations correlation could result if alleles accumulate within populations that increase adaptation to one environment and have no effect on performance in other environments. To answer this question, the contributions of various chromosomal blocks to the various character states expressed in each environment must be separated.

QTL mapping and the genetic architecture of specialization and assortative mating

We performed a QTL mapping experiment to partition the genetic architecture of differential

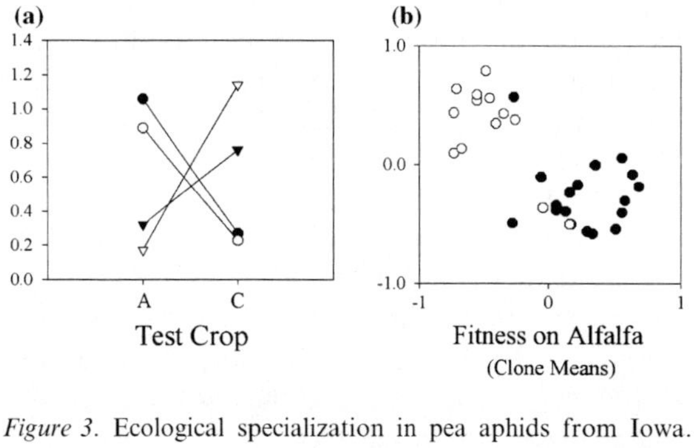

Figure 3. Ecological specialization in pea aphids from Iowa. (A) Population × environment interaction for two sets of clones collected from alfalfa (circles) and two collected from clover (triangles). (B) Scatterplot of adjusted clone mean fitness on each host for clones collected from alfalfa (solid circles) or from clover (open circles). Modified from Via (1991).

host plant use and host acceptance behavior into the effects of different chromosomal blocks. This experiment also allowed us to determine whether the type of facilitating network of genetic correlations seen in Figure 2 is observed in pea aphids.

The crosses
In 1995, we performed a single-pair reciprocal cross between one alfalfa specialist genotype and one clover specialist genotype. These genotypes were collected in 1993 in Lansing, NY and maintained in individual clonal culture on their natal host. These two clones were chosen after field testing in a reciprocal transplant because they typify the most specialized genotypes within the two races (Via, unpublished data). Since the genetic differentiation between races is much larger than the variability within races (e.g., Figure 3(B)), much of the genetic differentiation between the races is likely to have been captured in this single cross. In 1996, we mated two different F_1 genotypes from this cross to produce 200 F_2 progeny, which hatched in 1997. During meiosis in the F_1, crossing-over and recombination occurs between the parental genomes (F_1 have one unrecombined homolog from each specialized parent). Thus, each F_2 genotype bears a unique combination of chromosomal blocks from each of the parents. Given race-specific markers, we can identify the origin of each chromosomal segment, and correlate the possession of certain segments with variation in the phenotype. This is the essence of QTL mapping (review in Tanksley, 1993; Via & Hawthorne, 1998).

Phenotyping the F_2
Four phenotypic traits (fecundity on alfalfa, fecundity on clover, acceptance of alfalfa and acceptance of clover) were measured in replicate in the segregating F_2 population between September 1997 and June 2000 in a randomized block design (see Hawthorne & Via, 2001 for methods). Unlike progeny of a sexual species, these F_2 can be propagated parthenogenetically, permitting replication and estimation of the best linear unbiased predictor (BLUPs) for each F_2 genotype and character from the replicate trials of each genotype (SAS, PROC MIXED; Littel et al., 1996).

The correlations among the BLUPS for the four traits provide an estimate of the relevant genetic correlations as shown in Figure 2. These correlations (Figure 4) measure the segregating genetic covariance after one generation of recombination between the host-race genomes. Thus, any linkage disequilibrium (LD) between alleles caused by crossing divergent populations would have been reduced by 50%, while pleiotropy would remain constant. Thus, the observed correlations among BLUPs for the F_2 could be due to a combination of residual LD and pleiotropy. One way to test whether these correlations are due to pleiotropy or LD would be to carry the crosses into advanced hybrid generations and test for a decline in the correlation (as in Conner, 2002).

Construction of the linkage map
Amplified fragment length polymorphisms (AFLPs) were used to construct a linkage map for the pea aphid genome (Figure 5). Because AFLP are dominant markers, we constructed a separate map for each of the parental genomes, using markers that were recessive homozygotes in that parent. These two maps were aligned using seven sequence-tagged codominant markers generated from the AFLPs.

QTL were mapped separately for each of the four key traits [fecundity on alfalfa (FecA), fecundity on clover (FecC), acceptance of alfalfa (AccA), and acceptance of clover (AccC)], using composite interval mapping in QTL Cartographer (Basten et al., 1996). Using permutation tests in QTL Cartographer, 95% confidence intervals on QTL location were obtained. Directionality of the additive effect of each QTL was also determined using QTL Cartographer (Basten et al., 1996).

Results

Genetic correlations in the mapping population

The genetic correlations among the F_2 progeny clones mirror the pattern that is predicted to speed

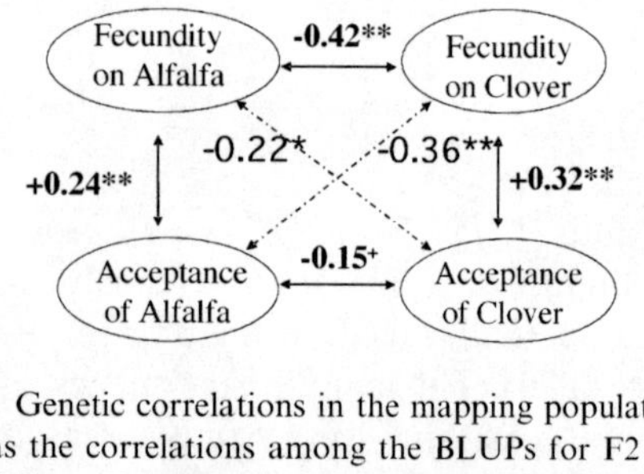

Figure 4. Genetic correlations in the mapping population, calculated as the correlations among the BLUPs for F2 progeny (modified from Hawthorne & Via, 2001).

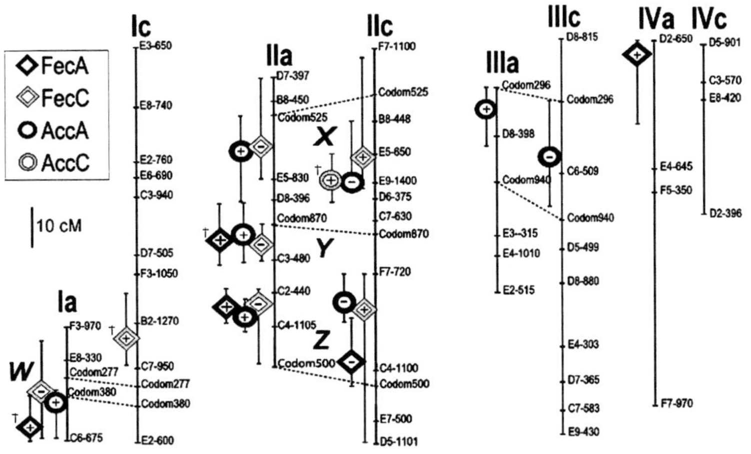

Figure 5. QTL map for four traits [fecundity on alfalfa (FecA)], fecundity on clover (FecC), behavioral acceptance of alfalfa (AccA), and acceptance of clover (AccC) in pea aphids (modified from Hawthorne & Via, 2001]. QTL are shown as symbols within the 95% confidence intervals for their location from permutation testing. QTL that are suggestive in composite interval mapping ($0.12 < p < 0.05$) are indicated by daggers. Signs inside the symbols indicate the directionality of QTL effect.

the evolution of specialization and reproductive isolation (Figures 2 and 4). Do these correlations reflect fundamental antagonisms of alleles at single or closely linked genes, or have associations among alleles at unlinked loci been built up by selection?

The pea aphid linkage map

Our map revealed four linkage groups, consistent with cytological observations of four chromosomes in pea aphids (Figure 5; Sun & Robinson, 1966). The AFLP markers on this framework map are separated by an average of 13 cM.

Mapped QTL for performance and acceptance behavior on alfalfa and red clover

If genetic correlations among these traits were caused only by LD of unlinked alleles, then QTL would be expected to be scattered across the genome. In contrast, our results suggest that many of the QTL for these key traits map together, appearing in several groups of two or more QTL each (Figure 5). Clusters *W* and *Y* are seen only on the alfalfa genome, while clusters *X* and *Z* may involve homologous QTL on both genomes. In each of these clusters, the directionality of the QTL effects matches the model of complementary correlated responses shown in Figure 2. Thus, selection on any one of these four key traits is expected to lead to correlated responses through pleiotropy or close linkage that could speed population divergence.

For example, if an individual were to inherit the chromosomal block between markers C2-440 and C4-1105 on linkage group II_a, it would be expected to inherit not only a QTL that increases fecundity on alfalfa, but also a QTL that decreases fecundity on clover, and a third QTL increasing the behavioral acceptance of clover. This cluster contains both the antagonistic allelic effects in performance in two environments that would produce a genetic tradeoff, and the correlated effects on performance and habitat choice that would lead to assortative mating.

In addition to these co-localized clusters of QTL, there are several independent QTL for the various traits (see the QTL for acceptance of alfalfa on linkage groups III_a and III_c, Figure 5). Such QTL contribute variation to the trait that is uncorrelated with that in any of the other traits, lowering the observed genetic correlation. This mixture of correlated and uncorrelated effects of alleles in the composite genetic correlation stands in contrast to the typical assumptions of equal allelic effects among loci made in many quantitative genetic models (e.g., Via & Lande, 1985).

In no case did the cumulative effects of the QTL that we discovered explain more than 50% of the variance among the F_2 clones. Thus, there are likely to be many undiscovered QTL of small effect that influence the phenotypic differences between the specialists in these four traits.

Discussion

Genetic correlations among ecologically important phenotypic traits can either facilitate or constrain the evolutionary dynamics of adaptation and speciation. We illustrate how a complementary pattern of genetic correlations among traits under divergent selection for resource use and those that influence habitat selection could facilitate the divergence of populations and speed the evolution of reproductive isolation under divergent selection in two environments.

We asked two questions in the QTL mapping analysis. First, is the negative genetic correlation across environments seen between pea aphid populations due to alleles with antagonistic effects in the two environments, as generally assumed in the discussion of genetic tradeoffs in specialization? Secondly, is the assortative mating that results from the habitat fidelity of specialists attributable to pleiotropy/close linkage, or to linkage disequilibrium between alleles at unlinked loci? Distinguishing between these alternatives allows us to determine whether the correlations may have been causally involved in facilitating specialization and reproductive isolation or whether they are the end products of divergent selection.

The results of our QTL mapping suggest that the genetic correlations among key traits in the mapping population are due in part to several clusters of closely linked or pleiotropic genes that affect several of the key character states, with additional uncorrelated variation contributed by QTL that affect only a single character state. Though we cannot distinguish close linkage from pleiotropy, the apparent QTL clustering does not support the hypothesis that the correlations are caused only by associations among alleles at unlinked loci that have accumulated under selection in the nearly reproductively isolated populations.

Given that a favorable network of genetic correlations could speed population divergence-with-gene-flow in sympatric populations (e.g., Rice & Hostert, 1993), it would be very useful to know how such a network could arise. Favorable effects on multiple character states could arise by pleiotropic mutation, but they could also occur by drift to favorable allelic combinations at closely linked genes, or even by a chance gene rearrangement that brings loci affecting key traits into physical adjacency. Once the appropriate pleiotropic alleles or allelic combinations are available, it could potentially spread rapidly under divergent selection, increasing even as the populations diverge.

As we analyze more cases of rapid population divergence-with-gene-flow or speciation, will we find additional examples of favorable networks of genetic correlations? If gene flow decreases as specialization increases due to a combination of pleiotropic or closely linked QTL with effects in the appropriate directions (see Figure 5), genetic correlations that arise through pleiotropic mutation or gene rearrangement could be a potent factor in initiating speciation. Perhaps many of the cases of sympatric divergence that can be observed today (see Via, 2001 for review) are those in which population divergence and reproductive isolation have evolved jointly under a genetic architecture of this type, permitting differentiation that is rapid enough to outrun the gene flow that might extinguish a slower process.

Further study of the mechanisms of genetic correlations among key traits involved in adaptation and reproductive isolation are likely to reveal important facets of the speciation process. For example, it would be fascinating to use QTL analyses of different taxa for a comparative study of the genetic architecture of divergence in a variety of ecological conditions, including sympatry and allopatry. Are genetic correlations among traits leading to the kind of complementary correlated responses that speed speciation (e.g., Figure 2) seen more often among sympatrically diverged taxa than among allopatrically diverged ones? That is certainly the hypothesis suggested by our work on pea aphids. By combining molecular approaches with quantitative genetics and genomics to address specific mechanistic hypotheses about how speciation occurs, the next decade promises to be an exciting time in speciation research.

Acknowledgments

We are grateful to Rodney Mauricio for organizing the symposium, and for inviting us to present this work. Andrea Badgely, Amy Bouck, Stasia

Skillman, Carrie Wicklund Wu and Christy Olson provided expert technical assistance. We thank Jeff Conner and Doug Schemske for comments on the manuscript, and NSF for support of our work on speciation.

References

Agrawal, A.A., 2000. Host range evolution: adaptation and tradeoffs in fitness of mites on alternative hosts. Ecology 81: 500–508.

Basten, C.J., B.S. Weir & Z.-B. Zeng, 1996. QTL Cartographer (Ver 1.13), Department of Statistics, North Carolina State University, Raleigh, NC.

Bradshaw, Jr., H.D., S.M. Wilbert, K.G. Otto & D.W. Schemske, 1995. Genetic mapping of floral traits associated with reproductive isolation in monkey flowers (*Mimulus*). Nature 376: 762–765.

Caillaud, C.M. & S. Via, 2000. Specialized feeding behavior influences both ecological specializtion and assortative mating in sympatric host races of pea aphids. Am. Nat. 156: 609–621.

Conner, J.K., 2002. Genetic mechanisms of floral trait correlations in a natural population. Nature 420: 407–410.

Diehl, S.R. & G.L. Bush, 1989. The role of habitat preference in adaptation and speciation, pp. 678 in Speciation and its Consequences, edited by D. Otte and J.A. Endler. Sinauer Associates, Sunderland, MA.

Eastop, V.F. 1973. Keys for the identification of *Acyrthosiphon* (Hemiptera: Aphididae). Bull. Br. Mus. Nat. Hist. B. Entomol. 26: 1–115.

Eberhart, S.A. & W.A. Russell, 1966. Stability parameters for comparing varieties. Crop Sci. 6: 36–40.

Falconer, D.S., 1952. The problem of environment and selection. Am. Nat. 86: 293–298.

Falconer, D.S. & T.F.C. Mackay, 1996. Introduction to Quantitative Genetics. 4th edn., Longman Sci. and Tech., Harlow, UK.

Fry, J.D., 1996. The evolution of host specialization, are tradeoffs overrated? Am. Nat. 148: S84–S107.

Fry, J.D., S.L. Heinsohn & T.F.C. Mackay, 1996. The contribution of new mutations to genotype–environment interaction for fitness in *Drosophila melanogaster*. Evolution 50: 2316–2327.

Futuyma, D.J. & G. Moreno, 1988. The evolution of ecological specialization. Annu. Rev. Ecol. Syst. 19: 201–233.

Hawthorne, D.J. & S. Via, 2001. Genetic linkage facilitates ecological specialization and reproductive isolation in pea aphids. Nature 412: 904–907.

Henderson, C.R., 1963. Selection index and the expected genetic advance, pp. 141–163 in Statistical Genetics and Plant Breeding, edited by W.D Hanson and H.F. Robinson Natl. Acad. Sci, Natl. Res. Council Publ. No. 982, Washington, DC.

Hill, W.G., 1970. Design of experiments to estimate heritability by regression of offspring on selected parents. Biometrics 26: 566–571.

Hill, W.G. & A. Robertson, 1966. The effect of linkage on limits to artificial selection. Genet. Res. 8: 269–294.

Jinks, J.L., 1954. The analysis of continuous variation in a diallel of *Nicotiana rustica* varieties. Genetics 39: 767–788.

Kempthorne, O., 1957. An Introduction to Genetic Statistics. John Wiley & Sons. NY.

Lande, R., 1975. The maintenanace of genetic variation by mutation in a polygenic character with linked loci. Genet. Res. 26: 221–235.

Lande, R., 1976. Natural selection and random genetic drift in phenotypic evolution. Evolution 30: 314–334.

Lande, R., 1979. Quantitative genetic analysis of multivariate evolution, applied to brain: body allometry. Evolution 33: 402–416.

Lande, R., 1980a. The genetic covariance between characters maintained by pleiotropic mutation. Genetics 94: 203–215.

Lande, R., 1980b. Genetic variation and phenotypic evolution during allopatric speciation. Am. Nat. 116: 463–479.

Lande, R., 1980c. Sexual dimorphism, sexual selection, and adaptation in polygenic characters. Evolution 34: 292–305.

Lande, R., 1981. Models of speciation by sexual selection on polygenic traits. Proc. Natl. Acad. Sci. USA 78: 3721–3725.

Lande, R., 1982. A quantitative genetic theory of life history evolution. Ecology 63: 607–615.

Lande, R. & S.J. Arnold., 1983. The measurement of selection on correlated characters. Evolution 37: 1210–1226.

Liu, B.H., 1997. Statistical Genomics: Linkage Mapping and QTL Analysis. CRC Press, NY. 612 pp.

Littel, R.C., G.A. Milliken, W.W. Stroup & R.D. Wolfinger, 1996. SAS System for Mixed Models. SAS Institute, Cary, NC.

Lynch, M. & B. Walsh, 1998. Genetics and Analysis of Quantitative Traits. Sinauer Press, Sunderland, MA.

Mather, K. & J.L. Jinks, 1982. Biometrical Genetics. 3rd edn. Chapman and Hall, NY.

Paterson, A.H. (Ed)., 1997. Molecular Dissection of Complex Traits. CRC Press, NY. 306 pp.

Provine, W.B., 1971. The Origins of Theoretical Population Genetics. University of Chicago Press, Chicago.

Rausher, M.D., 1988. Is coevolution dead? Ecology 69: 898–901.

Rice, W.R., 1987. Selection via habitat specialization, the evolution of reproductive isolation as a correlated character. Evol. Ecol. 1: 301–314.

Rice, W.R. & Hostert, E., 1993. Laboratory experiments on speciation: what have we learned in 40 years? Evolution 47: 1637–1653.

Robertson, A., 1959a. Experimental design in the evaluation of genetic parameters. Biometrics 15: 219–226.

Robertson, A., 1959b. The sampling variance of the genetic correlation coefficient. Biometrics. 15: 469–485.

Roff, D.A., 1997. Evolutionary Quantitative Genetics. Chapman and Hall, NY. 493 pp.

Schluter, D., 1998. Ecological causes of speciation, pp. 114–129 in Endless Forms: Species and Speciation, edited byHoward, D.J. & S.H. Berlocher, Oxford University Press, NY.

Schluter, D., 2001. Ecology and the origin of species. Trends Ecol. Evol. 16: 372–380.

Sun, R.Y. & A.G. Robinson, 1966. Chromosome studies of 50 species of aphids. Can. J. Zool. 44: 649–653.

Tanksley, S.D., 1993. Mapping polygenes. Annu. Rev. Genet. 27: 205–233.

Van Vleck, L.D. & C.R. Henderson, 1961. Empirical sampling estimates of genetic correlations. Biometrics 17: 359–371.

Via, S., 1987. Genetic constraints on the evolution of phenotypic plasticity, pp. 47–71 in Genetic Constraints on Adaptive Evolution, edited by Volker Loeschcke. Springer-Verlag.

Via, S., 1991. The genetic structure of host plant adaptation in a spatial patchwork: demographic variability among reciprocally transplanted pea aphid clones. Evolution 45: 827–852.

Via, S., 2001. Sympatric speciation in animals: the ugly duckling grows up. Trends Ecol. Evol. 16: 381–390.

Via, S. & D.J. Hawthorne, 1998. The genetics of speciation: Promises and prospects of QTL mapping. pp. 352–366 in Endless Forms: Species and Speciation, edited by D.J. Howard & S.H. Berlocher.

Via, S. & D.J. Hawthorne, 2002. Genetic architecture of ecological specialization and incipient speciation in divergent races of pea aphids. Am. Nat. 159(Suppl.): 576–588.

Via, S. & R. Lande, 1985. Genotype–environment interaction and the evolution of phenotypic plasticity. Evolution 39: 505–522.

Genetica (2005) 123: 157–170 © Springer 2005

Parallel genotypic adaptation: when evolution repeats itself

Troy E. Wood[1], John M. Burke[2] & Loren H. Rieseberg[1]
[1]*Indiana University, 1001 E. Third St, Jordan Hall 142, Bloomington, IN 47405, USA (Phone: +1-812-856-4996; Fax: +1-812-855-6705; E-mail: trowood@indiana.edu);* [2]*Vanderbilt University, Department of Biological Sciences, VU Station B 351634, Nashville, TN 37235, USA*

Received 1 January 2002 Accepted 13 February 2003

Keywords: adaptation, artificial selection, convergent evolution, experimental evolution, homoplasy, natural selection, parallel evolution

Abstract

Until recently, parallel genotypic adaptation was considered unlikely because phenotypic differences were thought to be controlled by many genes. There is increasing evidence, however, that phenotypic variation sometimes has a simple genetic basis and that parallel adaptation at the genotypic level may be more frequent than previously believed. Here, we review evidence for parallel genotypic adaptation derived from a survey of the experimental evolution, phylogenetic, and quantitative genetic literature. The most convincing evidence of parallel genotypic adaptation comes from artificial selection experiments involving microbial populations. In some experiments, up to half of the nucleotide substitutions found in independent lineages under uniform selection are the same. Phylogenetic studies provide a means for studying parallel genotypic adaptation in non-experimental systems, but conclusive evidence may be difficult to obtain because homoplasy can arise for other reasons. Nonetheless, phylogenetic approaches have provided evidence of parallel genotypic adaptation across all taxonomic levels, not just microbes. Quantitative genetic approaches also suggest parallel genotypic evolution across both closely and distantly related taxa, but it is important to note that this approach cannot distinguish between parallel changes at homologous loci versus convergent changes at closely linked non-homologous loci. The finding that parallel genotypic adaptation appears to be frequent and occurs at all taxonomic levels has important implications for phylogenetic and evolutionary studies. With respect to phylogenetic analyses, parallel genotypic changes, if common, may result in faulty estimates of phylogenetic relationships. From an evolutionary perspective, the occurrence of parallel genotypic adaptation provides increasing support for determinism in evolution and may provide a partial explanation for how species with low levels of gene flow are held together.

Introduction

Homoplasy, or the recurrence of similarity in distinct evolutionary lineages, occurs frequently in nature. Such similarities have been documented at practically every level of biological organization, from nucleotide/amino acid sequences (Stewart, Schilling & Wilson 1987) to large scale deletions (Downie & Palmer, 1992), whole genome duplications (Soltis & Soltis, 1991), and the acquisition of complex phenotypic characters such as succulent, spiny stems in the Euphorbiaceae and Cactaceae. There is even evidence of the repeated origin of animal and plant species (Soltis & Soltis, 1991; Rundle et al., 2000; reviewed in Levin, 2001). This list includes examples of both molecular and morphological homoplasy, which are generally thought to be the result of distinct evolutionary processes. Because it is unlikely that complex phenotypes would arise repeatedly via a stochastic process, morphological homoplasy is widely regarded to be the result of selection. In contrast, nucleotide sequences are limited in the number of ways that they can evolve, thus most instances of molecular homoplasy have been interpreted as the chance fixation of independently arising variants in diverging lineages (Doolittle, 1994; Wells, 1996).

Although morphological homoplasy is generally viewed as being driven by natural selection, many evolutionary biologists assume that the phenotypes of interest result from unique genetic changes. In some cases, they are clearly right: The evolution of spines in euphorbs and cacti results from the modification of non-homologous structures. In cases where homology is plausible, this view is perhaps best explained by the traditional acceptance of Fisher's infinitesimal model, in which quantitative traits are assumed to be controlled by an effectively infinite number of genes, each of very small effect (Fisher, 1930). Under this view, there should be numerous paths from any one phenotype to another. Thus, the likelihood that two lineages would independently accumulate changes at the same subset of underlying loci would be low. It has become increasingly clear, however, that continuous patterns of variation may sometimes be explained by the existence of a few major quantitative trait loci (QTLs) (Tanksley, 1993). Under this so-called oligogenic model of inheritance, the number of pathways from one phenotype to another is considerably more limited, increasing the likelihood that parallel phenotypic changes have a common genetic basis.

In organisms where connections between genotype and phenotype have been made, there is emerging evidence that molecular homoplasy is sometimes driven by natural selection. Unfortunately, our understanding of the genetic basis of all but the simplest traits in the simplest organisms is woefully incomplete. Thus, it is difficult to say with any certainty whether or not some of the more complex instances of morphological homoplasy have a common genetic basis. Here, we review the best examples of selection driving different lineages to the same phenotype through the fixation of independent changes at homologous loci. This pattern of evolution has several important implications. With respect to phylogeny reconstruction, it is widely recognized that homoplasy, regardless of the cause, can lead to inaccurate conclusions regarding the evolutionary history of taxa. Parallel selection responses at the genotypic level also suggest that adaptation may be a more deterministic process than previously believed, with genetic background effects and historical contingency playing a lesser role. If parallel changes prove to be common, they may provide a mechanism by which populations of a species can evolve collectively. Furthermore, such changes may increase the likelihood of the recurrent origin of taxa by allowing geographically isolated populations of the same species to independently invade a novel, unoccupied habitat.

Definitions

Historically, taxonomists have divided phenotypic homoplasy into two categories, parallelism and convergence. Parallel evolution is defined as 'the independent occurrence of similar changes in groups with a common ancestry and *because* they had a common ancestry' (Simpson, 1961, p. 103). In contrast, 'convergence is the development of similar characteristics separately in two or more lineages without a common ancestry pertinent to the similarity but involving adaptation to similar ecological status' (Simpson, 1961, pp. 78–79). As noted above, selection is believed to be the primary evolutionary force causing the recurrence in both situations.

The advent of DNA and protein sequencing necessitated a more precise definition of these terms. Molecular evolutionary biologists use parallelism and convergence in an analogous yet distinct manner. Nucleotide or protein sequence changes from the same ancestral state to the same derived state are called parallel changes, whereas changes from different ancestral states to a common derived state are considered convergent changes (Zhang & Kumar, 1997; Figure 1). Because our goal is to make an explicit connection between evolution at the phenotypic and genotypic levels, we need an operational definition that bridges the phenotypic and molecular views. Thus, we define parallel genotypic adaptation as the independent evolution of homologous loci to fulfill the same function in two or more lineages. Note that these changes need not be identical, just functionally equivalent. Under this definition, changes at non-homologous loci resulting in the same phenotype would be considered convergent (e.g., Chen Devries & Cheng, 1997), and fall outside the scope of this review.

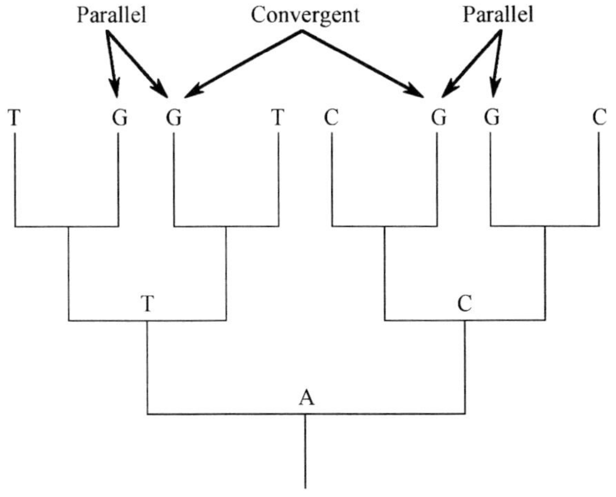

Figure 1. Parallelism versus convergence in molecular evolution. Character states at a single, homoplastic nucleotide site are mapped onto a gene tree. Parallelism refers to the independent evolution of the same derived state from a common ancestral state (the two Gs from T, or the two Gs from C). In contrast, convergence involves the evolution of the same derived state from different ancestral states (G derived independently from T and C). (After Zhang & Kumar, 1997)

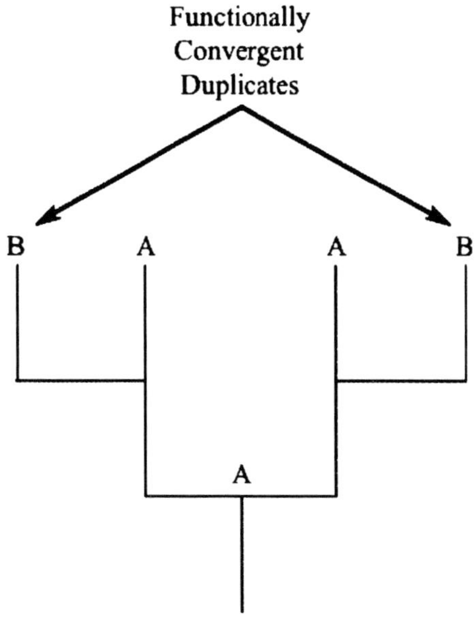

Figure 2. Convergent evolution of gene duplicates. The lateral branches leading to functional state B represent independent duplications of a homologous gene that fulfills function A. Functional state B evolved independently from changes in the duplicate copies. See text for details.

Another possibility involves the independent duplication of a homologous, ancestral locus (A) to yield two descendant loci (B) (Figure 2). In this case, the two independently derived loci are not technically homologous. However, because the two loci are direct descendants of true homologues, we consider cases in which such loci evolve to fulfill a common function to be examples of parallel genotypic evolution. The growing body of genomic data suggests that gene duplication is a common phenomenon (Lynch & Conery, 2000), and its importance in generating the raw material for adaptive evolution has been widely recognized (e.g., Haldane, 1932; Ohno, 1970). Thus, future analyses may reveal this process to be a common mode of parallel evolution.

Empirical evidence

Experimental evolution studies

The clearest evidence of parallel genotypic adaptation comes from artificial selection experiments in the lab or greenhouse (Table 1, Section A). The strength of this approach lies in the fact that researchers control both the relevant selective pressures acting upon and the evolutionary histories of the populations under study. The short generation time and relative ease of characterizing genetic variation in certain microbes makes them ideal organisms in which to study the genotypic response to uniform selective pressures. In general, these studies have revealed that selection pressures such as temperature or host shifts commonly lead to parallel genotypic adaptation (Table 1). Moreover, there is evidence that these phenotypic shifts often result from minor sequence changes; in some cases, one or a few nucleotide substitutions at a single locus accounted for the entire response to selection (Liao, Mckenzie & Hageman, 1986; Cunningham et al., 1997; Crill, Wichman & Bull, 2000). While these studies are intriguing, they have an obvious shortcoming – the dynamics of selection in these simple organisms might not be representative of adaptation in more complex organisms. In taxa with larger and more complex genomes, selective constraints due to genetic background effects or antagonistic pleiotropy may play a more important role.

Although our understanding of the molecular basis of selection response in higher organisms is incomplete, several studies in Table 1 document parallel evolution in eukaryotes. The best experimental evidence comes from a comparison of

Table 1. List of studies documenting parallel genotypic adaptation. The upper, middle, and lower panels include laboratory or greenhouse selection experiments, phylogeny-based studies, and genetic analyses of controlled crosses, respectively. See Appendix 1 for a summary of each study

Taxonomic Group(s)	Phenotype	Type of evidence	Reference
Maize & Cocklebur	Herbicide resistance	Amino acid substitution	Bernasconi et al., 1995
Human influenza A	Virulence	Amino acid substitution	Brown et al., 2001
Bacteriophage ΦX 174	Thermotolerance	Nucleotide substitution	Bull et al., 1997
Bacteriophage ΦX 174	Host shift	Amino acid substitution	Crill et al., 2001
Bacteriophage T7	Fitness	Deletion/nucleotide Substitution	Cunningham et al., 1997
Escherichia coli	Drug resistance	Amino acid substitution	Levin et al., 2000
Bacillus subtilis KNTase	Thermostability	Amino acid substitution	Liao et al., 1986
Annual Sunflower spp.	Fertility	Genome composition	Rieseberg et al., 1996
Arabidopsis thaliana	Fitness	Genome composition	Ungerer, 2000
Bacteriophage ΦX 174	Thermotolerance/host shift	Nucleotide substitution	Wichman, 1999
Flour Beetle	Pesticide resistance	Amino acid substitution	Andreev et al., 1999
Nematodes & Fungi	Pesticide resistance	Amino acid substitution	Elard et al., 1996
Coleopterans, Dipterans & Dictyopterans	Pesticide resistance	Amino acid substitution	ffrench-Constant, 1994
Arabidopsis thaliana	Flowering time	Deletion	Johanson et al., 2001
Wild Mice spp.	Immune response	Amino acid substitution	Jouvin-Marche et al., 1988
Human & Non-Human Primates	Blood groups	Nucleotide substitution	Kermarrec et al., 1999
Human & Old/New World Monkeys	Immune response	Nucleotide substitution	Kriener, 2000
Escherichia coli	Drug resistance	Nucleotide substitution	Low et al., 2001
Potato Virus X	Virulence	See Appendix 1	Malcuit et al., 2000
Human Immunodeficiency Virus (HIV)	Drug resistance	Amino acid substitution	Molla et al., 1996
Primates & Squid	Visual pigments	Amino acid substitution	Morris et al., 1993
Chimpanzee & Gorilla	Blood groups	Amino acid substitution	O'h Uigin et al., 1997
Human & Sooty Mangabey	Disease resistance	Deletion	Palacios et al., 1998
Escherichia coli	Virulence	Horizontal transfer	Reid et al., 2000
Escherichia coli	Thermotolerance	Duplication/deletion	Riehle et al., 2001
Cetaceans & Pinnipeds	Respiration	Amino acid substitution	Romero-Herrera et al., 1978
Human & Pea	Enzyme function	Amino acid substitution	Shafqat et al., 1996
Human, Marmoset & Squirrel Monkey	Visual pigments	Amino acid substitution	Shyue et al., 1995
Colobine Monkey, Ruminants & Hoatzin	Enzyme function	Amino acid substitution	Stewart et al., 1987; Zhang and Kumar, 1997
Human & Blind Cave Fish	Visual pigments	Amino acid substitution	Yokoyama and Yokoyama, 1990
Cowpea & Mung Bean	Seed weight	Comparative QTL mapping	Fatokun et al., 1992
Maize, Rice & Sorghum	Seed mass and dispersal	Comparative QTL mapping	Paterson et al., 1995
Silene vulgaris	Metal tolerance	Complementation test	Schat et al., 1996

resistance to acetolactate synthase inhibitors in naturally occurring cocklebur and two mutagenized maize lines (Bernasconi et al., 1995). Given that resistance in this case is based on a single enzyme, this result may not be predictive of the types of changes that underlie parallel phenotypic evolution in more complex traits. While there are very few studies that bear on this issue, Ungerer (2000) found that the frequency of QTL alleles governing life history traits responded uniformly to viability selection in replicate *Arabidopsis* populations, even when genetic background was varied. Similarly, working in sunflower, Rieseberg et al. (1996) showed that experimental hybrid lineages subjected to strong fertility selection converged on a common genomic composition. Because this fertility selection was primarily the result of selection for the recovery of viable gametes in interspecific hybrids, the underlying adaptive process is mechanistically distinct from classical examples of adaptation involving allelic substitution at a targeted locus. However, this study clearly demonstrates that parallel selection among lineages can yield remarkably similar genotypic responses. One weakness of conclusions drawn from these two studies is that they did not provide the necessary resolution to conclude that selection is acting on variation at homologous loci across populations. In addition, both of these studies relied on variation generated in crosses between different lineages, rather than on novel variation. They do, however, show that selection response at the genotypic level is repeatable across populations. Thus, given the appropriate genetic variation, we might expect the evolution of complex traits to mirror the findings from genetically simpler traits.

Phylogenetic studies

While experimental studies allow researchers to control the branching pattern of lineages and monitor their response to selection, parallel genotypic adaptation can be assessed in non-experimental systems as well. One approach is to use phylogenetic methods to infer the evolutionary history of the organisms of interest. This phylogeny can then be used to reconstruct the historical sequence of mutational changes in a nucleotide or protein sequence with known function. The advantage of this approach is that it can be applied to virtually any organism; thus, parallel evolution can be studied across vast taxonomic distances and in organisms that are not amenable to experimental manipulation. The main difficulty is that, in order to show that homoplasy is adaptive in origin rather than the result of chance fixation, the functional effects of a sequence change must be known, or at least inferred (Doolittle, 1994).

Once a relationship between genotype and phenotype has been established, the basic challenge is to demonstrate that shared sequence similarities are not simply the result of common ancestry. Because sequences that have evolved in parallel will show phylogenetic affinity, the detection of parallel genotypic adaptation can be problematic. Of course, if the adaptive change results from relatively few nucleotide substitutions, homoplasy may have only minor effects on phylogenetic inference. In other cases, where the ratio of informative sites to selectively advantageous substitutions is relatively low, the framework for these analyses should be based on independent phylogenetic data. Assuming that the structure of the resulting tree represents the true evolutionary history of the organisms, detecting homoplasy is as simple as mapping character states onto this tree (Figure 1). The phylogenetic approach can also be used within taxa to examine the pattern of evolution of a gene in a geographic context. For example, Andreev et al. (1999) used a phylogeny of alleles of *Resistance to dieldrin* to demonstrate that the same point mutation arose on multiple occasions in different populations of the red flour beetle, *Tribolium castaneum*.

The middle panel of Table 1 lists examples of parallel genotypic adaptation documented with phylogenetic methods. Although this set of studies includes examples from microorganisms, the taxonomic diversity represented clearly demonstrates that parallel genotypic adaptation occurs at all taxonomic levels. Once again, many of these examples involve minor sequence changes. In fact, parallel adaptation in four of these studies was based on a single amino acid substitution (Morris, Bowmaker & Hunt, 1993; Elard, Comes & Humbert, 1996; ffrench-Constant, 1996; Andreev et al., 1999).

While many of the traits listed would generally be viewed as complex, what the studies in Table 1 say about parallel evolution in simple versus complex traits is unclear. Part of the problem here stems from the definition of traits. For example, the spectral properties of visual pigments represent one aspect of color vision, which is clearly a complex trait (Yokoyama & Yokoyama, 1990; Morris, Bowmaker & Hunt, 1993; Shyue et al., 1995). Thus, parallel evolution of the genes encoding these pigments could be viewed as the parallel evolution of a highly complex trait. If, on the other hand, the trait is defined to be spectral tuning, then the trait of interest is Mendelian, no different from herbicide resistance in cocklebur and maize. The difficulty here lies in the fact that, from an evolutionary perspective, traits should be defined by what selection sees, not what the researcher sees. For example, if selection acts to increase the height of a hypothetical organism, parallel genotypic responses may be less likely than if selection acts on a specific component of height, such as cell number or cell size.

A number of the studies included in this section demonstrate sequence homoplasy for loci that have a known adaptive function, but the parallel changes themselves have not been demonstrated to be under selection. Thus, although an adaptive role for these changes is plausible, their functional significance has not been directly assessed (e.g., Romero-Herrera et al., 1978; Jouvin-Marche et al., 1988). Moreover, only two of the examples in this section (Stewart, Schilling & Wilson, 1987; Zhang & Kumar, 1997; Kriener, 2000) have been evaluated statistically. Unfortunately, the statistical model used to evaluate the role of selection in parallel sequence changes (Zhang & Kumar, 1997) is, out of necessity, naive to protein function. Because it uses a general evolutionary model to ascribe probabilities to changes between sequence states, this approach can lead to false positives. For example, if a given amino acid site is constrained on the basis of charge, it is free to evolve, but in a more limited number of ways. Therefore, the number of possible states can be far fewer than the model allows. In such cases, the test will be biased toward detecting significant parallelisms even though the changes may have occurred by chance. Ultimately, sequence changes need to be linked to a change in function to demonstrate unequivocally parallel genotypic adaptation.

Quantitative genetic studies

Another approach to detecting parallel genotypic adaptation in non-experimental systems involves quantitative genetic analysis. The most direct method is a complementation test, in which two lineages are crossed and the segregation patterns of their hybrid offspring are analyzed. If a shared, yet independently derived character state has a common genotypic basis, it will not segregate in the second (or later) generation(s). In contrast, if the character is determined by non-homologous loci, the hybrid progeny should exhibit significant phenotypic variation. An example of this approach is the work of Schat, Voour & Kuiper, (1996; Table 1), who demonstrated that metal tolerance in genetically isolated populations of *Silene* results from changes at homologous loci.

Comparative QTL mapping can also yield evidence for parallel genotypic responses. In this case, molecular markers are used to identify chromsomal regions underlying the trait(s) of interest in a segregating population (see Mauricio, 2001 for a review). In cases where homologous markers are shared across mapping populations, QTL positions can be compared between taxa. When QTLs map to the same marker intervals, the results are consistent with parallel genotypic adaptation. Although QTL methods have been applied to a wide variety of study organisms, there are only three good examples of parallel adaptation identified through this approach (Fatokun et al., 1992; Paterson et al., 1995; Hu et al., 2003; Table 1).

In all three of these cases, it is important to note that the effects of closely linked, but non-homologous loci cannot be discounted. Thus, like the map-based studies of Ungerer (2000) and Rieseberg et al. (1996) detailed above, conclusions regarding homology of the changes are premature. In addition, all three of the studies focus on domestication traits. Like the examples listed under experimental evolution above, these traits have evolved in response to strong artificial selection. Because artificially selected lineages are generally maintained in a controlled environment (e.g., lab, greenhouse, or agricultural setting), they are not necessarily subject to the same pleiotropic constraints as naturally evolving populations. Therefore, the relevance of these studies to the evolution of traits in the wild is tenuous (Coyne & Lande, 1985).

Evolutionary implications

Each of the studies reviewed here provides at least circumstantial evidence that parallel genotypic adaptation occurs at all taxonomic levels. This finding stands in stark contrast to the traditional view that parallel phenotypic evolution results from unique genetic changes. Given that a number of the traits listed above are simple (i.e., Mendelian), this result should not be surprising. After all, if a trait is controlled by a single gene, phenotypic evolution can involve changes in only that gene. As the complexity of an adaptation increases, the likelihood of its parallel recurrence should decrease. In other words, if there are numerous pathways connecting two phenotypic states, it is relatively unlikely that evolution will follow the same path twice. As stated above, however, there is a growing body of evidence that many quantitative traits are controlled oligogenically (Tanksley, 1993). In addition, apparently complex traits can often be decomposed into their component parts (e.g., color vision versus visual pigments; Morris, Bowmaker & Hunt, 1993; Shyue et al., 1995; Yokoyama & Yokoyama, 1990). If selection acts on these parts, rather than on their sum, the number of potential pathways will be fewer, which makes parallel genotypic adaptation even more likely. Finally, if the genetic variance–covariance matrices are similar across populations or taxa, then populations may be predisposed to adaptation along the path of least resistance, thereby leading to parallel genotypic adaptation (Endler, 1986; Schluter, 1996).

From a practical standpoint, perhaps one of the greatest concerns regarding homoplasy is the confounding effect it can have on phylogeny reconstruction. Because phylogenetic algorithms are designed to minimize homoplasy, shared character states that truly arose multiple times may be grouped together erroneously (Forey et al., 1992). However, a number of the studies reviewed here suggest that selection often targets only one or a few sites in a sequence (e.g., Andreev et al., 1999). Thus, even if a gene responds identically to selective pressures in evolutionarily distinct lineages, the majority of the sequence will track the branching patterns of the taxa. That is, if the selectively important changes are rare relative to the number of phylogenetically informative sites, the gene tree may still track the species tree. On the other hand, if the sequence changes represent a larger proportion of the informative sites, the resulting tree may be incongruent with the true phylogeny. For example, Kriener et al. (2000) examined sequence variation in certain alleles of the *DRB* gene family in monkeys and humans. Similarities among coding sequences were strong enough to cause a conflict between the exon-based tree and true organismal relationships. Because systematists are increasingly using multiple gene sequences to reconstruct phylogenies, these sorts of conflicts are less likely to lead to incorrect phylogenetic inferences.

From an evolutionary perspective, the occurrence of parallel genotypic adaptation suggests that adaptive evolution may be a more deterministic process than previously believed. Although some authors have argued that the most likely outcome of parallel selection in isolated populations is divergence (e.g., Wade & Goodnight, 1998; Goodnight, 2000; Levin, 2000), two studies in particular suggest that selection response at the genotypic level is repeatable across populations (Rieseberg et al., 1996; Ungerer, 2000). These studies, therefore, suggest that the effects of genetic background on selection response may have been overemphasized. If this turns out to be generally true, then parallel genotypic adaptation might provide a mechanism for both the collective evolution of populations within a species (Lande, 1983; Templeton, 1989) and the recurrent origin of taxa (reviewed in Levin, 2001).

Classical studies of gene flow have suggested that migration rates are too low to account for the apparent integration of species across their ranges (e.g., Ehrlich & Raven, 1969; Grant, 1980). If this were true, species would not be different from higher taxa, mere aggregates of the actual units of evolution (local populations or metapopulations). Recent work has revealed that the joint effects of selection and migration are, in general, sufficient to account for the integration of populations across a species range (Rieseberg & Burke, 2000). The studies reviewed above take this idea further, suggesting that local populations of a species subjected to similar selective pressures may arrive at the same genetical solutions. Another type of evidence supporting this idea comes from experimental selection studies in which populations subjected to parallel selection maintained reproductive compatibility, whereas those subjected to divergent selection often evolved incompatibilities (Rice & Hostert, 1993). The importance of parallel genotypic adaptation in species cohesion will vary with the relative rates of mutation and migration;

in cases where gene flow is limiting, parallel genotypic adaptation would be expected to play a more central role. In this context, it is interesting to note that many of the characters used to differentiate plant species are governed by one or two genes (Gottlieb, 1984; Hilu, 1983). Thus, traits used in species identification may be especially likely to evolve in parallel.

Just as parallel genotypic adaptation can help maintain species cohesion, the potential for recurrent evolution of key adaptations makes the repeated origin of taxa plausible. In general terms, this evolutionary process could allow local populations to independently invade a similar habitat. Because these lineages would share a common solution to a unique ecological challenge, they would be demographically exchangeable (sensu Templeton, 1989) for the same genetic reasons. Indeed, more and more evolutionary biologists are recognizing the importance of ecology in speciation (Schluter, 2001). Because different habitat types are often interspersed across the range of a species, the requisite ecological opportunities may occur frequently. An example of this process, albeit at the infraspecific level, would be metal tolerance in *Silene* (Schat, Voour & Kuiper 1996; Table 1). Given enough time, these independently derived populations may ascend to species status. Though not yet characterized genetically, threespine stickleback fishes are another possible example of recurrent divergence due to parallel genotypic adaptation (Rundle et al., 2000).

Taken together, the studies reviewed here provide evidence that parallel genotypic adaptation can occur in organisms ranging from microbes to plants to primates. Although the relevance of studies in microorganisms to adaptation in general has been questioned, this body of data suggests that Jacques Monod may have been right when he suggested that 'What is true for *E. coli* is true for elephants, only more so.' In some cases, the parallelisms spanned remarkably wide taxonomic distances – e.g., the independent evolution of ethanol-active ADH in pea plants and humans (Shafqat et al., 1996). Given that the genetic basis of most adaptations is still unknown, our understanding of the prevalence of parallel genotypic adaptation is still in its infancy. The advent of functional genomics should lead to a wealth of data connecting genotype to phenotype, allowing researchers to identify and compare the genetic mechanisms underlying adaptive traits in a variety of organisms.

Acknowledgements

We thank Jeff Demuth and Mark Welch for useful discussions of various aspects of this work. Our research on the genetics of adaptive evolution has been supported by grants from the National Institutes of Health, National Science Foundation, and United States Department of Agriculture.

References

Andreev, D., M. Kreitman, T.W. Phillips, R.W. Beeman & R.H. ffrench-Constant, 1999. Multiple origins of cyclodiene insecticide resistance in *Tribolium castaneum* (Coleoptera: Tenbrionidae). J. Mol. Evol. 48: 615–624.

Barlow, M. & B.G. Hall, 2002. Predicting evolutionary potential: *in vitro* evolution accurately reproduces natural evolution of the TEM β-lactamase. Genetics 160: 823–832.

Barlow, M. & B.G. Hall, 2003. Experimental prediction of the natural evolution of antibiotic resistance. Genetics 163: 1237–1241.

Bernasconi, P., A.R. Woodworth, B.A. Rosen, M.V. Subramanian & D.L. Siehl, 1995. A naturally occurring point mutation confers broad range tolerance to herbicides that target acetolactate synthase. J. Biol. Chem. 270: 17381–17385.

Brown, E.G., H. Liu, L. Chang Kit, S. Baird & M. Nesrallah, 2001. Pattern of mutation in the genome of influenza A virus on adaptation to increased virulence in the mouse lung: identification of functional themes. Proc. Natl. Acad. Sci. U.S.A. 98: 6883–6888.

Bull, J.J., M.R. Badgett, H.A. Wichman, J.P. Huelsenbeck, D.M. Hillis, A. Gulati, C. Ho & I.J. Molineux, 1997. Exceptional convergent evolution in a virus. Genetics 147: 1497–1507.

Chen, L., A.L. DeVries & C.-H.C. Cheng, 1997. Convergent evolution of antifreeze glycoproteins in Antarctic notothenioid fish & Arctic cod. Proc. Natl. Acad. Sci. U.S.A. 94: 3817–3822.

Coyne, J.A. & R. Lande, 1985. The genetic basis of species differences in plants. Am. Nat. 126: 141–145.

Crill, W.D., H.A. Wichman & J.J. Bull, 2000. Evolutionary reversals during viral adaptation to alternating hosts. Genetics 154: 27–37.

Cunningham, C.W., K. Jeng, J. Husti, M. Badgett, I.J. Molineux, D.M. Hillis & J.J. Bull, 1997. Parallel molecular evolution of deletions & nonsense mutations in bacteriophage T7. Mol. Biol. Evol. 14: 113–116.

Doolittle, R.F., 1994. Convergent evolution: the need to be explicit. Trends Biochem. Sci. 19: 15–18.

Downie, S.R. & J.D. Palmer, 1992. Use of chloroplast DNA rearrangements in reconstructing plant phylogeny, pp. 14–35 in Molecular Systematics of Plants, edited by P.S. Soltis, D.E. Soltis & J.J. Doyle. Chapman & Hall, N.Y.

Elard, L., A.M. Comes & J.F. Humbert, 1996. Sequences of B-tubulin cDNA from benzimidazole-susceptible and -resistant strains of *Teladorsagia cirumcincta*, a nematode parasite of small ruminants. Mol. Biochem. Parasitol. 79: 249–253.

Endler, J.A., 1986. Natural Seleciton in the Wild. Princeton University Press, Princeton.

Ehrlich, P.R. & P.H. Raven, 1969. Differentiation of populations. Science 165: 1228–1232.

Fatokun, C.A., D.I. Menancio-Hautea, D. Danesh & N.D. Young, 1992. Evidence for orthologous seed weight genes in cowpea & mung bean based on RFLP mapping. Genetics 132: 841–846.

ffrench-Constant, R.H., 1994. The molecular & population genetics of cyclodiene insecticide resistance. Insect Bicochem. Mol. Biol. 24: 335–345.

Forey, P.L., C.J. Humphries, I.L. Kitching, R.W. Scotland, D.J. Siebert & D.M. Williams, 1992. Cladistics. Oxford University Press, NY.

Goodnight, C.J., 2000. Quantitative trait loci & gene interaction: the quantitative genetics of metapopulations. Heredity 84: 587–598.

Gottlieb, L.D., 1984. Genetics & morphological evolution in plants. Am. Nat. 123: 681–709.

Grant, V., 1980. Gene flow and the homogeneity of species populations. Biol. Zentralbl. 99: 157–169.

Haldane, J.B.S., 1932. The Causes of Evolution. Longmans Green, London.

Hilu, K.W., 1983. The role of single-gene mutations in the evolution of flowering plants. Evol. Biol. 36: 97–128.

Hu, F.Y., D.Y. Tao, E. Sacks, B.Y. Fu, P. Xu, J. Li, Y. Yang, K. McNally, G.S. Khush, A.H. Paterson & Z.-K. Li, 2003. Convergent evolution of perenniality in rice & sorghum. Proc. Natl. Acad. Sci. U.S.A. 100: 4050–4054.

Johanson, U., J. West, C. Lister, S. Michaels, R. Amasino & C. Dean, 2001. Molecular analysis of FRIGIDA, a major determinant of natural variation in *Arabidopsis* flowering time. Science 290: 344–347.

Jouvin-Marche, E., A. Cuddihy, S. Butler, J.N. Hansen, W.M. Fitch & S. Rudikoff, 1988. Modern Evolution of a single-copy gene: the immunoglobulin Ck locus in wild mice. Mol. Biol. Evol. 5: 500–511.

Kermarrec, N., F. Roubinet, P.-A. Apoil & A. Blancher, 1999. Comparison of allele O sequences of the human & non-human primate ABO system. Immunogenetics 49: 517–526.

Kriener, K., C. O'hUigin, H. Tichy & J. Klein, 2000. Convergent evolution of major histocompatibility complex molecules in humans & New World monkeys. Immunogenetics 51: 169–178.

Lande, R., 1983. The response to selection on major & minor mutations affecting a metrical trait. Heredity 50: 47–65.

Levin, B.R., V. Perrot & N. Walker, 2000. Compensatory mutations, antibiotic resistance & the population genetics of adaptive evolution in bacteria. Genetics 154: 985–997.

Levin, D.A., 2000. The Origin, Expansion, & Demise of Plant Species. Oxford University Press, NY.

Levin, D.A., 2001. The recurrent origin of plant races & species. Syst. Bot. 26: 197–204.

Liao, H., T. McKenzie & R. Hageman, 1986. Isolation of a thermostable enzyme variant by cloning & selection in a thermophile. Proc. Natl. Acad. Sci. U.S.A. 83: 576–580.

Low, A.S., F.M. MacKenzie, I.M. Gould & I.R. Booth, 2001. Protected environments allow parallel evolution of a bacterial pathogen in a patient subjected to long-term antibiotic therapy. Mol. Microbiol. 42: 619–630.

Lynch, M. & J.S. Conery, 2000. The evolutionary fate & consequences of duplicate genes. Science 290: 1151–1155.

Malcuit, I., W. De Jong, D.C. Baulcombe, D.C. Shields & T.A. Kavanagh, 2000. Acquisition of multiple virulence/avirulence determinants by potato virus X (PVX) has occurred through convergent evolution rather than through recombination. Virus Genes 20: 165–172.

Mauricio, R., 2001. Mapping quantitative trait loci in plants: uses & caveats for evolutionary biology. Nat. Rev. Genet. 2: 370–381.

Molla, A., M. Korneyeva, Q. Gao, S. Vasavanonda, P.J. Schipper, et al., 1996. Ordered accumulation of mutations in HIV protease confers resistance to ritonavir. Nat. Med. 2: 760–766.

Morris, A., J.K. Bowmaker & D.M. Hunt, 1993. The molecular basis of a spectral shift in the rhodopsins of two species of squid from different photic environments. Proc. R. Soc. Lond. Ser. B Biol. Sci. 254: 233–240.

Ohno, S., 1970. Evolution by Gene Duplication. Springer, Berlin.

O'hUigin, C., A. Sato & J. Klein, 1997. Evidence for convergent evolution of A & B blood group antigens in primates. Hum. Genet. 101: 141–148.

Palacios, E., L. Digilio, H.M. McClure, Z. Chen, P.A. Marx, M.A. Goldsmith & R.M. Grant, 1998. Parallel evolution of CCR5-null phenotypes in humans & in a natural host of simian immunodeficiency virus. Curr. Biol. 8: 943–946.

Paterson, A.H., Y.-R. Lin, Z. Li, K.F. Schertz, J.F. Doebley, S.R.M. Pinson, S.-C. Liu, J.W. Stansel & J.E. Irivine, 1995. Convergent domestication of cereal crops by independent mutations at corresponding genetic loci. Science 269: 1714–1718.

Reid, S.D., C.J. Herbelin, A.C. Bumbaugh, R.K. Selander & T.S. Whittam, 2000. Parallel evolution of virulence in pathogenic *Escherichia coli*. Nature 406: 64–67.

Rice, W.R. & E.E. Hostert, 1993. Laboratory experiments on speciation: what have we learned in 40 years? Evolution 47: 1637–1653.

Riehle, M.M., A.F. Bennett & A.D. Long, 2001. Genetic architecture of thermal adaptation in *Escherichia coli*. Proc. Natl. Acad. Sci. U.S.A. 98: 525–530.

Rieseberg, L.H., B. Sinervo, C.R. Linder, M.C. Ungerer & D.M. Arias, 1996. Role of gene interactions in hybrid speciation: evidence from ancient & experimental hybrids. Science 272: 741–745.

Rieseberg, L.H. & J.M. Burke, 2001. The biological reality of species: gene flow, selection, & collective evolution. Taxon 50: 47–67.

Romero-Herrera, A.E., H. Lehmann, K.A. Josey & A.E. Friday, 1978. On the evolution of myoglobin. Philos. Trans. R Soc. Lond. Ser. B Biol. Sci. 283: 61–163.

Rundle, H.D., L. Nagel, J.W. Boughman & D. Schluter, 2000. Natural selection & parallel speciation in sympatric sticklebacks. Scinece 287: 306–308.

Saitou, N. & F. Yamamoto, 1997. Evolution of primate ABO blood group genes & their homologous genes. Mol. Biol. Evol. 14: 399–411.

Schat, H., R. Voous & E. Kuiper, 1996. Identical major gene loci for heavy metal tolerances that have independently evolved in different local populations & subspecies of *Silene vulagaris*. Evolution 50: 1888–1895.

Schluter, D., 1996. Adaptive radiation along genetic lines of least resistance. Evolution 50: 1766–1774.

Schluter, D., 2001. Ecology & the origin of species. Trends Ecol. Evol. 16: 372–380.

Shafqat, J., M. El-Ahmad, O. Danielsson, M.C. Martinez, B. Persson, X. Pares & H. Jornvall, 1996. Pea formaldehyde-active class III alcohol dehydrogenase: common derivation of the plant & animal forms but not of the corresponding ethanol-active forms (classes I & P). Proc. Natl. Acad. Sci. U.S.A. 93: 5595–5599.

Shyue, S.-K., D. Hewett-Emmett, H.G. Sperling, D.M. Hunt, J.K. Bowmaker, J.D. Mollon & W.-H. Li, 1995. Adaptive evolution of color vision genes in higher primates. Science 269: 1265–1267.

Simpson, G.G., 1961. Principles of Animal Taxonomy. New York, Columbia University Press.

Soltis, P.S. & D.S. Soltis, 1991. Multiple origins of the allotetraploid *Tragopogon mirus* (Compositae): rDNA evidence. Syst. Bot. 16: 407–413.

Stewart, C.-B., J.W. Schilling & A.C. Wilson, 1987. Adaptive evolution in the stomach lysozymes of foregut fermenters. Nature 330: 401–404.

Sucena, E., I. Delon, I. Jones, F. Payre & D.L. Stern, 2003. Regulatory evolution of *shavenbaby/ovo* underlies multiple cases of morphological parallelism. Nature 424: 935–938.

Tanksley, S.D., 1993. Mapping polygenes. Annual Rev. Genet. 27: 205–233.

Templeton, A.R., 1989. The meaning of species & speciation: a genetic perspective, pp. 3–27 in Speciation & Its Consequences, edited by D. Otte & J.A. Endler. Sunderland, MA.

Ungerer, M., (2000) Selection, genetic envrionment & adaptive evolution: an analysis of microevolutionary dynamics in experimental plant populations. Ph.D. dissert. Indiana Univ., Bloomington.

Wade, M.J. & C.J. Goodnight, 1998. Perspective: the theories of Fisher & Wright in the context of metapopulations: when nature does many small experiments. Evolution 52: 1537–1553.

Wells, R.S., 1996. Excessive homoplasy in an evolutionarily constrained protein. Proc. R. Soc. Lond. Ser. B Biol. Sci. 263: 393–400.

Wichman, H.A., M.R. Badgett, L.A. Scott, C.M. Boulianne & J.J. Bull, 2001. Different trajectories of parallel evolution during viral adaptation. Science 285: 422–424.

Yokoyama, R. & S. Yokoyama, 1990. Convergent evolution of the red- & green-like visual pigment genes in fish, *Astyanax fasciatus* & human. Proc. Natl. Acad. Sci. U.S.A. 87: 9315–9318.

Zhang, J. & S. Kumar, 1997. Detection of convergent & parallel evolution at the amino acid sequence level. Mol. Biol. Evol. 14: 527–536.

Appendix 1.

Brief summaries of studies

A. Experimental Evolution Studies

1. Barlow and Hall (2002, 2003), Mutations in genes in the TEM family of β-lactamases are known to confer resistance to the β-lactam antibiotics. The authors compared analyses of *in vitro* selection experiments targeting the TEM-1 gene to naturally occurring, resistant TEM-alleles. Nine substitutions have evolved multiple times in natural bacterial populations, and seven of these were recovered in the *in vitro* experiments. The authors (2003) also showed that mutagenized TEM-1 alleles conferred resistance to the relatively new antibiotic, cefepime. Resistant alleles contained two to six substitutions each, and many of these substitutions were shared across allelic variants. Thus, adaptation at this locus in response to antibiotic challenge is highly predictable.
2. Bernasconi et al. (1995), Bernasconi and colleagues examined the molecular basis of resistance to acetolactate synthase (ALS) inhibitors, which are commonly used as herbicides. The molecular basis of resistance was characterized in two field isolates of cocklebur and compared to experimentally mutagenized maize lines that also show resistance. Two different amino acid substitutions were responsible for resistance in the two cocklebur isolates. These mutations were identical to those conferring resistance in two mutagenized maize lines.
3. Brown et al. (2001), A clinical, mouse-naive isolate of human influenza A virus, A/HK/1/168, was selected for virulence in mice. This process resulted in three mutations identical to those characteristic of the virulent human H5N1 isolate A/HK/156/97, the strain that infected humans directly from birds in Hong Kong.
4. Bull et al. (1997), In this explicit test of parallel evolution, genomic sequence analysis of different lineages of bacteriophage ΦX174 challenged with high temperature revealed that over half of the substitutions were identical with substitutions in other lineages. The phages were grown on two different hosts, *Escherichia coli* C and *Salmonella typhimurium*, and some of the parallel changes were host-specific.

Appendix 1. (*Continued*)

5. Crill, Wichman and Bull (2001), Bacteriophage ΦX174 was grown alternately on its typical laboratory host, *Escherichia coli* C and a novel host, *Salmonella enterica*. Experimental adaptation to this novel host inhibited the phage's ability to grow on *E. coli* C. Two to three non-synonymous substitutions in the major capsid gene accounted for this inhibition, and when phages adapted to *S. enterica* were grown on *E. coli*, fitness recovery was based on reversions at these same sites.
6. Cunningham et al. (1997), Six bifurcating lineages of bacteriophage T7 were grown in the presence of the mutagen nitrosoguanidine. Every lineage evolved a ~1.5-kb deletion that fused the 0.3 and 0.7 genes, and this loss was associated with a gain in fitness. In addition, three different sets of parallel nonsense mutations, which produced identical ORFs in independent lineages and were under positive selection, resulted in truncation of the 0.7 gene product.
7. Levin, Perrot and Walker (2000), In the absence of an antibiotic challenge, antibiotic resistance often engenders a cost in the fitness of bacteria. In this study, two candidate genes (*rpsD* and *rpsE*) were sequenced from 24 independently derived, streptomycin resistant (*rpsL*) *Escherichia coli* strains known to be carrying compensatory mutations. For *rpsD*, there were three different single amino acid replacements and two instances of tandem duplications leading to the insertion of three or five amino acids. At *rpsE*, there were five different single base changes leading to four amino acid replacements. One of the non-synonymous changes occurred in five different strains. In no cases were there compensatory changes in both *rpsD* and *rpsE*.
8. Liao, McKenzie and Hageman (1986), In order to produce a thermostable enzyme, the authors transformed the thermophilic *Bacillus stearothermophilus* with a plasmid containing kanamycin nucleotidyltransferase (KNTase) from the mesophilic *B. subtilis* and subjected it to selection at 63°C. KNTases purified from variants that retained kanamycin resistance at 63°C shared a single amino acid replacement, Asp 80 to Tyr. Further selection at 70°C yielded another shared substitution, Thr 130 to Lys.
9. Riehle, Bennett and Long (2001), Six lines of *Escherichia coli* were adapted to 41.5°C and examined for duplications and deletions across their genomes. The authors detected five duplication/deletion events in three lines (no events were detected in the other three lines). Three of the events involved duplications at the same location in the genome, a region harboring four genes previously identified to be important in stress and starvation survival. In both instances examined, the duplications were coincident with increases in fitness.
10. Rieseberg et al. (1996), Rieseberg and colleagues analyzed the genomic composition of three experimental hybrid lineages derived from a cross between *Helianthus annuus* and *H. petiolaris*. As a result of fertility selection in the early generations, all three lineages converged on a common genomic composition. Moreover, this genomic structure was in accord with the recombinant genome of a natural hybrid species (*H. anomalus*) derived from the same two parental taxa. These findings suggest that selection plays a central role in the formation of hybrid species.
11. Ungerer (2000), Populations derived from a cross between the Niederzenz and Landsberg ecotypes of *Arabidopsis thaliana* were subjected to three generations of selection for increased viability. For QTL alleles governing life history traits, as well as other genomic regions, selection response was almost always uniform. The results of this work were consistent across different genetic backgrounds, suggesting that the selective value of an allele is not strongly influenced by variation in genetic background.
12. Wichman (1999), Replicates of two lines of bacteriophage ΦX174 were adapted to high temperature and a novel host and resultant populations were surveyed for genome-wide changes. Each replicate displayed over a dozen nucleotide changes that reached high frequency, and half the substitutions in one line also arose in the second line. In total, six nucleotide changes and one 27–bp deletion arose in parallel. All of these changes were determined to be adaptive, and the order of occurrence of these changes varied between the lineages. This result suggests that their selective value is independent of genetic background. An important antithetical point is that the parallel changes were not those with the largest beneficial effect.

B. Phylogenetic Studies

13. Andreev et al. (1999), In the flour beetle, *Tribolium castaneum*, point mutations in the gene *Resistance to dieldrin (Rdl)* confer resistance to cyclodiene pesticides. Resistance results from a point mutations resulting in replacement of Ala 302 by Ser. Of 141 strains examined, 24 contained resistant individuals. A phylogeny of resistant alleles inferred from a 694–bp stretch of *Rdl* that contains the codon for Ala 302, resolved six distinct clades. The pattern of nucleotide variation in this region is better explained by multiple (parallel) independent origins of the resistant genotype.
14. Elard, Comes and Humbert (1996), The authors demonstrate that resistance to benzimidazole (BZ) antihelmentics is conferred by a substitution at residue 200 (Phe to Tyr) in beta-tubulin (a precursor to the structural microtubules) in the nematode *Teladorsagia circumcincta*. A review of the literature shows that this same substitution is associated with BZ resistance in two other nematode species and two of four fungi examined.

Appendix 1. (*Continued*)

15. ffrench-Constant (1994), A survey of *Rdl* sequences (see #11) from a wide range of insects (Coleopterans, Dipterans and Dictyopterans) resistant to cyclodiene revealed that all these lineages share the same point muation, the replacement of Ala 302 by Ser.
16. Johanson et al. (2001), The *FRIGIDA* locus (*FRI*) has been shown to be a major determinant of flowering time in *Arabidopsis*. A majority of *Arabidopsis* early-flowering ecotypes (i.e., those that do not require vernalization to flower early) contain one of two deletions that cause a frame shift in the ORF of *FRI*, suggesting that this phenotype has arisen at least twice.
17. Jouvin-Marche et al. (1988), Sequence analysis of the immunoglobulin kappa light-chain constant region gene (Ck) sampled from five wild mouse species suggests that parallel evolution of sequences is common at this single-copy locus. Of 47 codons with at least one substitution, 21 of these changes are most likely the result of parallel evolution. Thirteen of these 21 changes result in amino acid substitution. In two cases, parallelism is exhibited at the amino acid level only.
18. Kermarrec et al. (1999), Human and non-human primates share the ABO histoblood group system. This system is based on a single locus encoding a galactosyltransferase, which modifies the O antigen and whose specificity determines the blood group. *O* alleles are null-recessives resulting from a deletion, and their non-functional products do not affect the O antigen. Molecular phylogenetic analysis of human and non-human primate *O* alleles established that these alleles are the result of four independent silencing mutations. The large coalescence times of these alleles at intermediate frequencies suggests that balancing selection (Saitou & Yamamoto, 1997) governs the dynamics of this locus, but the selective value of the silent *O* alleles is unknown.
19. Kriener (2000), Some alleles in the *DRB* gene family in Old and New World monkeys resemble human *DRB1*03* and *DRB3* sequences in their second exon. Phylogenetic analyses based on the flanking intron sequences grouped genes in a taxon-specific fashion (i.e., gene and species trees were congruent). In contrast, the exon-based tree conflicts with taxonomic groupings (i.e., gene and species trees were incongruent). In other words, exon sequences with similar motifs grouped together, even though the flanking intron sequences suggest that the sequences had separate evolutionary histories. The authors found statistical support for the hypothesis that the sequence similarities among these diverse lineages were selected independently, allowing them to reject the hypothesis of common ancestry.
20. Low et al. (2001), Low and colleagues isolated multiple, independently derived strains of β-lactam resistant *Escherichia coli* from the infected kidney cysts of a single patient. Resistance resulted from one to three nucleotide substitutions in the promoter region of the *ampC* locus (four variable sites total), which led to an increase in expression of the AmpC enzyme. Two of the resistant strains carried the same set of three substitutions. Because the strains carried the same basic *ompC* sequence, which is often highly variable among strains, their results are consistent with an initial infection by a single *E. coli* strain, followed by the acquisition of resistance with different cysts.
21. Malcuit et al. (2000), Potato (*Solanum tuberosum*) has evolved two distinct modes of resistance to potato virus X (PVX): one controlled by the N genes (Nx and Nb), and one governed by the Rx genes (Rx1 and Rx2). For each of these host genes, PVX has a single determinant that specifies virulence (i.e., breaks resistance) or avirulence. While this study does not pinpoint the substitutions responsible for these determinants, a genomic phylogeny of strains variable for these determinants revealed that the Nb-resistance breaking factor (located in ORF2 of the viral genome) has evolved on five separate occasions. Alternatively, the topology could be the result of seven independent losses.
22. Molla et al. (1996), The evolution of resistance at the HIV protease gene was monitored in 48 patients treated with the protease inhibitor, ritonavir. While there was variation among sequences in resistant lineage, the authors pinpointed nine amino acid changes that resulted from drug selection. For example, mutation at site 82 (V to A or F) was always associated with the evolution of resistance and associated mutations at four other sites occurred in more than one half of the sequences analyzed. Moreover, multiple mutations consistently accumulated in an ordered fashion.
23. Morris et al. (1993), The absorbance maxima (max) of the rhodopsin visual pigments of squid species have been shown to be correlated with their maximum depth distribution – species that inhabit deeper waters have lower maxima. In this study, the authors show that the 5 nm spectral shift in rhopdopsin maxima between *Alloteuthis subulata* (max depth of 200 m) and *Loligo forbesi* (360 m) is associated with a substitution of phenylalanine by serine at residue 270. This residue is homologous to site 277 in primate cone visual pigments, a site that is important in spectral tuning in primates (Neitz et al., 1991 and Williams et al., 1992).

Appendix 1. (*Continued*)

24. O'hUigin, Sato and Klein (1997) Sequences of introns 5 and 6 of the *ABO* gene were analyzed to distinguish between parallel evolution and trans-species inheritance of polymorphism at this locus. Four substitutions and one indel separate human *A*, *B*, and *O* variants from chimpanzee *A* and gorilla *B* alleles. There is no phylogenetic support for trans-species inheritance, thus the authors conclude that the chimpanzee *A* and gorilla *B* alleles evolved in parallel with the human *A* and *B* alleles, respectively. Note that cloning and homology assessment demonstrated that the *A* and *B* alleles are distinguished by the same four amino acid residues (sites 176, 234, 265 and 267) within humans and between the chimpanzee *A* and gorilla *B* alleles. In a similar study, Saitou and Yamamoto (1997) hypothesize that *B* alleles have evolved at least three times from an ancestral *A* form.
25. Palacios et al. (1998), The transmembrane receptor, CCR5, serves as a cellular gateway for the entry of HIV-1 and all strains of SIV. Humans homozygous for a null allele of *CCR5*, which has a 32-bp deletion, are highly resistant to HIV-1. A novel 24-bp deletion allele of *CCR5* was discovered in sooty mangabeys (*Cercocebus torquatus atys*), a host of SIV, at an appreciable frequency. This allele is expressed, but its encoded protein is not transported to the cell surface, and thus monkeys homozygous for this allele are expected to be resistant to SIV infection.
26. Reid et al. (2000), The authors constructed a phylogeny of enteropathogenic *Escherichia coli* strains based on six housekeeping genes. The phylogenetic distribution of mobile elements that confer virulence suggests that the high virulence of certain lineages is a derived (not ancestral) state. More importantly, the phylogeny supports the parallel gain and loss of specific mobile virulence elements. For example, the chromosomal acquisition of the LEE pathogenicity island, a critical first step in the evolution of pathogenicity, occurred at least twice. In addition, a plasmid-borne haemolysin and phage-encoded Shiga toxins were acquired in parallel in distinct lineages.
27. Romero-Herrera et al. (1978), Phylogeny reconstruction of vertebrate myoglobin sequences revealed that 139 of 278 mutations, corresponding to 39 of 83 variable sites, occurred in parallel. Although the adaptive significance of these changes is unclear, myoglobin function is likely to be under strong selection in diving mammals. Certain changes that arose independently in cetaceans and pinnipeds are also intriguing: 54 Asp and 122 Glu in both harbour seal and dolphin, 83 Asp in sea lion and dolphin, 121 Ala and 152 His in harbour seal, dolphin and porpoise.
28. Shafqat et al. (1996), Shafqat and colleagues examined the interrelationships of formaldehyde-active and ethanol-active alcohol dehydrogenase (*ADH*) in plants and animals. Their results indicate that the plant and animal forms of formaldehyde-active (class III) *ADH* share a common ancestor. In contrast, the ethanol-active (classes P and I) forms are derived from independent duplications of the class III enzyme-encoding loci within each lineage, followed by functional convergence. These forms are characterized by parallel changes at four of the thirteen substrate binding amino acid residues. See also Fliegmann and Sandermann (1997).
29. Shyue et al. (1995), Color vision is governed by two genes in the New World marmosets and squirrel monkeys, one of which is X-linked. Both marmosets and squirrel monkeys have evolved multiple alleles at the X-linked locus, each encoding photopigments with distinct spectral sensitivities. Consequently, heterozygous females are trichromatic. Phylogenetic analysis supports the independent evolution of these multi-allelic systems. In addition, a comparison of the amino acid sequences of the X-linked loci in New World monkeys and humans (which have two such loci) reveals parallel changes at three sites that are believed to be critical for spectral tuning.
30. Stewart, Schilling and Wilson (1987), Zhang and Kumar (1997), The digestive system of colobine monkeys, ruminants, and the avian hoatzin all involve the recruitment of lysozyme expression (lysozyme *c*) in the stomach, where it serves as a bacteriolytic enzyme. Phylogenetic analysis revealed that two amino acid sites evolved in parallel across taxa, supporting the hypothesis that these substitutions were the result of positive selection.
31. Yokoyama and Yokoyama (1990), Red- and green-like visual pigment genes of the blind cave fish, *Astyanax fasciatus*, were compared to their homologous counterparts in humans. Like humans, this species of fish has one red-like pigment gene and multiple green-like pigment genes. A phylogeny of these genes allowed the authors to infer the direction of evolution of amino acid sequences. The results of this analysis point to independent origins of the red pigments, from a green ancestor, in human and fish by identical amino acid substitutions at two, or possibly three, critical positions.

Appendix 1. (*Continued*)

C. Quantitative Genetic Studies

32. Fatokun et al. (1992), The most important yield trait in both cowpea (*Vigna unguiculata*) and mung bean (*V. radiata*) is seed weight, thus this trait has been the target of selection during the independent domestication of both of these species. Fatokun and colleagues identified QTLs with major effects on seed weight in both species. Furthermore, they used orthologous RFLP markers to demonstrate that the QTL with the greatest magnitude maps to the same marker interval in both species.
33. Hu et al. (2003), Rhizomatousness was mapped in an F2 population derived from a cross between *Oryza sativa* and *O. longistaminata*. Two key loci were identified, each having strong affects on several rhizome traits. Each of these QTLs is coincident in map position to a major QTL affecting rhizome growth in *Sorghum propinquum*, a wild congener of domesticated sorghum.
34. Paterson et al. (1995), Paterson and coworkers mapped agronomically important traits in rice, maize and sorghum, which diverged up to 65 million years ago. A significant portion of QTLs underlying seed mass and seed dispersal (i.e., shattering versus non-shattering) show correspondence among rice, maize and sorghum. QTLs for daylength-insensitive flowering also map to corresponding regions in rice, maize, sorghum, wheat and barley, suggesting that artificial selection resulted in parallel changes at a single ancestral locus.
35. Schat, Voous and Kuiper (1996), Schat and colleagues crossed individuals from four geographically isolated, zinc tolerant *Silene vulgaris* populations *inter se* and to a non-tolerant line. One of the tolerant lines exhibited an intermediate level of tolerance. The segregation patterns in F2 and F3 families fit a major genes model of inheritance, and the authors concluded that tolerance was governed by two additive genes. All three highly tolerant populations appear to be homozygous tolerant at both loci, while the intermediate population possesses only one tolerant allele. Because the tolerant populations are geographically isolated, it is unlikely that tolerance in these populations resulted from common descent. In addition, copper and cadmium tolerance are controlled by two loci that correspond among all tolerant populations examined.
36. Sucena et al. (2003), In the *Drosophila virilis* species group, the loss of thin trichomes on the dorsal cuticle of first-instar larvae has evolved in parallel in three distinct lineages. Sucena et al. examine controlled crosses and gene expression patterns to demonstrate that all three instances of trichome loss are the result of regulatory changes affecting the *shavenbaby/ovo* gene.

Genetica (2005) 123: 171–179 © Springer 2005

Hybridization as a source of evolutionary novelty: leaf shape in a Hawaiian composite

Stacy Jørgensen[1,2] & Rodney Mauricio[3]
[1]*Department of Geography, University of Georgia, Athens, GA 30602, USA;* [2]*Current address: Department of Geography, University of Hawai'i at Mānoa, Honolulu, HI 96822, USA (Phone: +808-956-7526; Fax: +808-956-3512; E-mail: jorgy@hawaii.edu);* [3]*Department of Genetics, University of Georgia, Athens, GA 30602, USA*

Received 1 May 2002 Accepted 4 December 2003

Key words: adaptation, Asteraceae, AFLP, gene complexes, Hawaiian Islands, leaf shape, *Lipochaeta tenuifolia*, *Lipochaeta tenuis*, morphological variation, hybridization

Abstract

Hybridization is increasingly recognized as a significant creative force in evolution. Interbreeding among species can lead to the creation of novel genotypes and morphologies that lead to adaptation. On the Hawaiian island of O'ahu, populations of two species of plants in the endemic genus *Lipochaeta* grow at similar elevations in the northern Wai'anae Mountains. These two species represent extremes of the phenotypic distribution of leaf shape: the leaves of *Lipochaeta tenuifolia* individuals are compound and highly dissected while leaves of *L. tenuis* are simple. Based primarily on leaf shape morphology, a putative hybrid population of *Lipochaeta* located at Pu'u Kawiwi was identified. Individuals in this population exhibit a range of leaf shapes intermediate in varying degrees between the leaf shapes of the putative parental species. We analyzed individuals from pure populations of *L. tenuifolia*, *L. tenuis* and the putative hybrids using 133 AFLP markers. Genetic analysis of these neutral markers provided support for the hybrid origin of this population. The correlation between genetic background and leaf morphology in the hybrids suggested that the genome of the parental species with simple leaves might have significantly contributed to the evolution of a novel, compound leaf morphology.

Introduction

The diverse flora and fauna of remote island chains have been studied by evolutionary biologists for many decades (e.g., Darwin, 1859; Mayr, 1942; Carson, 1996; Grant & Grant, 1996). Geographic isolation and founder-mediated speciation have historically been emphasized as the driving forces behind adaptive radiation on these islands (e.g., Weller, Sakai & Straub, 1996). However, there has long been interest in the role of interbreeding among species, or hybridization, as a creative force in evolution (Anderson & Stebbins, 1954; Lewontin & Birch, 1966). Hybridization is increasingly recognized as an evolutionary force that can lead to adaptation through the creation of novel genotypes and morphologies (Rieseberg, 1995; Arnold, 1997).

Despite its recognition as a recurrent process in the diversification of flowering plants, the importance of hybridization as a general mechanism of evolution driving speciation and adaptation has been and remains unclear (Heiser, 1973; Levin, 1979; Rieseberg, 1991). Many workers have pointed to the fact that early-generation hybrids often exhibit significant reductions in viability and fertility (Barton & Hewitt, 1980; Templeton, 1981), thought to be caused by the disruption of coadapted gene complexes (Dobzhansky, 1951; Mayr, 1963) or by the introduction of maladapted genes (Waser & Price, 1991). Additionally, hybridization may result in the creation of morphologically intermediate offspring, adapted to neither parental habitat and outcompeted by non-hybrid individuals (Arnold & Hodges, 1995).

Given these findings, it is perhaps not unexpected that the role of hybridization in speciation on islands has historically been considered minor (Humphries, 1979; Ganders & Nagata, 1984; Francisco-Ortega, Jansen & Santos-Guerra, 1996). In fact, contemporary examples of hybridization in the Hawaiian flora, for example, appear to be rare, presumably because the allopatric distribution of species prevents pollen flow (Mayer, 1991). However, there are reasons to suspect that hybridization may, indeed, play a role in plant speciation on oceanic islands. For example, within the Hawaiian flora, a high rate of fertility is often observed in artificially induced interspecific and intergeneric hybrids (Carr, 1995). Examples include a number of groups within the Asteraceae: *Bidens* (Gillet & Lim, 1970), *Tetramolopium* (Lowrey, 1986), and the silversword alliance (Carr & Kyhos, 1981), which are known to hybridize freely in the few locations where different species co-occur (Caraway, Carr & Morden, 2001). Furthermore, non-concordance between nuclear- and organelle-derived phylogenies of groups such as the silversword alliance (Baldwin, Kyhos & Dvorák, 1990) and the Drosophilidae (DeSalle & Giddings, 1986) are generally interpreted as indicative of a role for hybridization in the diversification of these groups. These findings, along with the general lack of post-zygotic genetic barriers to hybridization among congeners, makes the fact that hybridization has been generally discounted as a factor in adaptive radiation in island settings surprising (Crawford, Whitkus & Stuessy, 1987).

In this study, we examined a putative example of natural hybridization in plants from the Hawaiian Islands. On the island of O'ahu, two species of plants in the Hawaiian endemic genus *Lipochaeta* (family Asteraceae) grow in the northern Wai'anae Mountains: *Lipochaeta tenuifolia* and *L. tenuis*. Both species are found at similar elevations in mesic forest, with *L. tenuifolia* found in the extreme northern portion of the mountain range and *L. tenuis* known from locations to the south. Individual species of *Lipochaeta* have diverged in a number of vegetative and floral traits, including leaf shape. *Lipochaeta tenuifolia* and *L. tenuis* represent the extremes in the genus with regard to leaf shape: the leaves of *L. tenuifolia* are compound and highly dissected, while the leaves of *L. tenuis* are simple. A population of *Lipochaeta* in the northern Wai'anae Mountains has been hypothesized to be of hybrid origin because individuals within the population possess a variety of leaf morphologies intermediate between those characteristic of *L. tenuifolia* and *L. tenuis* (J. Lau, Hawai'i Natural Heritage Program, pers. comm.). Our primary objective in this study was to use genetic markers to test the hypothesis that the population of *Lipochaeta* in the northern Wai'anae Mountains is of hybrid origin. Furthermore, within this putative hybrid population, we were interested in identifying correlations between leaf shape and the parental origin of our genetic markers.

Materials and methods

Study species

Lipochaeta DC (Asteraceae) is an endemic Hawaiian genus of about 20 species of primarily suffruticose perennials (Wagner, Herbst & Sohmer, 1990); two sections, based on morphology and cytology (*Lipochaeta*, $n = 26$, four-petaled disk florets; *Aphanopappus*, $n = 15$, five-petaled disk florets), are recognized within the genus. Artificial hybrids can be induced in crosses within and between sections (Rabakonandrianina, 1980), and between *Lipochaeta* and *Wollastonia biflora* ($n = 15$), the presumed progenitor of *Lipochaeta* (Rabakonandrianina & Carr, 1981). Although the exact relationship between the two sections is unclear, section *Lipochaeta* likely arose from a hybridization event involving a member of section *Aphanopappus* and another member of the genus *Wollastonia* (Gardner, 1977; Chumley et al., 2002).

Members of section *Aphanopappus* ($n = 14$, of which 11 are extant) are distributed in a classic adaptive radiation pattern; all but two species are single-island endemics (Wagner & Robinson, 2001). Individual species have diverged in vegetative and floral morphology including leaf shape, growth habit, and the color, number, and size of ray florets. Natural hybridization within the group appears to be uncommon (Gardner, 1979) but not unknown (Wagner Herbst & Sohmer, 1990). Heretofore, reports of natural hybridization within *Lipochaeta* were based solely on morphological descriptions of intermediacy rather than the genetic criteria we are employing.

Field sampling and laboratory techniques

Individuals were sampled from naturally growing populations of *L. tenuifolia*, *L. tenuis*, and the putative hybrid population, which was assumed to be composed entirely of hybrid individuals (Figure 1); sample sizes were five, three, and 13 individuals, respectively. Two leaves were collected per individual and placed in plastic bags with desiccating silica gel. Each individual collected in the hybrid population was assigned to a leaf shape class (Figure 2): 1, *L. tenuis*-type, deltate; 2, deltate with basal lobes; 3, deltate with several distinctive lobes; 4, deltate with numerous lobes and some further dissection of lobes; 5, very highly dissected with numerous lobes and sub-lobes, but less dissected than the parental species *L. tenuifolia*.

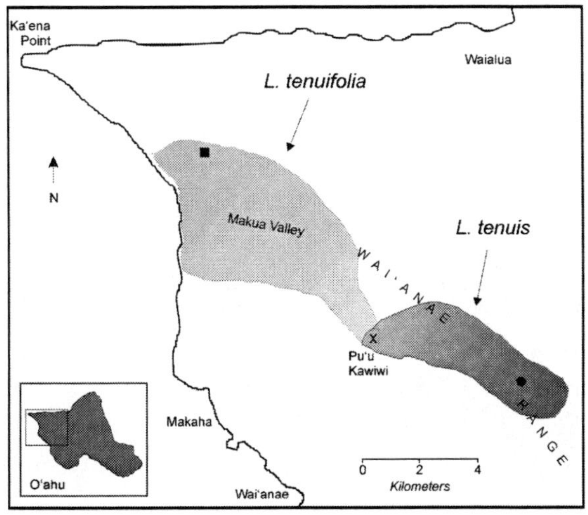

Figure 1. Distribution of *Lipochaeta tenuifolia* and *L. tenuis* and the location of a putative hybrid population in the northern Wai'anae Mountains, O'ahu. The locations of the populations sampled from the parental taxa are indicated by a closed square for *L. tenuifolia* and a closed circle for *L. tenuis*; the hybrid population is indicated by a ×. Species distributions were extrapolated from occurrences in the Hawai'i Natural Heritage Program database.

Leaves were crushed by vortexing with ball bearings (Colosi & Schaal, 1993), and total genomic DNA was extracted according to a standard phenol-chloroform procedure (Sambrook, Fritsch & Maniatis, 1989). Following phenol extraction, DNA was precipitated with ethanol and resuspended in deionized water to an approximate concentration of 50 ng/μl. Amplified fragment length polymorphism (AFLP) fragments (Vos et al., 1995) were detected using standard kits available from Applied Biosystems (ABI). A restriction-ligation was conducted with the enzymes *Eco*RI and *Mse*I and enzyme-specific ligators from the preselective amplification kit (ABI part # 402004). Following ligation, two rounds of PCR were conducted. During preselective amplification, a single nucleotide was added to the 3′ end of the primers; the preselective product was diluted to serve as the template for the subsequent selective amplifications. During selective amplification, two additional nucleotides were added to the primers, and the *Eco*RI primer was fluorescently labeled to permit fragment detection. Six *Eco*RI-*Mse*I primer-pair combinations were used for selective amplification (listed by the additional nucleotides added): ACA-CAT, ACA-CTT, ACG-CTG, ACT-CTG, ACC-CAT, and, AGG-CTT.

Fragments were separated by electrophoresis using 4.75% polyacrylamide gels on an ABI 377 sequencer. A ROX-500 fluorescently labeled size standard was loaded with each sample during electrophoresis to permit fragment-size determination. The software package GeneScan® (version 3.1, Applied Biosystems) was used to visualize the gels and determine fragment size by interpolating to the ROX-500 standard (ABI #401734) loaded

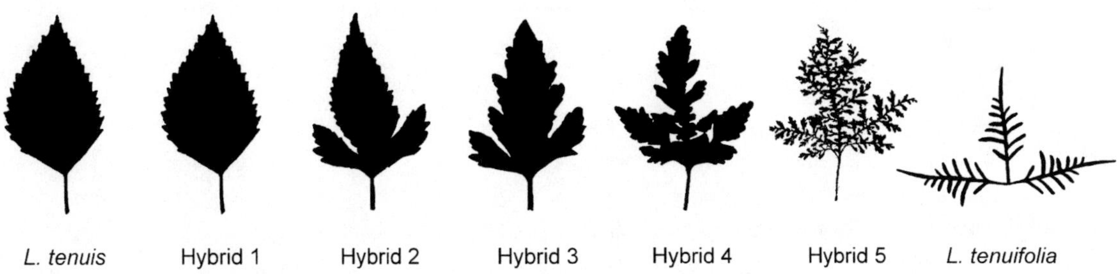

Figure 2. Variation in leaf shape among *L. tenuifolia*, *L. tenuis* and their putative hybrids.

with each sample, which permitted the analysis of fragments between 70 and 450 bp. Each differentially sized fragment was considered a single gene locus, and individuals were scored by the presence or absence of the indicated fragment.

Data analysis

We analyzed the AFLP data to quantify genetic diversity in the parental and putative hybrid populations and to examine individual plant genotypes for correlations among fragments and overall genetic similarity among individuals. Three assumptions were necessary for these analyses: (1) Mendelian segregation of polymorphic fragments, (2) allelic identity of same-size fragments, and (3) the existence of a single dominant (amplified) and recessive (null) allele at each locus. The calculation of standard measures of genetic diversity and structure required the additional assumption of Hardy–Weinberg proportions within populations (Travis, Maschinshi & Keim, 1996). Genetic diversity within each of the three groups was assessed by the percentage of polymorphic loci (P) and heterozygosity (H). A locus was considered polymorphic if its associated fragment did not occur in every individual analyzed. Heterozygosity at each locus was estimated from the equation $H = 1 - [(1 - q)^2 + q^2]$ where q^2 is the frequency of individuals in which a fragment was absent; total heterozygosity was calculated as the mean heterozygosity among loci.

When a large number of loci are examined, there are likely to be non-independent associations among loci. Traditional analyses of genetic structure, which are based on a locus-by-locus approach, are unlikely to reveal the effects of such associations or linkages (Edwards, 2003); also, traditional analyses of genetic structure require *a priori* divisions into groups. Therefore, the relationships among individuals sampled from the three populations were analyzed via principal components analysis (PCA). This analysis was selected because the components generated by the analysis will reflect correlations among fragments in their presence or absence (i.e., non-independence) and because divisions into groups are not required (Wiley, 1981; Caraway, Carr & Morden, 2001). All loci were used for the analysis; however, only those individuals/samples for which all six primer-pair combinations were resolved were included in the PCA analysis; calculations were conducted with the software package PC-ORD (McCune & Mefford, 1999).

Results

The six primer pairs yielded 133 AFLP fragments among all individuals. Well over half (61%) of the fragments were shared by the parental species (Table 1). Four unique fragments (i.e., also absent from hybrids) were detected in each *L. tenuifolia* and *L. tenuis*. Fixed differences between the parental species were detected at only two loci; in both cases, the fragments were present in *L. tenuis* and absent in *L. tenuifolia*. Twenty-two fragments were detected in only *L. tenuifolia* and the hybrids, and nine were shared by only *L. tenuis* and the

Table 1. Summary of AFLP markers analyzed in *Lipochaeta tenuifolia*, *L. tenuis*, and their putative hybrids. One hundred thirty-three markers were detected among all sampled individuals

AFLP markers	*L. tenuifolia*	*L. tenuis*	*L. tenuifolia* × *L. tenuis*
Total number[a]	107	94	124
Constant markers	17	47	15
Polymorphic markers	90	47	109
Shared by both parental species	81	81	–
Constant in both parental species	12	12	–
Shared by parent and hybrid	102	89	–
Absent in other parent	22	9	–
Unique to species or hybrid	4	4	13

[a]Number of fragments present in at least one individual of the group.

hybrids. A single fragment was shared by the parental species but was absent in the hybrids. In contrast, 13 fragments detected in the hybrids were absent from both parental species.

The number of polymorphic markers varied substantially between the parental species. Ninety (84%) of the fragments detected in *L. tenuifolia* were polymorphic, while only 47 (50%) polymorphic fragments occurred in *L. tenuis*. The putative hybrids possessed the greatest number (109) of polymorphic fragments. Heterozygosity, as averaged across all 133 loci, also was greatest in the putative hybrid population ($H = 0.30$). Heterozygosity in *L. tenuifolia* was, at $H = 0.24$, almost twice the level observed in *L. tenuis* ($H = 0.13$) (Table 2).

The three groups largely segregated into discrete groups along the first two principal component axes (Figure 3), which accounted for 27 and 18% of the variance observed in the total data set, respectively. Individuals of the two parental species largely segregated from individuals from the hybrid population along the first principal component. Notably, this division was not complete: individuals of the most highly dissected leaf shape (hybrid 5) clearly segregated with individuals of the simple-leafed parent. Individuals of the two parental species segregated from one another along the second principal component. Again, hybrid 5 individuals, which are morphologically most similar to *L. tenuifolia*, segregated with *L. tenuis*.

Table 2. Sample sizes, percent polymorphic loci, and heterozygosity calculated in *L. tenuifolia*, *L. tenuis*, and their hybrids determined from 133 AFLP loci

Species	N	P	P'	H_e
L. tenuifolia	5	67.7	84.1	0.238
L. tenuis	3	35.3	50.0	0.131
Hybrids	13	82.0	87.9	0.300

The percentage of polymorphic loci was calculated using all fragments (P) and only those fragments actually occurring within each group (P').

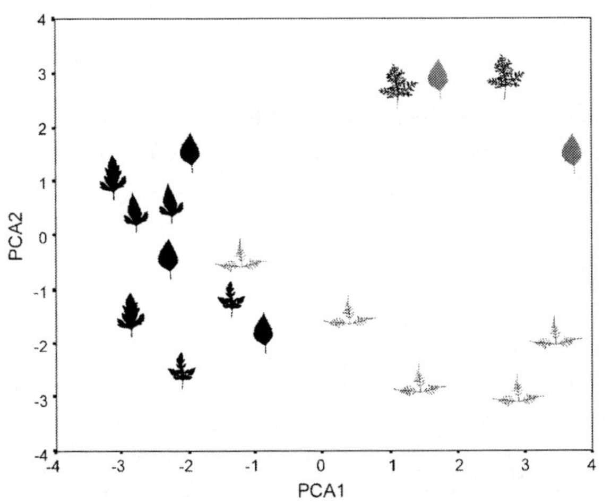

Figure 3. PCA of AFLP data using all scored fragments. Individuals are depicted by their leaf shape as shown in Figure 2; hybrids are shown in black, while individuals of the parental species are shown in gray.

Discussion

Evidence for hybridization

The AFLP data presented here strongly suggest a *L. tenuifolia* × *L. tenuis* hybrid origin for the population at Puʻu Kawiwi. As would be expected in a hybrid population, the Puʻu Kawiwi population contained a mix of the AFLP fragments detected in the parental taxa (Rieseberg, 1991); in fact, virtually all the fragments detected in the parental species were also found in the hybrid population.

Only 13 of the 133 fragments detected in the putative hybrids were absent from both parental species, although it is likely that sampling error could explain this discrepancy. Only one population each was sampled from the parental species, and these populations were located well away from Puʻu Kawiwi; it is possible that populations of *L. tenuifolia* and *L. tenuis* closer to the hybrid population might contain these fragments. The failure to detect these fragments could also indicate another species of *Lipochaeta* has been involved in the formation of the hybrid population. Two other species of diploid *Lipochaeta* are known from extreme northwestern Oʻahu locations; however, these species are known from coastal (*L. integrifolia*) and lowland (*L. remyi*) locations fairly removed from the mesic forest locales of *L. tenuifolia* and *L. tenuis*.

In addition to possessing fragments from both the parental species, the greater percentage of polymorphic loci and higher levels of heterozygosity found in the putative hybrid population are also consistent with the hybrid origin hypothesis. Although there were virtually no apparent fixed differences between the parental species, *L. tenuifolia* and *L. tenuis* have diverged in their allele frequencies at many loci. Crosses of the two parental taxa would result in a greater number of polymorphic loci, more even allele frequencies and therefore higher levels of heterozygosity in populations consisting of hybrid individuals.

Fixed differences between parental species would have made possible a determination of whether the population at Pu'u Kawiwi consists of early or late generation hybrids, backcrossed individuals or some combination of these crosses. For example, fixed differences between the silversword alliance members *Dubautia ciliolata* and *D. scabra* allowed Caraway, Carr and Morden (2001) to conclude that many individuals in a hybrid population from lava flows on the island of Hawai'i represented later generation backcrosses to *D. ciliolata*. The lack of fixed differences between *L. tenuifolia* and *L. tenuis*, precludes this analysis, however. Genetically, most individuals in the population appear intermediate or equally similar to *L. tenuifolia* and *L. tenuis*, which would seem to argue for a large occurrence of F_1 individuals. However, the varying leaf morphologies found in the population and the genetic identities of the hybrid individuals with the most highly dissected leaf pattern are inconsistent with this explanation.

Morphology and genetics uncoupled?

The hybrid individuals show a variety of intermediate leaf morphologies that are distinctly different from those of the parental species. In an F_1 hybrid population, a single, intermediate leaf morphology would be expected (Rieseberg, 1991) if loci contributing to leaf shape act additively. Later generation hybrid crosses or backcrosses could generate a variety of leaf forms as segregation occurs among loci. Overall, the hybrid individuals were genetically intermediate to the parental species, but there was variation in the degree of genetic similarity to the parental species with regard to the various leaf morphologies. Most strikingly, those individuals with the most highly dissected leaf morphologies, that is, most resembling *L. tenuifolia*, were genetically very similar to *L. tenuis*. In other words, a *L. tenuifolia*-like leaf morphology was present with a *L. tenuis*-like genetic background. Obviously, this conclusion is tempered by the very small number of hybrid five individuals we were able to sample from this small, natural population. However, cautiously taking the result at its face value, it suggests that genes from a simple-leafed parent, segregating in novel hybrid genomes, might play a role in generating a highly dissected leaf shape.

Such uncoupling of genetics and morphology is not unusual in hybrids. For example, present-day varieties of cultivated cotton are tetraploid, but are derived from two distinct diploid parental species (Jiang et al., 1998). Surprisingly, QTL that contribute to fiber quality were found to come from the diploid parent species that possesses no spinnable fiber on its seeds, suggesting a non-additive interaction between the two parental genomes affecting seed fiber quality. As in the cotton example, our present study illustrates that the merger of genomes with divergent evolutionary histories can produce 'unique avenues' for selection (Anderson & Stebbins 1954; Jiang et al., 1998; Wright et al., 1998).

Backcrossing to the *L. tenuis* parent could explain how individuals within the hybrid population have become genetically almost identical to that parental species. Although their status as 'pure' may be questionable, populations identified as *L. tenuis* do occur near Pu'u Kawiwi; pollen flow from these populations is a likely mechanism of backcrossing. Although individuals genetically similar to *L. tenuis* could theoretically arise by later generation crosses among hybrids (i.e., not involving backcrossing), this mechanism seems unlikely given the very small hybrid population size (tens of individuals). Only a very small percentage of late-generation filial hybrids would randomly end up with a predominantly *L. tenuis* genetic make-up, and there is no reason to expect that all these individuals would possess the dissected leaf morphology similar to that of *L. tenuifolia*. In fact, one would predict such advanced generation hybrid individuals to possess an external morphology virtually indistinguishable from *L. tenuis*. It is highly unlikely, then, that the pairing of the external morphology of *L. tenuifolia* with the genetic background of *L. tenuis* would arise by chance alone, making selection the best explanation for this pattern.

In fact, different classes of hybrids may have varying levels of fitness (Arnold & Hodges, 1995), with selection often favoring the native phenotype (Nagy, 1997). Based on this prediction and given populations of *L. tenuis* near Pu'u Kawiwi, hybrid individuals with entire leaves would be expected to possess the highest levels of fitness. However, native phenotypes do not always possess the highest fitness, and there are scenarios under which a non-native phenotype could be selected. For example, Nagy (1997) examined a variety of morphological traits, including leaf shape, petal shape, and petal color, in F_2 individuals created by crossing individuals from two subspecies of the annual plant *Gilia capitata* occurring in coastal and inland habitats in California. For all traits except leaf shape, native phenotypes were favored; the inland leaf shape, with fewer lobes or dissection, was favored at both locations.

Adaptive significance of leaf shape

Leaf shape itself has long been recognized as a trait of adaptive significance (e.g., Raschke, 1960; Givnish, 1979). In particular, leaf dissection appears correlated with environmental characteristics, with highly dissected leaves often favored in dry, sunny habitats, because the leaves are less likely to become overheated (Gurevitch, 1988). In addition to having significance with regard to the abiotic environment, leaf shape has also been shown to have adaptive significance with regard to interspecific interactions. For example, differences in leaf shape among closely related species with similar geographic ranges may be a response to avoid predation by herbivorous insects (Gilbert, 1975). Rausher (1978) demonstrated that females of the pipevile swallowtail butterfly *Battus philenor* discriminated between broad- and narrow-leaved *Aristolochia* when searching for specific plants on which to oviposit.

It is worth noting that hybrid individuals of *Lipochaeta* with the most highly dissected leaves were clearly morphologically distinct from both of the typical parental leaf morphologies. Furthermore, that such variation in leaf shape occurs between the parental species, which occur in similar habitats (i.e., mid-elevation mesic forest), suggests a selective pressure other than simple environmental conditions. Arthropods comprise over 75% of the Hawaiian fauna, and many are highly host specific (Roderick & Gillespie, 1998). Co-evolution with arthropods has been suggested as an important factor in the diversification of the silversword alliance (Roderick, 1997). If leaf shape in *Lipochaeta* is, in part, driven by herbivory a novel leaf shape might have a selective advantage over either parental phenotype. Concordant with this hypothesis is the leaf shape variety of *Lipochaeta* present when multiple diploid species occur on a single island. For example, on Kaua'i *L. fauriei* (entire, deltate), *L. waimeaensis* (entire, elongated), and *L. micrantha* (highly dissected) all have very different leaf morphologies, and these morphologies are not consistent with the general predictions based on the physical environment alone: *Lipochaeta waimeaensis* occurs on dry, exposed slopes within Waimea Canyon while *L. micrantha* is a forest species.

Although the diversity of leaf shape in the hybrid population seems remarkable, the genetic basis of transition between simple and compound leaves is well understood (Sinha, 1997). In fact, the transition between simple and compound leaves in the hybrid population is remarkably similar in appearance to induced mutants in leaf morphology known in the cultivated tomato, *Solanum esculentum* (Kessler et al., 2001). In the tomato model system, whether a plant makes complex, divided leaves or simple ones is controlled by KNOTTEDI-like (KNOXI) homeobox genes (Bharathan & Sinha, 2001). This group of genes is found in most plants; they are switched on in the leaves of all plants with complex leaves but are inactive in plants with simple leaves (Bharathan et al., 2002). A single gene, *PHANTASTICA* (PHAN) controls whether a leaf is pinnate or palmate (Kim et al., 2003). Although the genetic basis of leaf shape seems remarkably simple considering the complexity of the phenotype, even simple leaves can begin development as 'complex' primordia (Bharathan et al., 2002). Certainly, the molecular genetic studies of leaf shape illustrate that small genetic changes can lead to the generation of great morphological diversity. It seems likely that an analysis of KNOXI gene expression in *Lipochaeta*, and in the hybrid population specifically, would yield interesting results.

Conclusions

DNA markers are powerful tools for the confirmation of hybridization within plant species, and, in fact, are necessary to assess the contribution of each parental taxon to the hybrid population. The leaf morphologies of *L. tenuifolia* and *L. tenuis* represent the ends of a continuum found within the genus, and hybrids between the two species yield individuals with a variety of intermediate morphologies. In fact, the variety of leaf morphologies found in the *L. tenuifolia* × *L. tenuis* hybrid population at Pu'u Kawiwi is indicative of later generation hybrids or backcrosses. The genetic composition of the hybrid individuals could not be predicted from their vegetative morphology. Further studies of this hybrid population should include controlled crosses between the parental taxa; these crosses could yield important information about the number of genes controlling leaf morphology and whether epistatic interactions among loci may affect leaf morphology.

Acknowledgements

We thank K. Kawelo, J. Lau and J. Rohrer for assistance with field collections. M. Arnold, A. Bouck and K.C. Parker provided helpful comments that improved the manuscript. This research was supported by grants from the National Science Foundation (9901022) and the Biogeography Specialty Group of the Association of American Geographers to SJ and National Science Foundation (0129191) to RM.

References

Anderson, E. & G.L. Stebbins, 1954. Hybridization as an evolutionary stimulus. Evolution 8: 378–388.

Arnold, M.L, 1997. Natural Hybridization and Evolution. Oxford University Press, New York.

Arnold, M.L. & S.A. Hodges, 1995. Are natural hybrids fit or unfit relative to their parents? Trends Ecol. Evol. 10: 67–71.

Baldwin, B.G., D.W. Kyhos & J. Dvorák, 1990. Chloroplast DNA evolution and adaptive radiation in the Hawaiian silversword alliance (Asteraceae-Madiinae). Ann. MO. Bot. Gard. 77: 96–109.

Barton, N.H. & G.M. Hewitt, 1980. Hybrid zones and speciation, pp. 109–145 in Evolution and Speciation. Essays in Honor of M.J.D. White, edited by W.R. Atchley & D.S. Woodruff. Cambridge University Press, Cambridge.

Bharathan, G. & N.R. Sinha, 2001, The regulation of compound leaf development. Plant Physiol. 127: 1533–1538.

Bharathan, G., T.E. Goliber, C. Moore, S. Kessler, T. Pham & N.R. Sinha, 2002. Homologies in leaf form inferred from KNOX1 gene expression during development. Science 296: 1858–1860.

Caraway, V., G.D. Carr & C.W. Morden, 2001. Assessment of hybridization and introgression in lava-colonizing Hawaiian *Dubautia* (Asteraceae: Madiinae) using RAPD markers. Am. J. Bot. 88: 1688–1694.

Carr, G.D., 1995. A fully fertile intergeneric hybrid derivative from *Argyroxiphium sandwicense* ssp. *macrocephalum* × *Dubautia menziesii* (Asteraceae) and its relevance to plant evolution in the Hawaiian Islands. Am. J. Bot. 82: 1574-1581.

Carr, G.D. & D.W. Kyhos, 1981. Adaptive radiation in the silversword alliance (Compositae-Madiinae) I. Cytongenetics of spontaneous hybrids. Evolution 35: 543–556.

Carson, H.L., 1996. Pacific Basin biotas and evolutionary theory, pp. 7–17 in The Origin and Evolution of Pacific Island Biotas, New Guinea to Eastern Polynesia: Patterns and Processes, edited by A. Keast & S.E. Miller. SPB Academic Publishing, Amsterdam.

Chumley, T.W., S.C. Keeley, J. Panero & R.K. Jansen, 2002. A phylogeny of *Lipochaeta* (Asteraceae) inferred from the internal and external transcribed spacers of nuclear ribosomal DNA. Supplement to Am. J. Bot. 89(6).

Colosi, J.C. & B.A. Schaal, 1993. Tissue grinding with ball bearings and a vortex mixer. Nucl. Acids Res. 21: 1051–1052.

Crawford, D.J., R. Whitkus & T.F. Stuessy, 1987. Plant evolution and speciation on oceanic islands, pp. 183–199 in Differentiation Patterns in Higher Plants, edited by K. Urbanska. Academic Press, London.

Darwin, C., 1859. On the Origin of Species by Means of Natural Selection. John Murray, London.

DeSalle, R. & L.V. Giddings, 1986. Discordance of nuclear and mitochondrial DNA phylogenies in Hawaiian *Drosophila*. Proc. Natl. Acad. Sci. USA 83: 6902–6906.

Dobzhansky, T., 1951. Genetics and the Origin of Species, Third edition. Columbia University Press, N.Y.

Edwards, A.W.F., 2003. Human genetic diversity: Lewontin's fallacy. BioEssays 25: 798-801.

Francisco-Ortega, F., R.K. Jansen & A. Santos-Guerra, 1996. Chloroplast DNA evidence of colonization, adaptive radiation, and hybridization in the evolution of the Macaronesian flora. Proc. Natl. Acad. Sci. USA 93: 4085–4090.

Ganders, F.R. & K.M. Nagata, 1984. The role of hybridization in the evolution of *Bidens* on the Hawaiian Islands, pp. 179–194 in Plant Biosystematics, edited by W.F. Grant. Academic Press, Toronto.

Gardner, R.C., 1977. Chromosome numbers and their systematic implications in *Lipochaeta* (Compositae: Heliantheae). Am. J. Bot. 64: 810–813.

Gardner, R.C., 1979. Revision of *Lipochaeta* (Compositae: Heliantheae) of the Hawaiian Islands. Rhodora 81: 291–339.

Gilbert, L.E., 1975. Ecological consequences of a coevolved mutualism between butterflies and plants, pp. 210–240 in Coevolution of Animals and Plants, edited by L.E. Gilbert & P.H. Raven. University of Texas Press, Austin.

Gillet, G.W. & E.K.S. Lim, 1970. An experimental study of the genus *Bidens* (Asteraceae) in the Hawaiian Islands. Univ. Calif. Publ. Bot. 56: 1–63.

Givnish, T.G., 1979. On the adaptive significance of leaf form, pp. 375–401 in Topics in Plant Population Biology, edited by O.T. Solbrig, S. Jain, G.B. Johnson & P.H. Raven. Columbia University Press, New York.

Grant, P.R. & B.R. Grant, 1996. Speciation and hybridization in island birds. Philos. Trans. R. Soc. Lond. B Biol. Sci. 351: 765–772.

Gurevitch, J., 1988. Variation in leaf dissection and leaf energy budgets among populations of *Achillea* from an altitudinal gradient. Am. J. Bot. 75: 1298–1306.

Heiser, C.B., 1973. Introgression re-examined. Bot. Rev. 39: 347–366.

Humphries, C.J., 1979. Endemism and evolution in Macaronesia, pp. 171–199 in Plants and Islands, edited by D. Bramwell. Academic Press, New York.

Jiang, C.X., R.J. Wright, K.M. El-Zik & A.H. Paterson, 1998. Polyploid formation created unique avenues for response to selection in *Gossypium* (cotton). Proc. Natl. Acad. Sci. USA 95: 4419–4424.

Kessler, S., M. Kim, T. Pham, N. Weber & N.R. Sinha. 2001. Mutations altering leaf morphology in tomato. Int. J. Plant Sci. 162: 475–492.

Kim, M., S. McCormick, M. Timmermans & N. Sinha, 2003. The expression domain of *PHANTASTICA* determines leaflet placement in compound leaves. Nature 424: 438–443.

Levin, D.A., 1979. Introduction, pp.1–6 in Hybridization: An Evolutionary Perspective, edited by D.A. Levin. Dowden, Hutchinson, and Ross, Stroudsburg.

Lewontin, R.C. and L.C. Birch, 1966. Hybridization as a source of variation for adaptation to new environments. Evolution 20: 315–336.

Lowrey, T.K., 1986. A biosystematic revision of Hawaiian *Tetramolopium* (Compositae: Astereae). Allertonia 4: 203–265.

Mayer, S. S., 1991. Artificial hybridization in Hawaiian *Wikstroemia* (Thymelaeaceae). Am. J. Bot. 78: 122–130.

Mayr, E., 1942. Systematics and the Origin of Species. Columbia University Press, New York.

Mayr, E., 1963. Animal Species and Evolution. Harvard University Press, Cambridge, Massachusetts.

McCune, B. & M.J. Mefford, 1999. PC-ORD. Multivariate Analysis of Ecological Data. Version 4.01. MjM Software, Gleneden Beach, Oregon.

Nagy, E.S., 1997. Selection for native characters in hybrids between two locally adapted plant subspecies. Evolution 51: 1469–1480.

Rabakonandrianina, E., 1980. Intrageneric relationships and the origin of the Hawaiian endemic genus *Lipochaeta* (Compositae). Pac. Sci. 34: 29–39.

Rabakonandrianina, E. & G. D. Carr, 1981. Intergeneric hybridization, induced polyploidy, and the origin of the Hawaiian endemic *Lipochaeta* from *Wedelia* (Compositae). Am. J. Bot. 68: 206–215.

Raschke, K., 1960. Heat transfer between the plant and the environment. Annu. Rev. Plant Physiol. 11: 111–126.

Rausher, M.D., 1978. Search image for leaf shape in a butterfly. Science 200: 1071–1073.

Rieseberg, L.H., 1991. Homoploid reticulate evolution in *Helianthus* (Asteraceae): evidence from ribosomal genes. Am. J. Bot. 78: 1218–1237.

Rieseberg, L.H., 1995. The role of hybridization in evolution: old wine in new skins. Am. J. Bot. 82: 944–953.

Roderick, G.K., 1997. Herbivorous insects and the Hawaiian silversword alliance: coevolution or cospeciation? Pac. Sci. 51: 440–449.

Roderick, G. K. & R. G. Gillespie, 1998. Speciation and phylogeography of Hawaiian terrestrial arthropods. Mol. Ecol. 7: 519-531.

Sambrook, J, E.F. Fritsch & T. Maniatis, 1989. Molecular Cloning: A Laboratory Manual. Cold Spring Harbor Laboratory Press, New York.

Sinha, N., 1997. Simple and compound leaves: reduction or multiplication? Trends Plant Sci. 2: 396–402.

Templeton, A.R., 1981. Mechanisms of speciation – a population genetic approach. Annu. Rev. Ecol. Syst. 12: 23–48.

Travis, S. E., J. Maschinshi & P. Keim, 1996. An analysis of genetic variation in *Astragalus cremnophylax* var. *cremnophylax*, a critically endangered plant, using AFLP markers. Mol. Ecol. 5: 735–745.

Vos, P., R. Hogers, M. Bleeker, M. Reijans, T. Van de Lee, M. Hornes, A. Frijtens, J. Pot, J. Peleman, M. Kuiper & M. Zabeau, 1995. AFLP: a new technique for DNA fingerprinting. Nuc. Acids. Res. 23: 4407–4414.

Wagner, W.L. & H. Robinson. 2001. *Lipochaeta* and *Melanthera* (Asteraceae: Heliantheae subtribe Ecliptinae): establishing their natural limits. Brittonia 53: 539–561.

Wagner, W.L., D.R. Herbst & S.H. Sohmer, 1990. Manual of the Flowering Plants of Hawai'i. University of Hawaii Press and Bishop Museum Press, Honolulu.

Waser, N.M. & M.K. Price, 1991. Outcrossing distance effects in *Delphinium nelsonii*: pollen loads, pollen tubes, and seed set. Ecology 72: 171–179.

Weller, S.G., A.K. Sakai & C. Straub, 1996. Allozyme diversity and genetic identity in *Schiedea* and *Alsinidendron* (Caryophyllaceae: Alsinoideae) in the Hawaiian Islands. Evolution 50: 23–34.

Wiley, E.O., 1981. Phylogenetics. The Theory and Practice of Phylogenetic Systematics. John Wiley and Sons, New York.

Wright, R.J., P.M. Thaxton, K.M. El-Zik & A.H. Paterson, 1998. D-subgenome bias of *Xcm* resistance genes in tetraploid *Gossypium* (cotton) suggests that polyploid formation has created novel avenues for evolution. Genetics 149: 1987–1954.

Genetica (2005) 123: 181–189 © Springer 2005

Discovery and utilization of QTLs for insect resistance in soybean

H. Roger Boerma & David R. Walker
Center for Applied Genetic Technologies, The University of Georgia, 111 Riverbend Road, Athens, GA 30602, USA (Phone: 706 542-0927; Fax: 706 583-8120; E-mail: rboerma@uga.edu)

Received 6 June 2003 Accepted 30 March 2004

Key words: Bt, insect resistance, linkage drag, molecular markers, QTL, soybean

Abstract

Insect resistance in soybean has been an objective in numerous breeding programs, but efforts to develop high yielding cultivars with insect resistance have been unsuccessful. Three Japanese plant introductions, PIs 171451, 227687 and 229358, have been the primary sources of insect resistance alleles, but a combination of quantitative inheritance of resistance and poor agronomic performance has hindered progress. Linkage drag caused by co-introgression of undesirable agronomic trait alleles linked to the resistance quantitative trait loci (QTLs) is a persistent problem. Molecular marker studies have helped to elucidate the numbers, effects and interactions of insect resistance QTLs in the Japanese PIs, and markers are now being used in breeding programs to facilitate transfer of resistance alleles while minimizing linkage drag. Molecular markers also make it possible to evaluate QTLs independently and together in different genetic backgrounds, and in combination with transgenes from *Bacillus thuringiensis*.

Abbreviations: Bt – *Bacillus thuringiensis*; IRQTL – insect resistance QTL; LG – linkage group; PI – plant introduction; QTL – quantitative trait locus; RFLP – restriction fragment length polymorphism; SSR – simple sequence repeat.

Introduction

Modern agriculture is characterized by monocultural cropping, often over vast areas of land. This practice results in agroecosystems in which crop plants are highly vulnerable to pathogens and insect pests, which can spread easily from one field to another. Although many insect pests can be effectively controlled through cultural practices and/or the application of pesticides, environmental and economical concerns, along with the appearance of pesticide-resistant insect populations have made a heavy reliance on pesticides undesirable. Integrated pest management (IPM) is a more holistic approach to pest control. The goal of IPM is to integrate various cultural, chemical, and genetic approaches to controlling pests, and to use pesticides only when pest populations approach economic thresholds of damage tolerance. Plant resistance to the most important pests and pathogens is viewed as an important component of IPM, and is therefore an objective in many crop breeding programs. This article reviews what is currently known about the genetics of insect resistance in some soybean [*Glycine max* (L.) Merr.] germplasm which has been studied and used in breeding programs since the late 1960s.

The cultivated soybean is a member of the Leguminosae family, and is thought to have originated in northern and central China (Probst & Judd, 1973). Soybean is one of the major crop species in North America, South America, and Eastern Asia, where it was first cultivated at least 3000 years ago. Soybean is important for human nutrition in Asia, but is grown primarily as an oil crop and source of protein-rich meal for poultry and livestock feeds in the Western Hemisphere, where it is a major agricultural commodity in the United States, Brazil, and Argentina.

Soybean is a host for 36 insect species in North America, but only eight of these are of major importance (Lambert & Tyler, 1999). Five of the eight major pests feed exclusively on foliage, two exclusively on fruit forms, and one on both fruit forms and foliage. Damage to seeds by chewing or piercing and sucking insect pests can cause abortion or deformation of seeds, thus reducing both the weight and quality of the mature seeds. Soybean can tolerate up to 40% defoliation prior to the onset of fruiting, and 30% after fruiting with little or no yield loss (Lambert & Tyler, 1999). The effect of insect feeding damage on yield is reduced when environmental conditions (particularly soil moisture) favor foliage regrowth after insect feeding pressure subsides.

The most serious insect pests are in the orders Coleoptera, Lepidoptera, and Heteroptera (Lambert & Tyler, 1999). Major lepidopteran pests in North America include corn earworm [*Helicoverpa zea* (Boddie)], soybean looper [*Pseudoplusia includens* (Walker)], and velvetbean caterpillar [*Anticarsia gemmatalis* (Hübner)]. Larvae of all three species are foliage feeders, but corn earworm also feeds on reproductive structures and developing seeds. Soybean looper is noteworthy in that it has an unusually high tolerance to a variety of insecticides, and the ability to develop resistance to many pesticides rapidly. Velvetbean caterpillar is also a major pest in the soybean producing regions of Southern Brazil and Northern Argentina, and is a particularly voracious foliage feeder. In the United States, these insects cause the most damage in the Southeast and Delta regions because of the long growing season and their proximity to tropical regions where soybean looper and velvetbean caterpillar overwinter.

Three modalities of plant insect resistance have been described (Painter, 1951; Kogan & Ortman, 1978), and all three exist in soybean. *Antixenosis*, or non-preference, involves a morphological or biochemical trait that affects insect behavior to discourage oviposition, colonization, or feeding. *Antibiosis* involves a negative effect on insect growth, development, and/or reproduction following ingestion of plant tissue. Examples would include toxins and antinutrients such as certain proteinase inhibitors. Phytoalexins produced by soybean and other plants can be involved in either or both types of resistance, so antibiosis and antixenosis should not be viewed as discrete modes of resistance. The third mode is *tolerance*, which refers to the ability to tolerate a moderate amount of damage without appreciable yield loss.

Insect resistance in soybean

Most of the elite cultivars grown in North America are descendents of a small group of progenitor genotypes (Gizlice, et al., 1994). These ancestors consisted of plant introductions (PIs) or early-generation progeny of PIs that exhibited desirable agronomic qualities when grown under North American environmental conditions. Although there may have been some degree of selection based on response to natural infestations by certain insects, agronomic performance and seed composition traits were the primary criteria for selection. This narrow genetic base severely limited genetic diversity, and consequently, the number of alleles conditioning resistance to various pests and pathogens within the elite breeding populations used for cultivar improvement.

Evaluations of maturity group VII and VIII PIs from the USDA Soybean Germplasm Collection in the late 1960s identified three Japanese PIs resistant to the Mexican bean beetle [*Epilachna varivestis* (Mulsant)] (Van Duyn et al., 1971, 1972). 171451 ('Kosamame') had been collected in Kanagawa, Japan, 229358 ('Soden-daizu') from an unspecified location, and 227687 ('Miyako White') from Okinawa (USDA-ARS Germplasm Resources Information Network; http://www.ars-grin.gov/npgs/searchgrin.html). These PIs exhibit both antixenosis and antibiosis resistance to a number of soybean insect pests, including soybean looper, velvetbean caterpillar, cabbage looper [*Trichoplusia ni* (Hübner)], corn earworm, tobacco budworm [*Heliothis virescens* (Fabricius)], bean leaf beetle [*Cerotoma trifurcata* (Forster)], and the striped blister beetle [*Epicauta vittata* (Fabricius)] (Clark et al., 1972; Hatchett et al., 1976; Kilen et al., 1977; Luedders & Dickerson, 1977; Lambert & Kilen, 1984). The PIs also show resistance to some soybean pests from Taiwan, including the lepidopterans beet armyworm [*Spodoptera exigua* (Hübner)] (Family Noctuidae), *Porthesia taiwana* (Shiraki) (Family Liparidae), and *Orgyia* sp.

(Family Lymantridae), and two Scarabaeidae coleopterans, *Anomala cupripes* (Hope) and *A. expansa* (Bates) (Talekar et al., 1988).

171451, 227687, and 229358 differ in their relative resistance to some pest species. By intermating the three PIs and analyzing resistance in the progenies, Kilen and Lambert (1986) found that each possessed at least one unique resistance gene. Talekar et al. (1988) reported that the level of antibiosis resistance of the three PIs to four Asian insects varied, with 227687 most resistant to *S. exigua*, 171451 most resistant to *P. taiwana* and *Orgyia* sp., and 229358 most resistant to *A. cupripes*. Resistance of a particular PI to the adult and larval forms of the same insect can also vary. Oviposition by corn earworm moths was lower on 171451 than on PI 227687, suggesting a higher level of antixenosis towards adult females (Clark et al., 1972). However, PI 227687 plants showed a lower level of pod damage in the same experiments, suggesting a higher level of antibiosis towards larvae. Some resistance mechanisms in a PI may be effective against both lepidopteran and coleopteran pests, whereas others appear to be order-specific. Smith and Brim (1979a) tested the corn earworm leaf-feeding resistance of four F_3 lines derived from PI 229358 which had been previously selected for resistance to Mexican bean beetle. They found that one line showed a high incidence of resistance towards corn earworm, whereas another line had no significant resistance. Among PI 171451-derived backcross populations with high levels of Mexican bean beetle resistance, few of the progeny showed resistance to corn earworm (Smith & Brim, 1979b).

Efforts to transfer insect resistance from PIs 171451, 227687, and 229358 to elite soybean lines have been hindered by quantitative inheritance of resistance and the poor agronomic qualities of the PIs (Boethel, 1999). In 1987, breeding programs in 10 states were using one or more of the PIs in crosses to elite cultivars, and insect resistance remained a breeding objective in nine states in 1998 (Lambert & Tyler, 1999). Studies by Sisson et al. (1976) showed that inheritance of resistance to Mexican bean beetle was quantitative. All three PIs are low yielding and are susceptible to some important diseases and nematodes (Lambert & Kilen, 1984). Other problematic traits that contribute to low yield from the PIs include premature dehiscence of pods and a tendency to lodge (Kilen & Lambert, 1986). Tight linkages between resistance alleles at quantitative trait loci (QTLs) associated with insect resistance and inferior alleles at nearby agronomic or other resistance trait loci result in linkage drag, which refers to the inadvertent co-selection of an undesirable allele genetically linked to a desirable one (Boethel, 1999). The combination of linkage drag and quantitative inheritance has been a major obstacle to soybean breeders, and has made it very difficult to develop agronomically competitive cultivars with good insect resistance. Although three insect-resistant cultivars ('Crockett,' 'Lyon,' and 'Lamar') and >40 breeding lines have been released since 1969, none of them possesses both the resistance level of the PI donor parent and the yield performance of existing elite cultivars (Boethel, 1999; Lambert & Tyler, 1999). As a result, these cultivars have never been popular with producers.

Transfer of resistance from unadapted insect-resistant germplasm has also been restricted by the expense and difficulty of conducting phenotypic assays to evaluate insect resistance in segregating breeding populations. Selection in early generations is particularly problematic because it is difficult to obtain reliable data on the resistance of single plants. Delaying selection until families can be assayed, however, wastes resources on planting and assaying many lines that do not have the desired level of resistance. In addition, it is seldom possible to assay a breeding population for resistance to more than one insect pest. The quantitative nature of insect resistance and agronomic traits, and requirements for resistance to other pests and pathogens means that large populations are necessary to ensure recovery of lines possessing most of the desired traits. The challenge to soybean breeders can be appreciated if one considers that in addition to a good agronomic performance, a new cultivar may have to show resistance to up to 12 diseases (including multiple pathogen races or biotypes), and five nematode species (including six biotypes) (Lambert & Tyler, 1999).

DNA marker investigations of soybean insect resistance

DNA markers have proven a useful tool for investigating the genetics of insect resistance in soybean, and for marker-assisted selection (MAS) of insect-resistant individuals in breeding populations.

To map insect resistance QTLs (IRQTLs), populations derived from a cross between resistant and susceptible parents are tested for non-random associations between phenotype and the genotype at a marker locus. Statistically significant associations suggest linkage between the marker and a gene associated with resistance. QTLs can thus be identified and analyzed in a Mendelian fashion to determine their relative contribution to the phenotype (Tanksley et al., 1989). Genetic studies using classical techniques have identified >250 soybean loci since Piper and Morse's discovery of the *T* locus for pubescence color in 1910. In comparison, over 300 QTLs associated with various traits have been identified in soybean using molecular markers since 1990 (Orf et al. 2003). Yencho et al. (2000) listed 233 insect resistance QTLs that have been mapped in six different crop species. Although DNA marker technology is powerful, it nevertheless has limitations in detecting QTLs with relatively small effects (i.e., 'modifier genes'). Of the soybean QTLs reported in the literature, at least 162 appear to condition >10% of the variation in phenotype, and only a small fraction of the total have actually been confirmed. DNA markers linked to important genes or QTLs can be used for MAS, thereby reducing the need for phenotype-based selection. Tagging IRQTLs with markers also makes it possible to study them in different genetic backgrounds.

Rector et al. (1998, 1999, 2000) used restriction fragment length polymorphisms (RFLPs) to identify IRQTLs segregating in three populations developed by crossing the susceptible cultivar Cobb to PIs 171451, 227687 and 229358. Assays were conducted on $F_{2:3}$ lines to find IRQTLs associated with antixenosis and antibiosis resistance to corn earworm. Antixenosis in two of the populations was measured as percent defoliation in field plots. A greenhouse antixenosis assay was used to measure defoliation in the Cobb × PI 227687 population. Antibiosis was evaluated using a no-choice Petri plate assay to measure weight gain of larvae feeding on detached leaves.

The corn earworm IRQTLs identified by Rector et al. (1998, 1999, 2000), and in a follow-up mapping study by D. Hulburt (personal communication) are shown in Table 1. The percentage of phenotypic variance explained by the genotype at a particular IRQTL (R^2) was calculated to estimate

Table 1. Corn earworm IRQTLs[a] with resistance alleles contributed by Cobb, PI 171451, PI 227687, and/or PI 229358

Linkage group	Mode of action	Cobb	PI 171451	PI 227687	PI 229358
A1	Antibiosis	16	–	–	–
B2	Antibiosis	–	–	12–20	–
B2	Antixenosis	–	–	17	–
C1	Antixenosis	–	?	11–12	–
D1b	Antixenosis	–	–	–	10
E	Antibiosis and antixenosis	–	–	26 20	–
F	Antibiosis	–	–	12	–
F	Antibiosis	33	–	–	–
F	Antixenosis	20	–	–	–
G	Antibiosis	–	?	?	19
H	Antixenosis	–	19	9	16
J	Antibiosis	19	–	–	–
M	Antibiosis and antixenosis	– –	28 37	– –	22 37
O	Antixenosis	19	–	–	–

[a]Based on Rector et al. (1998, 1999, 2000); Narvel et al. (2001) and D. Hulburt, unpublished data.
Mode of action (antibiosis or antixenosis) is indicated and percent of phenotypic variance explained by each IRQTL (R^2) is shown as a percentage under the soybean genotype possessing the resistance allele. Question marks mean that the effect of a QTL in a particular population has not yet been determined.

the relative contribution of that IRQTL. With the exception of the IRQTLs on molecular linkage group (LG) A1, LG F, LG J, and LG O, the allele contributed by the PI parent was superior to the one from Cobb.

A major IRQTL on LG M (IRQTL-M) is associated with antixenosis ($R^2 = 0.37$) and antibiosis ($R^2 = 0.22$–0.28) in both PI 229358 and in PI 171451 (Rector et al., 1998, 1999, 2000). Narvel et al. (2001) re-mapped this QTL and other IRQTLs in the Cobb × PI 229358 population with simple sequence repeat (SSR) markers, and then conducted a retrospective analysis of an 82-cM region surrounding IRQTL-M in 15 cultivars and breeding lines to determine how many of these carry PI alleles at the QTL. These lines and cultivars had been selected phenotypically for resistance to coleopteran and/or lepidopteran pests, and were developed in six independent breeding programs using various selection and breeding methods (bulk, pedigree and backcross). In some programs, lines had been selected for resistance to Mexican bean beetle and corn earworm, while others had been selected for resistance to soybean looper and velvetbean caterpillar. Most of the lines and cultivars had PI 229358 as their resistant ancestor, but 171451 was listed as the resistant progentitor of the cultivar Crockett and one of the breeding lines. Graphical genotypes for the 15 lines and cultivars show that at least 13 of them carry a PI allele at the SSR marker Satt536, which maps about 0.5 cM from the estimated location of the antixenosis/antibiosis IRQTL-M (Figure 1). The fact that many of the lines had been selected for resistance to Mexican bean beetle suggests that IRQTL-M affects resistance to this coleopteran pest. In the two lines that had a non-PI allele at Satt536, the origin of the allele at Satt220, one of the markers flanking Satt536, could not be determined, so if IRQTL-M resides in the Satt220–Satt536 interval, these lines may also have the PI allele at IRQTL-M. Work is currently underway to fine-map the region around IRQTL-M, with the ultimate objective of cloning this QTL (Shuquan Zhu, personal communication). This will resolve whether IRQTL-M is a single locus with pleiotropic effects, or multiple loci that co-segregate. It will also be possible to determine whether PI 171451 and PI 229358 carry the same resistance allele(s) at IRQTL-M.

Rector et al. (2000) detected another antibiosis QTL ($R^2 = 19\%$) on LG G (IRQTL-G) in the Cobb × PI 229358 population. RFLP markers around IRQTL-G were monomorphic in the PI 171451-derived population, so it was not possible to determine whether the QTL affected resistance. The cultivar Crockett and a related breeding line supposedly descended from PI 171451 were found to have an allele at the SSR nearest IRQTL-G indicating that their true progenitor was PI 229358, and other DNA marker evidence also supported this hypothesis. The origin of IRQTL-G DNA could not be determined in any of the remaining 13 lines and cultivars due to monomorphic banding patterns at the nearest marker locus.

A corn earworm antixenosis IRQTL was identified on LG H (IRQTL-H) in all three PI-derived mapping populations used by Rector et al. (1998, 1999). IRQTL-H accounted for portions of phenotypic variance ranging from 9% in the PI 227687-derived population to 19% in the PI 171451-derived population. The detection of IRQTL-H in three independent populations provided confirmation of its antixenosis resistance, and suggests that it probably had adaptive value in the different environments where the three PIs originated. Nevertheless, among the 15 cultivars and breeding lines that Narvel et al. (2001) analyzed, only two carried PI alleles at a marker close to IRQTL-H. Both of these lines came from a program in which soybean looper and velvetbean caterpillar had been used to select for insect resistance, suggesting that IRQTL-H may associated with resistance to other lepidopteran pests. It is not known whether the three Japanese PIs carry the same allele at IRQTL-H.

The IRQTLs discovered by Rector et al. (1998, 1999, 2000) accounted for most of the genotypic variance for corn earworm resistance in the Cobb × PI 229358 and Cobb × PI 171451 populations, but a substantial amount of the genotypic variance observed in the Cobb × PI 227687 population remained unexplained by the identified QTLs. When soybean SSR markers became available in abundance, they were used to fill gaps in the RFLP map generated from the Cobb × PI 227687 population (D. Hulburt, personal communication). Antibiosis IRQTLs were identified on LGs A1, B2, E, and F (Table 1). The resistance

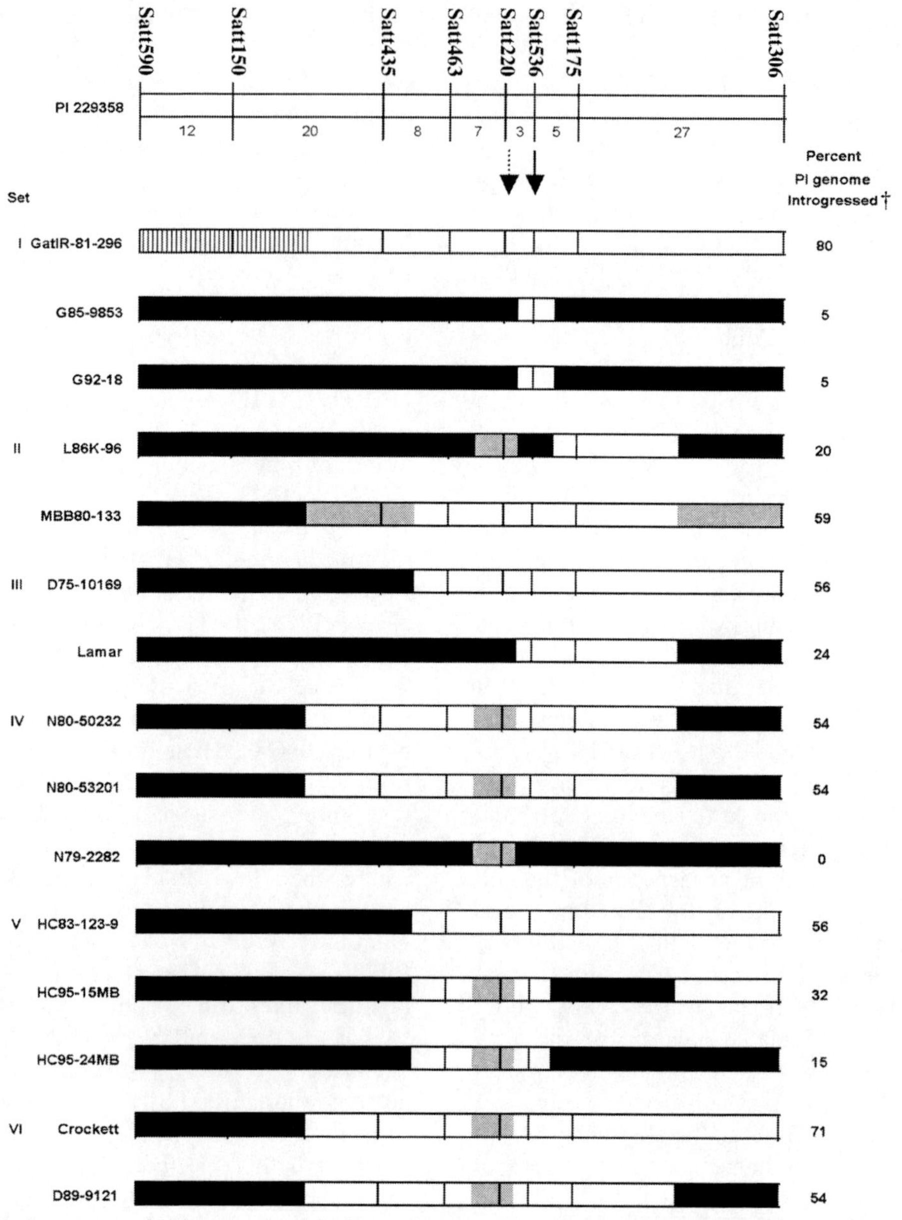

Figure 1. Graphical genotypes of 15 insect-resistant soybean cultivars and breeding lines in an 82-cM region encompassing IRQTL-M on molecular LG M. Cultivars and lines are grouped into sets based on the six different breeding programs in which they were developed. The most likely positions for the antixenosis (dotted arrow) and antibiosis (solid arrow) QTL(s) are indicated. The bar at the top representing PI 229358 shows the order and approximate genetic distances (cM) between SSR markers. Genomic segments are coded according to origin of the alleles at a marker locus, with crossovers portrayed as having occurred midway between markers. White segments indicate PI 229358 origin and black segments indicate non-PI origin. Vertical lines show that a genotype was heterogeneous for a locus, and gray represents regions in which a marker locus was uninformative (i.e. monomorphic). The approximate percentage of PI 229358 genome introgressed was estimated from informative markers (Figure from Narvel et al., 2001, used by permission from Crop Science).

alleles at all of the IRQTLs except the one on LG A1 came from PI 227687. The QTL on LG E (IRQTL-E) ($R^2 = 26\%$) is of particular interest because it mapped to a position 0.4 cM from the *Pb* gene, which conditions trichome tip morphology (http://soybase.agron.iastate.edu/).

The proximity of IRQTL-E to the *Pb* locus and the fact that PI 227687 has sharp-tipped trichomes, whereas Cobb has blunt-tipped trichomes, suggested that the *Pb* locus might actually be the IRQTL-E. This hypothesis was investigated by conducting antixenosis and antibiosis assays on near-isolines (NILs) of 'Clark' and 'Harosoy' that differed for trichome tip morphology (D. Hulburt, personal communication). Corn earworm larvae feeding on detached leaves from the sharp-tipped NILs of both cultivars consumed less tissue and weighed less (indicative of antixenosis and antibiosis, respectively) than larvae fed leaves from the blunt-tipped NILs. Defoliation by beet armyworm and soybean looper on the sharp-tipped NILs was also lower, though the only significant difference in weight gain in these two species was found for beet armyworm fed leaf tissue from one pair of NILs. These data support the hypothesis that IRQTL-E may be the *Pb* locus, and this is the first case we are aware of in which a morphological or biochemical trait has been convincingly associated with an IRQTL mapped with molecular markers in soybean.

Trichome density might also contribute to the resistance of PI 227687 resistance to some insects. Johnson and Hollowell (1935) reported that soybean genotypes with pubescence were less susceptible to damage by the potato leafhopper [*Empoasca fabae* (Harris)] than glabrous genotypes. Talekar, et al., (1988) analyzed trichome density among the three Japanese PIs and a susceptible control line, and found that PI 227687 was the only one of the three PIs that had a trichome density higher than the susceptible control. In other experiments with pubescent and glabrous NILs, the lines with dense pubescence were more resistant to the larvae of corn earworm, velvetbean caterpillar, and soybean looper, though oviposition by adult females was actually higher on plants with dense trichomes (Lambert et al., 1992).

The IRQTL mapping studies also identified loci at which resistance alleles originated from Cobb (Rector et al., 1999, 2000; D. Hulburt, personal communication). Although Cobb is susceptible relative to the Japanese PIs, it is not unusual for certain alleles contributed by a parent to have an effect opposite that expected from the phenotype (De Vicente & Tanksley, 1993). The relatively large effect of some Cobb alleles on resistance relative to that of the PI alleles is, however, surprising (Table 1). An antixenosis IRQTL on LG F ($R^2 = 20\%$) was detected in the Cobb × PI 171451 population (Rector, 1999), and another on LG O ($R^2 = 19\%$) has been identified in the Cobb × PI 227687 population (D. Hulburt, personal communication). At a different location on LG F, a major IRQTL explained 33% of the variance for antibiosis in the Cobb × PI 227687 population, while another antibiosis IRQTL on LG J explained 19% of the variance in the Cobb × PI 229358 population (Rector et al., 2000). These results show that useful insect resistance alleles exist in elite germplasm, and could therefore be transferred to other elite lines with minimal linkage drag. The results also suggest that the failure to detect some of the IRQTLs in a certain population could be explained if the allele from Cobb at those loci also conditioned a similar level of resistance.

Other unidentified IRQTLs probably exist, but these could not be detected with the mapping populations, markers, and pest species used to IRQTL detection may be difficult or impossible in regions of the genome where markers are either scarce or monomorphic with respect to parents used to generate the mapping population. Furthermore, IRQTLs with relatively small contributions ($R^2 < 0.10$) are difficult to identify because the risk of identifying false positives is high in the small mapping populations (<200 individuals) used in many mapping studies. Finally, some IRQTLs may not be involved in resistance to pest species other than corn earworm.

Although resistance assays may be designed to allow identification of IRQTLs associated with either antixenosis or antibiosis, these are not discrete modalities, so care must be taken in assuming that an IRQTL exclusively effects one type of resistance or the other (Smith, 1989). For example, a gene conditioning a trait that induces larvae to spend time searching for a different feeding site would be classified as antixenotic, yet the time and effort spent searching instead of feeding could indirectly result in a lower larval weight. In other cases, an IRQTL with purely antibiotic effects against one pest may also have antixenotic effects against a different pest.

Pyramids of IRQTLs and a Bt transgene

The value of pyramiding IRQTLs in resistance gene pyramid with a *cry1Ac* transgene from *Bacillus thuringiensis* (Bt) has been investigated in growth chamber and field studies (Walker et al., 2002, 2004). Some native gene/transgene pyramids could ameliorate two shortcomings of Bt-derived insect resistance. First, the Cry protein produced by a single Bt transgene will only protect the host plant from one, or at most two classes of insects. For example, the Cry1Ac toxin is lethal to many lepidopteran pests, but is non-toxic to coleopteran pests. If a native gene such as IRQTL-M conditioned resistance to Mexican bean beetle, then combining that gene with the Bt transgene could broaden resistance of Bt transgenic plants to include coleopteran pests that are insensitive to Cry1Ac toxins. Second, several insect pests have demonstrated the ability to develop resistance to Cry toxins, so effective strategies are needed to manage resistance to Bt (Roush, 1997). Populations of the diamondback moth [*Plutella xylostella* (L.)] have already developed resistance to Bt toxins in several parts of the world where Bt preparations are routinely applied to cruciferous crops (Tabashnik et al., 1997). Walker et al. (2002, 2004) found that soybean lines carrying the PI 229358 allele at IRQTL-M in addition to a *cry1Ac* transgene were better protected against defoliation by corn earworm and soybean looper than related transgenic lines lacking the PI 229358 allele. Additional experiments to investigate weight gain of tobacco budworm larvae from Cry1Ac-resistant and Cry1Ac-sensitive strains demonstrated that larvae fed leaves of plants with both a *cry1Ac* transgene and the IRQTL-M resistance allele gained weight more slowly than larvae fed leaves from transgenic plants lacking the IRQTL-M resistance allele (Walker et al., 2004). In related lines, some with and some without the Bt transgene, the PI 229358 allele at IRQTL-H, the presence of the PI allele did not improve the level of resistance.

The nature of IRQTLs is that even the ones with the largest effects seldom account for dramatic differences in the level of resistance observed at the single plant level. In contrast, a single Bt transgene can provide almost complete control of some sensitive insect species because the high toxicity of the expressed protein. Resistance of this type would appear to be more qualitative than quantitative. Despite the effectiveness of transgene-derived resistance towards certain insects, however, this technology has restrictions associated with proprietary issues in addition to biological limitations. It is therefore important to continue investigating native insect resistance genes in soybean and other crops to better evaluate their potential to increase and/or broaden resistance in both transgenic and non-transgenic cultivars.

References

Boethel, D.J., 1999. Assessment of soybean germplasm for multiple insect resistance, pp. 101–129 in Global Plant Genetic Resources for Insect-Resistant Crops, edited by S.L. Clement & S.S. Quisenbury. CRC Press, Boca Raton, FL.

Clark, W.J., F.A. Harris, F.G. Maxwell & E.E. Hartwig, 1972. Resistance of certain soybean cultivars to bean leaf beetle, striped blister beetle, and bollworm. J. Econ. Entomol. 65: 1669–1672.

De Vicente, M.C. & S.D. Tanksley, 1993. QTL analysis of transgressive segregation in an interspecific tomato cross. Genetics 134: 585–596.

Gizlice, Z., T.E. Carter & J.W. Burton, 1994. Genetic base for North American public soybean cultivars released between 1947 and 1988. Crop Sci. 34: 1143-1151.

Hatchett, J.H., F.G. Bland & E.E. Hartwig, 1976. Leaf-feeding resistance to bollworm and tobacco budworm in three soybean plant introductions. Crop Sci. 16: 277–280.

Johnson, J.W. & E.A. Hollowell, 1935. Pubescence and glabrous characters of soybeans as related to resistance to injury by the potato leafhopper. J. Agric. Res. 51: 371–381.

Kilen, T.C., J.H. Hatchett & E.E. Hartwig, 1977. Evaluation of early generation soybeans for resistance to soybean looper. Crop Sci. 17: 397–398.

Kilen, T.C. & L. Lambert, 1986. Evidence for different genes controlling insect resistance in three soybean genotypes. Crop Sci. 26: 869–871.

Kogan, M. & E.E. Ortman, 1978. Antixenosis – a new term proposed to define Painter's 'nonpreference' modality of resistance. Bull. Ent. Soc. Am. 24: 175–176.

Lam, W.F. & L.P. Pedigo, 2001. Effect of trichome density on soybean pod feeding by adult bean leaf beetles (Coleoptera: Chrysomelidae). J. Econ. Entomol. 94: 1459–1463.

Lambert, L. & T.C. Kilen, 1984. Multiple insect resistance in several soybean genotypes. Crop Sci. 24: 887–890.

Lambert, L. & J. Tyler, 1999. Appraisal of insect-resistant soybeans, pp. 131–148 in Economic, Environmental, and Social Benefits of Insect Resistance in Field Crops, edited by J.A. Webster & B.R. Webster. Thomas Say Publ., Entomol. Soc. of America, Lanham, MD.

Luedders, V.D. & W.A. Dickerson, 1977. Resistance of selected soybean genotypes and segregating populations to cabbage looper feeding. Crop Sci. 17: 395–396.

Narvel, J.M., D.R. Walker, B.G. Rector, J.N. All, W.A. Parrott & H.R. Boerma, 2001. A retrospective DNA marker assessment of the development of insect resistant soybean. Crop Sci. 41:1931–1939.

Orf, J.H., B.W. Diers & H.R. Boerma. 2003. Genetic improvement: conventional and molecular-based strategies, pp. 417–450 in Soybeans: Improvement, Production, and

Uses, edited by H.R. Boerma & J.E. Specht. Amercian Society of Agronomy, Madison, WI.

Painter, R.H., 1951. Insect resistance in crop plants, Univ. Press of Kansas, Lawrence, KS.

Probst, A.H. & R.W. Judd, 1973. Origin, U.S. history, development and world distribution, pp. 1–15 in Soybeans: Improvement, Production, and Uses, edited by B.E. Caldwell. Am. Soc. Agron., Madison, WI.

Rector, B.G., J.N. All, W.A. Parrott & H.R. Boerma, 1998. Identification of molecular markers associated with quantitative trait loci for soybean resistance to corn earworm. Theor. Appl. Genet. 96: 786–790.

Rector, B.G., J.N. All, W.A. Parrott & H.R. Boerma, 1999. Quantitative trait loci for antixenosis resistance to corn earworm in soybean. Crop Sci. 39: 531–538.

Rector, B.G., J.N. All, W.A. Parrott & H.R. Boerma, 2000. Quantitative trait loci for antibiosis resistance to corn earworm in soybean. Crop Sci. 40: 233–238.

Roush, R., 1997. Managing resistance to transgenic crops, pp. 271–294 in Advances in Insect Control: The Role of Transgenic Plants, edited by N. Carozzi & M. Koziel. Taylor & Francis, Ltd., London.

Sisson, V.A., P.A. Miller, W.V. Campbell & J.W. Van Duyn, 1976. Evidence of inheritance of resistance to the Mexican bean beetle in soybeans. Crop Sci. 16: 835–837.

Smith, C.M., 1989. Introduction, pp. 3–14 in Plant Resistance to Insects: A fundamental approach. John Wiley & Sons, New York.

Smith, C.M. & C.A. Brim, 1979a. Field and laboratory evaluations of soybean lines for resistance to corn earworm leaf feeding. J. Econ. Entomol. 72: 78–80.

Smith, C.M. & C.A. Brim, 1979b. Resistance to the Mexican bean beetle and corn earworm in soybean genotypes derived from 227687. Crop Sci. 19: 313–314.

Tabashnik, B.E., Y.-B. Liu, T. Malvar, D.G. Heckel, L. Masson, V. Ballester, F. Granero, J.L. Ménsua & J. Ferré, 1997. Global variation in the genetic and biochemical basis of diamondback moth resistance to *Bacillus thuringiensis*. Proc. Natl. Acad. Sci. USA 94: 12780–12785.

Talekar, N.S., H.R. Lee & Suharsono, 1988. Resistance of soybean to four defoliator species in Taiwan. J. Econ. Entomol. 81: 1469–1473.

Tanksley, S.D., N.D. Young, A.H. Paterson & M.W. Bonierbale, 1989. RFLP mapping in plant breeding: new tools for an old science. Bio. Technol. 7: 257–264.

Van Duyn, J.W., S.G. Turnipseed & J.D. Maxwell, 1971. Resistance in soybeans to the Mexican bean beetle: I. Sources of resistance. Crop Sci. 11: 572–573.

Van Duyn, J.W., S.G. Turnipseed & J.D. Maxwell, 1972. Resistance in soybeans to the Mexican bean beetle: II. Reactions of the beetle to resistant plants. Crop Sci. 12: 561–562.

Walker, D., H.R. Boerma, J. All & W. Parrott, 2002. Combining *cry1Ac* with QTL alleles from PI 229358 to improve soybean resistance to lepidopteran pests. Mol. Breeding 9: 43–51.

Walker, D.R., J.M. Narvel, H.R. Boerma, J.N. All & W.A. Parrott, 2004. A QTL that enhances and Broadens Bt insect resistance in soybean. Theor. Appl. Genet. In press.

Yencho, G.C., M.B. Cohen & P.F. Byrne, 2000. Applications of tagging and mapping insect resistance loci in plants. Ann. Rev. Entomol. 45: 393–422.

Genetica (2005) 123: 191–196 © Springer 2005

Polyploidy, evolutionary opportunity, and crop adaptation

Andrew H. Paterson
Plant Genome Mapping Laboratory, University of Georgia, Athens, GA 30602, USA (Phone: +1-706-583-0162; Fax: 1-706-583-0160; E-mail: paterson@uga.edu)

Received 1 January, 2003 Accepted 17 April, 2003

Key words: diversity, gene dosage, genome duplication, QTL, transgressive variation

Abbreviations: LG – linkage group; QTL – quantitative trait locus; RFLP – restriction fragment length polymorphism; WUE – water use efficiency

Abstract

The finding that even the smallest of plant genomes has incurred multiple genome-wide chromatin duplication events, some of which may predate the origins of the angiosperms and therefore shape all of flowering plant biology, adds new importance to the molecular analysis of polyploidization/diploidization cycles and their phenotypic consequences. Early clues as to the possible phenotypic consequences of polyploidy derive from recent QTL mapping efforts in a number of diverse crop plants of recent and well-defined polyploid origins. A small sampling examples of the role(s) of polyploidy in conferring crop adaptation from human needs include examples of (1) dosage effects of multiple alleles in autopolyploids, and (2) 'intergenomic heterosis' conferring novel traits or transgressive levels of existing traits, associated with merging divergent genomes in a common allopolyploid nucleus. A particularly interesting manifestation of #2 is the evolution of complementary alleles at corresponding ('homoeologous') loci in divergent polyploid taxa derived from a common ancestor. Burgeoning genomic data for both botanical models and major crops offer new avenues for investigation of the molecular and phenotypic consequences of polyploidy, promising new insights into the role of this important process in the evolution of botanical diversity.

Background

Polyploidy permeates virtually all of angiosperm biology. While it has long been apparent that many angiosperm taxa had undergone one or more chromosomal duplication events in their evolutionary history, early hints (McGrath et al., 1993; Kowalski et al., 1994) of chromosomal duplication even in the smallest of angiosperm genomes were recently borne out (Blanc et al., 2000; Paterson et al., 2000; Arabidopsis Genome Initiative, 2000; Vision, et al. 2000) by analysis of the completed *Arabidopsis* sequence. The finding that one period of chromatin duplication (perhaps a single event) predates most the divergence of most dicots from a common ancestor, and another event may predate the monocot–dicot divergence (Bowers et al. 2003), implies that most if not all angiosperm lineages may have been shaped by a few common ancient polyploidization events, then further modified by additional recent events.

While polyploidy as traditionally defined appears to be roughly equally prevalent in cultivated and non-cultivated plants (Hilu, 1993), analysis of crop plant genomes offers opportunities to study many phenotypic consequences of polyploidy in a manner that combines applications-oriented research with investigation of phenomena that may be fundamental to botanical evolution. Polyploidy is far less abundant in animals than plants, arguably due in part to the need in animals for monosomic sex-determining chromosomes. Consequences of polyploidy in plants may include a much higher rate of gene loss, and more rapid apparent decay of synteny than in animals (Bowers et al., 2003). Several recent studies associate non-linear phenotypic effects with the additive or even less-than-additive (Eckhardt, 2001). merger of two or more genomes with divergent evolutionary histories in a common nucleus. In this chapter, a tiny sampling of cases that have been investigated in my lab are reviewed, then I suggest how emerging research opportunities may yield new insights into the phenotypic consequences of polyploidy.

Case studies

Non-linear dosage effects of corresponding ('homoeologous') alleles in sugarcane, an autopolyploid

Autopolyploid genomes, containing many different homologous chromosomes that can pair and recombine in most or all possible combinations, have been under-explored at the molecular level due to their special problems in genetic and molecular analysis. The importance of autopolyploidy is highlighted by its prominence among cultivated crops, including sugarcane (8–18x), sugar beet (3x), ryegrass (4x), bermuda grass (3–4x), cassava (4x), potato (4x), alfalfa (4x), red clover (4x), Grande Naine banana (3x), apple cultivars (3x), and many ornamentals. It is noteworthy that many of these crops are cultivated for vegetative products and are vegetatively propagated, autopolyploidy often being associated with reduced seed production.

Sugarcane is a classical example of a complex autopolyploid genome. Cultivated sugarcane varieties have about 80–140 chromosomes, comprising 8–18 copies of a basic $x = 8$ or $x = 10$ (Irvine, 1999). Most chromosomes of cultivated sugarcane appear to be largely derived from *Saccharum officinarum*–however, *in situ* hybridization data suggest that about 10% may be derived from *S. spontaneum* (D'Hont et al., 1995).

Like other vegetatively propagated plant species, cultivated sugarcane (*Saccharum* spp. hybrids) and its wild relatives are highly heterozygous. Pure inbred lines do not exist due to the difficulty of self pollination and the random pairing of multiple homologous chromosomes. The segregating populations used in genetic studies are first-generation progenies from crosses between two cultivated varieties, or cultivated varieties and wild species. Genetic mapping uses the subset of DNA polymorphisms that show simplex segregation ratios, and these 'single-dose' markers can also be employed to locate QTLs. However, the fundamental complexity of autopolyploid genetics resulting from heterozygosity and lack of preferential pairing is further complicated by the fact that economically important traits such as sugar content are complex industrial traits, influenced by variation in carbon fixation, photosynthate partitioning into sucrose, transportation and accumulation of sucrose in harvestable biomass, and extractability of sucrose from biomass.

We have used a detailed genetic map to analyze the inheritance of numerous traits in two interspecific F_1 populations (Ming et al., 2001). For example, 36 significant associations between variation in sugar content and unlinked loci detected by 31 different probes were found. The 36 sugar content QTLs correspond to only eight non-overlapping regions of the sorghum genome, with single homologous genomic regions accounting for three QTLs in three cases, and two QTLs in five cases. In a subset of four of these cases, single DNA probes detected sugar content QTLs at each of two or more unlinked loci, making it possible to investigate whether the dosage (zero, one, or two 'copies') of the chromosomal region(s) containing the favorable allele(s) had non-additive (i.e. non-linear) effects on phenotype. Considering sugar content, all four cases showed non-linear tendencies suggesting less-than-additive effects, but in only one case (CSU0428b, dM) did the regression line have a significant non-linear (in this case, quadratic) component. Other traits for which significant effects were linked to larger numbers of loci detected by common probes provided a test of higher dosages. For example, two DNA probes each detected three loci associated with plant height, and another two DNA probes each detected four loci associated with plant height. In all four cases, the regression lines showed less-than-additive gene action, with significant ($p < 0.05$) quadratic trends in three cases, and a significant quartic trend in one case.

Multiplex segregation at QTL loci may be partly responsible for the phenotypic buffering that is argued by many to be one factor in the success of autopolyploid crops. Detecting this type of phenotypic buffering provides strategic information for marker-assisted selection in autopolyploid crops. Although diagnostic DNA markers enable us to pyramid multiple QTLs in a polyploid, incorporating any one copy of the multiple alleles may obtain most of the desired effect in the breeding population.

Non-additive gene action in multiple dose QTLs may also provide evolutionary opportunities. If a single copy of a gene/QTL is physiologically sufficient, the extra copies are free to collect mutations, often becoming non-functional, but perhaps occasionally resulting in a distinctive new function which improves fitness.

An important future investigation regards the contribution of multi-locus QTL genotypes to stability of performance across different environments. Sugar content is a trait of relatively high heritability – however, a role of multiple-dose QTLs in enhancing environmental stability would be of potentially great importance for less heritable traits.

Unique evolutionary opportunities associated with merging divergent genomes in a common allopolyploid nucleus

The evolution of the genus *Gossypium* (cotton) has included a very successful experiment in polyploid formation, one that fosters investigation of the consequences of re-uniting divergent genomes in a common nucleus after millions of years of divergence. World cotton commerce of about $20 billion annually is dominated by improved forms of two (among 5 extant) 'AD' tetraploid ($2n = 4x = 52$) species, *G. hirsutum* L. and *G. barbadense* L. Tetraploid cottons are thought to have formed about 1–2 million years ago, in the New World, by hybridization between a maternal Old World 'A' genome taxon resembling *G. herbaceum* ($2n = 2x = 26$), and paternal New World 'D' genome taxon resembling *G. raimondii* (Wendel, 1989) or *G. gossypioides* (Wendel et al., 1995), both $2n = 2x = 26$. The antiquity of this New World event precludes human involvement in polyploid formation.

Two aspects of the cotton 'experiment' are considered further below, in the context of 'intergenomic heterosis' arising from re-joining of the A and D genomes in a common tetraploid nucleus.

A non-fiber producing ancestral genome accounts for the majority of phenotypic variation in fiber attributes of modern cottons

Wild A-genome diploid and AD-tetraploid *Gossypium* taxa each produce spinnable fibers that were a likely impetus for domestication. Domesticated tetraploid cottons existed in the New World by 3500–2300 BC, and have been widely distributed by humans throughout the world's warmer latitudes. Domesticated A-genome diploids existed in the Old World by 2700 BC, and one (of only two extant) species, *G. arboreum*, remains intensively bred and cultivated in Asia.

Although the seeds of D-genome diploids are pubescent, none produce spinnable fibers. There is no evidence that domestication of D-genome *Gossypium* taxa has ever been attempted, although their geographic distribution overlaps that of several wild tetraploids.

Intense directional selection by humans has consistently produced AD-tetraploid cottons that have superior yield and/or quality characteristics than do A-genome diploid cultivars. Selective breeding of *G. hirsutum* (AADD) has emphasized maximum yield, while *G. barbadense* (AADD) is prized for its fibers of superior length, strength, and fineness. Side-by-side trials of 13 elite *G. hirsutum* genotypes and 21 *G. arboreum* diploids (AA) adapted to a common production region (India) show average seed cotton yield of 1135 (± 90) kg/ha for the tetraploids, a 30% advantage over the 903 (± 78) kg/ha of the diploids, at similar quality levels (Anonymous, 1997). Such an equitable comparison cannot be made for *G. barbadense* and *G. arboreum*, as they are bred for adaptation to different production regions. However, the fiber of 'extra-long-staple' *G. barbadense* tetraploids, representing ~5% of the world's cotton, commands a premium price due to ~40% higher fiber length (ca. 35 mm), strength (ca. 30 g per tex or more), and fineness over leading A-genome cultivars, at similar yield levels. Obsolete *G. barbadense* cultivars reportedly had up to 100% longer fibers (50.8 mm; Niles and Feaster, 1984) than modern *G. arboreum* (25.5 ± 1.6 mm; Anonymous, 1997).

A detailed RFLP map made in my lab has been used to determine the chromosomal locations and subgenomic (A versus D) distributions of QTLs segregating in at least four different crosses between high-fiber-quality *G. barbadense* cultivars, and high-yielding *G. hirsutum* cultivars (both AADD). Results are summarized in Table 1. The D subgenome, from the non-fiber-producing ancestor, generally accounts for more genetic variation in fiber traits of *G. barbadense* and *G. hirsutum* than does the A subgenome, from the fiber-producing ancestor. Not only do these data clearly demonstrate the role of the non-fiber-producing D subgenome in cotton fiber traits, but they suggest that the D-subgenome may even play the larger role (of the two subgenomes) in the inheritance of fiber characteristics of modern cottons.

While the molecular and evolutionary basis of these findings remains to be demonstrated, we can falsify a few alternative hypotheses, and speculate about some possible mechanisms. The D-subgenome bias of fiber QTLs is not explained by differences in either recombinational or physical size, or by levels of genetic variation (as reflected by DNA marker alleles) in the two subgenomes. Curiously, although extensive correspondence in the locations of QTLs has been found in other genomes diverged by up to 65 million years (Paterson et al., 1995), there have been few cases of correspondence between fiber QTLs in the A and D-subgenomes, thought to have diverged from a common ancestor only about 10 million years ago. The A-subgenome, in which fiber evolution preceded polyploid formation, has a much longer history of selection (albeit largely natural) for formation of an elongated seed epidermal fiber that presumably contributes to dispersal. (It is noteworthy that formation in the New World, of the polyploid between native D genome taxa and Old World A genome taxa, clearly required long-distance dispersal of the A-genome ancestor – Wendel, 1989). By contrast, the D-subgenome may not have come under selection for such a trait until after polyploid formation. One albeit speculative notion that has been suggested (Jiang et al., 1998) is that natural or human selection for fiber attributes of tetraploid cotton may have conferred a relatively greater likelihood that mutations at D-subgenome loci confer a fitness advantage for this trait – by virtue of a multi-million year history of natural selection for the trait in the A subgenome. Formal testing of this hypothesis will require cloning and characterization of the evolutionary history of a sampling of the determinants of this important trait, work that is underway in many labs using a variety of approaches that at a minimum include candidate gene evaluation, analysis of discrete mutants, and dissection of genomic regions containing QTLs.

Table 1. Subgenomic distribution of QTLs conferring fiber yield and quality components

	A	D	Uncertain
Jiang et al. (1998)	4	11	0
Saranga et al. (2001)	26	22	0
Paterson et al. (2002)	34	45	0
Chee et al. in prep[a]	29	38	1
TOTAL	93	116	1

[a] Subject to revision.

Evolution of complementary alleles at corresponding loci in divergent polyploid taxa

A second investigation of the cotton genome focused on response to water deficit. Water loss from a plant (transpiration) is an unavoidable consequence of photosynthesis, whereby the energy of solar radiation is used for carbon fixation. About one-third of the world's arable land suffers from chronically inadequate supplies of water for agriculture, and in virtually all agricultural regions, yields of rain-fed crops are periodically reduced by drought (Boyer, 1982). In this study, we crossed two superior genotypes (in terms of adaptation to water deficit) of different species to investigate the similarities and differences in how these species had become adapted to this important abiotic stress. Specifically, we crossed GH cv. Siv'on with GB cv. F-177, each of which had the highest WUE among cultivars of their species grown in the test environment in Israel (Saranga et al., 1998).

Among a total of 161 QTLs detected for the 16 measured traits (Saranga, Menz et al., 2001), the polyploidy of cotton was especially well reflected by two cases in which corresponding 'homoeologous' loci on each of the two different subgenomes appeared to account for common sets of traits. The *G. hirsutum* allele at a QTL on chromosome 6 (the A-subgenome) was associated with lower leaf osmotic potential, lower canopy temperature, and higher seed-cotton yield than the *G. barbadense* allele in the water-limited environment. At the homoeologous location on

Chr. 25, the *G. barbadense* allele conferred both lower OP and higher SC than the *G. hirsutum* allele. A second case of QTLs on homoeologous regions involved *G. hirsutum* (Chr. 22) and *G. barbadense* (LG D05) alleles that each conferred higher carbon isotope ratio ($\delta^{13}C$) under the water-limited treatment and lower chlorophyll content under the water-limited or both treatments. The discovery that each of two homoeologous locations account for genetic variation in the same phenotypes suggests that subsequent to polyploid formation in cotton, new functionally significant mutations (alleles) appear to have arisen at each of the two homoeologous loci (or nearby linked loci).

The finding that the *G. hirsutum* allele is favorable at some loci and the *G. barbadense* allele at other loci shows that subsequently to polyploid formation, these different lineages have taken very different evolutionary paths. Moreover, recombination of favorable alleles from each of these species may form novel genotypes that are better-adapted to arid conditions than either of the parental species. The 'genomic exploration' of other accessions of these species, or other wild tetraploid cottons (*G. tomentosum, G. darwinii, G. mustelinum*) may yield still additional valuable alleles, and is being actively pursued by crop breeders.

Looking ahead

Even based on this tiny sampling, it seems clear that the many polyploidization events that characterize angiosperm evolution (Bowers et al. 2003) appear to add a unique dimension to the means by which plants can adapt. The availability of complete sequence for one plant, *Arabidopsis thaliana*, has been key to realizing the true extent of gene duplication in plants, and perhaps also hints at some possible molecular mechanisms that may contribute to phenotypic evolution. For example, one of the more surprising findings (at least to this author) of the analysis of ancient duplication in *Arabidopsis* was the extent of gene loss. Many of the advantages postulated to be associated with polyploidy are contingent on the presence of two somewhat redundant copies of a gene – yet for the most recent duplication of *Arabidopsis*, most authors agree that fewer than 30% of genes retain a 'homoeolog' (syntenic duplicate). The notion of polyploidy as a 'buffer', and the rapid pace of 'diploidization' in some taxa (Eckhardt, 2001), seem at least superficially incongruous. Information from many additional taxa, together with more information about the extent to which the consequences of polyploidy are general, or peculiar to individual genes and gene families, will be especially important in better understanding of the consequences of polyploidy for both angiosperm diversity and agricultural productivity.

Acknowledgements

The author thanks many past and present collaborators, postdocs, students, and staff who contributed in many ways to the three case studies presented, the US Department of Agriculture, US National Science Foundation, International Consortium for Sugarcane Biotechnology, and Texas and Georgia Cotton Commissions for financial support, and R. Mauricio for motivating the writing of this chapter.

References

Arabidopsis Genome Initiative, 2000. Analysis of the genome sequence of the flowering plant *Arabidopsis thaliana*. Nature 408: 796–815.

Anonymous, 1997. Zonal Coordinators Annual Report of All-India Coordinated Cotton Improvement Project.

Blanc, G., A. Barakat, R. Guyot, R. Cooke & M. Delseny, 2000. Extensive duplication and reshuffling in the *Arabidopsis* genome. Plant Cell 12: 1093–1101.

Bowers, J.E., B.A. Chapman, J. Rong & A.H. Paterson, 2003. Unravelling angiosperm chromosome evolution by phylogenetic analysis of chromosomal duplication events. Nature 422: 433–438.

Boyer, J., 1982. Plant productivity and environment. Science 281: 443–448.

D'Hont, A., P.S. Rao, P. Feldmann, L. Grivet, N. Islamfaridi, P. Taylor & J.C. Glaszmann, 1995. Identification and characterization of sugarcane intergeneric hybrids, *Saccharum officinarum* x *Erianthus arundinaceus*, with molecular markers and DNA in situ hybridization. Theor. Appl. Genet. 91: 320–326.

Eckhardt, N., 2001. A sense of self: the role of DNA sequence elimination in allopolyploidization. Plant Cell 13: 1699–1704.

Hilu, K.W., 1993. Polyploidy and the evolution of domesticated plants. Am. J. Bot. 80: 1494–1499.

Irvine, J.E., 1999. *Saccharum* species as horticultural classes. Theor. Appl. Genet. 98: 186–194.

Jiang, C.X., R.J. Wright, K.M. El-Zik & A. H. Paterson, 1998. Polyploid formation created unique avenues for response to selection in Gossypium (cotton). Proc. Natl. Acad. Sci. USA 95: 4419–4424.

Kowalski, S.P., T.H. Lan, K.A. Feldmann & A.H. Paterson, 1994. Comparative mapping of *Arabidopsis thaliana* and *Brassica oleracea* chromosomes reveals islands of conserved gene order. Genetics 138: 499–510.

Martin, W., A. Gierl & H. Saudler, 1989. Molecular evidence for pre-Cretaceous angiosperm origins. Nature 339: 46–48.

McGrath, J.M., M.M. Jancso & E. Pichersky, 1993. Duplicate sequences with a similarity to expressed genes in the genome of *Arabidopsis thaliana*. Theor. Appl. Genet. 86: 880–888.

Ming, R., S.C. Liu, P.H. Moore, J.E. Irvine & A.H. Paterson, 2001. QTL analysis in a complex autopolyploid: genetic control of sugar content in sugarcane. Genome Res. 11: 2075–2084.

Niles, G.A. and C.V. Feaster, 1984. Breeding, pp. 202–229 in Cotton edited by R.J. Kohel & C.F. Lewis. American Society of Agronomy, Madison, WI, USA.

Paterson, A.H., Y.R. Lin, Z.K. Li, K.F. Schertz, J.F. Doebley, S.R.M. Pinson, S.C. Liu, J.W. Stansel & J.E. Irvine, 1995. Convergent domestication of cereal crops by independent mutations at corresponding genetic loci. Science 269: 1714–1718.

Paterson, A.H, J.E. Bowers, M.D. Burow, X. Draye, C.G. Elsik, C.X. Jiang, C.S. Katsar, T.H. Lan, Y.R. Lin, R. Ming & R.J. Wright 2000. Comparative genomics of plant chromosomes. Plant Cell 12: 1523–1539.

Paterson, A.H., Y. Saranga, M. Menz, C.X. Jiang & R. Wright, 2002. QTL Analysis of genotype x environment interactions affecting cotton fiber quality. Theor. Appl. Genet. 106: 384–396.

Saranga, Y., I. Flash & D. Yakir, 1998. Variation in water-use efficiency and its relation to carbon isotope ratio in cotton. Crop Sci. 38: 782–787.

Saranga, Y., M. Menz, C.X. Jiang, R.J. Wright, D. Yakir & A.H. Paterson, 2001. Genomic dissection of genotype x environment interactions conferring adaptation of cotton to arid conditions. Genome Res. 11: 1988–1995.

Vision, T., D.G. Brown & S.D. Tanksley, 2000. The origins of genomic duplications in *Arabidopsis*. Science 290: 2114–2117.

Wendel, J.F., 1989. New World tetraploid cottons contain Old World cytoplasm. Proc. Natl. Acad. Sci. USA 86: 4132–4136.

Wendel, J., A. Schnabel & T. Seelanan, 1995. Bidirectional interlocus concerted evolution following allopolyploid speciation in cotton (*Gossypium*). Proc. Natl. Acad. Sci. USA 92: 280–284.

Genetica (2005) 123: 197–204 © Springer 2005

Quantitative trait loci and the study of plant domestication

Jeffrey Ross-Ibarra
Department of Genetics, University of Georgia, Athens, GA 30602, USA (Phone: +1-706-542-0290; Fax: +1-706-542-3910; E-mail: jri@uga.edu)

Received 8 January 2004 Accepted 26 March 2004

Key words: adaptation, artificial selection, domestication, QTL

Abstract

Plant domestication ranks as one of the most important developments in human history, giving human populations the potential to harness unprecedented quantities of the earth's resources. But domestication has also played a more subtle historical role as the foundation of the modern study of evolution and adaptation. Until recently, however, researchers interested in domestication were limited to studying phenotypic changes or the genetics of simple Mendelian traits, when often the characters of most interest – fruit size, yield, height, flowering time, etc. – are quantitative in nature. The goals of this paper are to review some of the recent work on the quantitative genetics of plant domestication, identify some of the common trends found in this literature, and offer some novel interpretations of the data that is currently available.

Abbreviations: DRT – domestication related trait; QTL – quantitative trait locus.

Introduction

Plant domestication ranks as one of the most important developments in human history, giving human populations the potential to harness unprecedented quantities of the earth's resources. But domestication has also played a more subtle historical role as the foundation of the modern study of evolution and adaptation. Darwin explicitly identified domestication as the basis for his ideas of natural selection and evolution (Darwin, 1899), and many of his ideas about how natural selection might function are based on keen observations of the human-mediated selection of domesticated plants and animals. In fact, Darwin had good reason to look to domestication for an understanding of adaptation in nature. Unlike most natural cases of adaptation, studies of plant domestication have the potential to identify what selection pressures populations have responded to and infer how selection may have acted. Moreover, it has often been possible to pinpoint the geographic and phylogenetic origin of domesticates, thus allowing direct comparisons of descendents with their (usually extant) ancestors.

With only rare exceptions (e.g. Anderson et al., 1991; Dudley & Lambert, 1992; Cowie & Jones, 1998; Visser et al., 1998; Grant & Grant, 2002), studies of adaptation are restricted by the inability to observe selection in action over a meaningful period of time; the resulting changes are frequently the only clues biologists have with which to infer the processes involved in adaptation. Though focusing on domesticates alleviates many of the difficulties inherent in the study of adaptation, until recently researches interested in domestication were limited to studying phenotypic changes or the genetics of simple Mendelian traits, when often the characters of most interest – fruit size, yield, height, flowering time, etc. – are quantitative in nature.

The last 15 years, however, have seen an outpouring of data on the genetic basis of quantitative traits. Dozens, if not hundreds, of articles have investigated the number, location, and effects of the chromosomal regions responsible for the phenotypic variation observed among organisms in the natural world. Whether for expediency or scientific curiosity, much of this research has focused on quantitative variation in crop plants, and a number of studies have specifically investigated traits thought to have been important in domestication. Two recent reviews highlight several of the major patterns that have emerged from the growing body of quantitative mapping studies in domesticated plants (Paterson, 2002; Frary & Doganlar, 2003) including the number, effect, and distribution of the quantitative trait loci (QTL) underlying domestication related traits (DRT), as well as similarities across species in the QTL involved in the domestication process. In the last two years, however, several new studies have helped to flesh out the patterns recognized by these reviews. These data reinforce many of the conclusions of earlier reviewers, but also allow us to extrapolate beyond the patterns recognized by those authors.

I will begin with a brief discussion of the major patterns present in QTL mapping studies of domesticated plants. Many of these trends have been recognized previously (Paterson, 2002; Frary & Doganlar, 2003), and I will instead focus on extending the analysis of these trends, adding information from the recent literature and suggesting some novel interpretations of the data currently available.

Major patterns

Distribution of QTL

Perhaps the most widely cited pattern to emerge from QTL mapping studies in domesticated plants has been the clustering of QTL. Most mapping studies have found that QTL are not randomly or even uniformly distributed throughout the genome, but occur in apparently linked clusters in certain regions of the chromosome (Cai & Morishima, 2002; Paterson, 2002). The few studies that fail to find extensive clustering (e.g. Hashizume, Shimamoto & Hirai, 2003) tend to suffer from methodological problems that severely constrain the power of these studies to detect QTL. In spite of the strong empirical support for this pattern, its genetic basis (i.e. tight physical linkage or pleiotropic effects) and its significance in terms of adaptation remain open to debate.

Size and number of QTL

To many biologists, one of the most surprising finds of QTL studies has been the number of loci controlling many quantitative traits. QTL analysis allows the determination of a lower bound on the number of genes that control a given trait. And while classical quantitative genetic theory attributes continuous variation in nature to the small, additive effects of a nearly infinite number of genes, many studies of traits associated with domestication have found that much of the phenotypic variation can be explained by a few loci of relatively large effect. Though methodological problems – marker density, sample size, crossing scheme, etc. – can cloud the interpretation of these data (Beavis, 1994; Mauricio, 2001), the claim that most DRT are controlled by few loci of large effect seems to hold true for many studies across a variety of taxa. Counterexamples (Burke et al., 2002) do exist however, and the reasons for differences in effect size across studies or taxa are not completely clear. One difficulty in comparing QTL across studies has been the definition of 'major effect,' since transgressive variation among the progeny can decouple absolute morphological change from percent of phenotypic variance explained by a QTL.

QTL homology

The central theme of Frary and Doganlar's (2003) review is the similarity of QTL location and identity across taxa. Extensive synteny among QTL of major effect for DRT has been well established in the grass family (Paterson et al., 1995), and recent work has extended these findings to the Solanaceae, revealing similarities in QTL number and location across several genera of the family (Doganlar et al., 2002, Frary et al., 2003b). This similarity of genic and phenotypic character variation across a wide array of taxa seems to corroborate Vavilov's (1922) 'law of homologous series in variation,' – the assertion that character variation found in one taxa should exist in related or similar taxa.

Interpretations

Genetic basis of DRT

Loci of major effect are commonly found in mapping studies of DRT. The size of effect of a QTL is usually determined by the amount of phenotypic variation it explains. The percent variation explained by a QTL, however, does not necessarily correlate with the heritability of a given trait, nor with the absolute amount of change a gene effects (Burke et al., 2002). While there is good reason to interpret results evidencing QTL of major effect with some caution (Beavis, 1994; Mauricio, 2001; Paterson, 2002), the overall pattern is too common to ignore. Classic theory suggests that quantitative traits should be controlled by many genes of small effect, and that, more often than not, mutations of large effect would be deleterious in nature (Lande, 1983). This contrasts with reviews of phenotypic evolution in plants, which offer results similar to those reported in mapping studies: Hilu (1983) and Gottlieb (1984) both point to the important role of mutations of large effect. Similarly, recent theoretical advances find fault with the Neo-Darwinian dogma, suggesting an adaptive role for mutations of large phenotypic effect (Orr & Coyne, 1992; Orr, 1998a, 2003). On finding no QTL of large effect for DRT in crosses between wild and domesticated sunflower, Burke et al. (2002) make the argument that 'domestication may have occurred more readily without requiring the fortuitous occurrence of multiple major mutations.' While this may be true if adaptation under artificial selection depends solely on novel mutations, theory suggests that the opposite could occur if selection acts on standing genetic variation: selection will fix single alleles of large effect much faster than it could fix a multitude of small alleles (Barton & Keightley, 2002). Loci of large effect can then be later modified by selection acting on other genes (Hillman & Davies, 1990), which could well lead to distributions of allele effects quite similar to those seen in empirical mapping studies.

In addition to measuring the size of effect of QTL, mapping studies can elucidate the mode of action of the loci. Given that random mutation is more likely to inactivate a functional gene than to modify it or create a new function, it has been argued that the majority of DRT should be recessive. Many domesticated characters are in fact recessive (Ladizinsky, 1985; Lester, 1989), and both of the so-called 'domestication genes' which have been successfully cloned are essentially recessive (Doebley, Stec & Hubbard, 1997; Frary et al., 2000). Data from Burke et al. (2002) contradict this idea, showing no evidence for a predominance of recessive types among the alleles from domesticated sunflower. Other mapping studies show mixed results. Some show few or no recessive alleles in the domesticates (Paterson et al., 1991; Peng et al., 2003), yet other crosses find recessive alleles to be frequent (Doganlar et al., 2002; Xiong et al., 1999). Burke et al. (2002) actually argue that a lack of recessive alleles should have made the domestication of sunflower simpler. Again, however, if adaptation depends predominantly on standing variation rather than novel mutations, theory suggests that recessive alleles for DRT would be more likely to be fixed than nonrecessive ones (Orr & Betancourt, 2001). Until more data – especially on the relative importance of novel mutations and existing genetic variation – is available, however, it does not seem possible to make any general conclusions about the significance of the mode of action of QTL involved in crop domestication.

Tempo of domestication

Several lines of evidence suggest that the traditional Neo-Darwinian view of gradual change under domestication is no longer a tenable hypothesis. Paterson (2002) discusses the issue in some detail, arguing that the size of QTL, the existence of QTL clusters that could act as coadapted gene complexes, the coincidence of QTL across taxa, and the relative ease with which domesticates can lose DRT and become feral or weedy all support a relatively fast or punctuational tempo of domestication. Mathematical models of domestication based on empirical estimates of selection coefficients support his conclusion, estimating that domestication could take as little as 20–100 years (Hillman & Davies, 1990). Analysis of nucleotide variation in maize corroborates this conclusion, concluding that the current patterns of diversity are consistent with domestication having taken as little as ten years in very small populations (Eyre-Walker et al., 1998). Population bottlenecks, such as those suggested by the data in

Eyre-Walker et al., have long been thought to play an important role in plant domestication (Ladizinsky, 1985), and recent work (Ross-Ibarra, 2004) is consistent with the prediction that elevated levels of drift in such small populations would select for increased recombination (Otto & Barton, 2001). Finally, work in maize has provided a molecular model for rapid evolutionary change in domesticates, linking changes in the regulation of a single gene to major shifts in branching and inflorescence structure (Wang et al., 1999).

QTL distribution and adaptation

As mentioned above, nonrandom distribution of QTL has been a nearly ubiquitous finding in mapping studies of DRT. Most authors are careful to note that these clusters can be interpreted in at least two ways: either multiple genes are actually clustered together in linked groups, or the same genes are identified as QTL for several different traits (pleiotropy). The latter explanation seems probable for many of the reports of QTL for similar or correlated traits such as fruit weight and yield in peppers (Rao et al., 2003) or color shade and intensity in eggplant (Doganlar et al., 2002). Yet many studies have nonetheless found clustering of QTL for traits that do not seem likely to be pleiotropic effects of a single gene: Cai and Morishima (2002) mapped QTL relating to mineral tolerance, heading behavior, germination speed, and anther length all to a very short interval on one of the 12 chromosomes of rice, and similar clusters of seemingly unrelated QTL have been reported in a variety of species (Koinange et al., 1996; Poncet et al., 2000; Bres-Patry et al., 2001; Baum et al., 2003; Huang et al., 2003). As crossing strategies and mapping technologies improve, continued efforts at fine-scale mapping of QTL clusters (e.g. Takeuchi et al., 2003) combined with the development of new statistical analyses (e.g. Varona et al., 2004) should enable researchers to better distinguish between pleiotropy and linkage.

Many authors have made some variation of an adaptive argument for the observed presence of QTL clusters. Koinange et al. (1996) adopt the explanation of Pernes (1983) that, in allogamous plants, selection against recombinant hybrids between wild and cultivated plants will lead to the clustering of QTL for DRT in tightly linked groups, and computer simulations (Le Thierry D'Ennequin et al., 1999) of wild to crop gene flow during domestication seem to support this argument. Theoretical work has similarly shown that maladaptive gene flow creates positive associations among beneficial alleles in the reference population, thus selecting for increased linkage or decreased recombination (Lenormand & Otto, 2000). Cai and Morishima (2002) ascribe clustering of QTLs to Grant's (1981) concept of 'multifactorial linkages,' or weak linkages brought about by the random distribution of multiple factors throughout the genome. These linkages are then somehow preserved by selection for coadapted gene complexes, perhaps via a process similar to that of Pernes (1983). Poncet et al. (1998) proposed that linked clusters of QTL for DRT would become fixed more rapidly in a population than unlinked genes, through a type of 'reciprocal' hitchhiking effect.

There is, however, no a priori reason to believe that the clustering of genes is caused or maintained by strong selection. Westerbergh and Doebley (2002) analyzed the genetic basis of quantitative traits between two wild species of maize. Applying Orr's (1998b) QTL sign test, they conclude that phenotypic differences between the species can be best explained by neutral drift or temporal fluctuation in the direction of selection. Yet, in spite of an apparent lack of strong directional selection for any of the traits studied, Westerbergh and Doebley's linkage map shows the familiar pattern of clustered QTL. Furthermore, Pernes' (1983) hypothesis predicts a lack of clustering in selfing species, a result that is not supported by data gathered for common bean (Koinange et al., 1996), eggplant (Doganlar et al., 2002), rice (Thomson et al., 2003), soybean (Wang et al., 2004) or wheat (Peng et al., 2003), all predominantly selfing species.

Different interpretations of the pattern are entirely possible, however. It is well known that genes are not uniformly distributed throughout the genome, but that chromosomes usually contain both gene-rich and gene-poor regions (Gill et al., 1996; Ware & Stein, 2003; Aert et al., in press) I argue that QTL for DRT are found more often than not in tight clusters simply because all genes, more often than not, are found clustered together – the pattern does not require any adaptive explanation peculiar to domestication. Peng et al. (2003), for example, note that each of their seven domestication syndrome factors (clusters of QTL for DRT) land squarely in one of these gene-rich regions of the wheat genome. Gene rich regions have also been shown to be 'hot spots' of recombination – Gill et al. (1996) found that 1 cM of genetic distance on a barley linkage map corresponds to approximately 120 kb in gene rich regions but to more than 22 Mb of DNA in areas of low gene density. While increased recombination might make linkage seem less likely in gene-dense regions, the comparatively small size of these regions means that genes within clusters could nonetheless be fairly tightly linked – genes in part of the *bz* gene cluster in maize are separated by less than 0.1 cM (Fu, Zheng & Dooner, 2001). If tight linkage were selected for during domestication, one might expect to find genes for DRT in regions of low density and low recombination. Furthermore, a recent comparison of the literature on recombination rates in domesticated plants suggests that domestication actually selects for an increase in recombination rate (Ross-Ibarra, 2004), a finding that is in good concordance with theory on the evolution of recombination (Otto & Barton, 1997, 2001). It is even conceivable that genes are clustered together for precisely the opposite reason that Pernes (1983) and others suspected – there might well be a selective advantage for genes that occur in regions of high recombination.

The argument could even be taken a step further, turning the logic of Pernes (1983) and Le Thierry D'Ennequin et al. (1999) on its head: both theory and simulation show that maladaptive gene flow should select for decreased recombination, yet revision of the empirical data available reveals that recombination has actually increased, suggesting that maladaptive gene flow was not of great impact during the domestication of most crop plants. Indeed, Poncet et al. (1998) claim that the relatively high levels of gene flow currently observed between wild and cultivated pearl millet have not adversely affected cultivation.

Direction of effects

Given the strong directional selection associated with domestication and the presumed genetic basis of morphological variation, it is not surprising to find QTL whose effect are in the direction of the domesticated trait. In fact one would expect the domesticated allele to increase seed size, fruit sweetness, quantity of seed produced, or whatever other DRT was under investigation. This is in fact what is generally found: in a review of QTL effects in domesticated taxa, Rieseberg et al. (2002) found that the vast majority of QTL for DRT are in the direction expected, suggesting a central role for directional selection in their differentiation.

Not all QTL for DRT show this trend, however. Burke et al. (2002) discovered a large number of QTL of the opposite direction expected in a mapping study of domesticated sunflower. They suggest that negative QTL could become established in domesticates via hitchhiking selecting on other linked QTL, and they interpret the existence of multiple positive QTL in the wild species as evidence consistent with the idea of multiple domestications of sunflower. Evidence from studies of other purported multiple domesticates is not entirely convincing: bean (Koinange et al., 1996), pearl millet (Poncet et al., 1998), barley (Pillen, Zacharias & Leon, 2004) and rice (Xiao et al., 1998; Xiong et al., 1999) show similar evidence of beneficial alleles in their wild progenitors, but a mapping study in peppers finds only very few of these alleles (Rao et al., 2003). Moreover, numerous studies of crops not thought to be of recurrent origin report alleles of varying direction in both the wild and domesticated parents (Doebley et al., 1990; Fulton et al., 1997; Johnson et al., 2000; Doganlar et al., 2002; Peng et al., 2003).

Unless the genetic basis of DRT is thought to have originated completely by novel mutations that arose during the process of domestication, the genetic variation present in the wild progenitor of a cultivated plant would have to include some agriculturally beneficial alleles. Given the equivocal evidence available and the improbability of successful domestication relying entirely on novel mutations, the most likely conclusion is that the pattern of cryptic allelic variation observed by Burke et al. (2002) is probably not a result of multiple domestications but instead quite possibly a common feature of domestication in general.

Conclusions

We have clearly come a long way towards a more concrete understanding of the genetic basis of domestication, and current data allow for many intriguing speculations as well. Equally clear, however, is the fact that we still have a long way to go. The patterns that we have thus far observed suggest questions that we do not yet have the data to answer, and future studies are sure to raise as many new questions as they answer old ones. Much is still lacking in the way of basic data: one has only to compare a list of the most important agricultural crops to the (much shorter) list of domesticated plants for which we have some idea of the genetic basis of quantitative DRT to get an idea of how much work is still ahead. Students of domestication should see this not as a disheartening lack of data but instead as a great opportunity to more fully understand a process that has not only been key in our own history, but key to our conceptualization of evolution as well.

Acknowledgements

I would like to acknowledge R. Mauricio and an anonymous reviewer for useful discussion and the University of Georgia and the US National Institute of Health for financial support.

References

Aert, R., L. Sági & G. Volckaert, in press. Gene content and density in banana (*Musa acuminata*) as revealed by genomic sequencing of BAC clones. Theor. Appl. Genet.

Anderson, W.W., J. Arnold, D.G. Baldwin, A.T. Beckenbach, C.J. Brown, S.H. Bryant, J.A. Coyne, L.G. Harshman & W.B. Heed, 1991. Four decades of inversion polymorphism in *Drosophila pseudoobscura*. Proc. Natl. Acad. Sci. USA 88: 10367–10371.

Barton, N.H. & P.D. Keightley, 2002. Understanding quantitative genetic variation. Nat. Rev. Genet. 3: 11–21.

Baum, M., S. Grando, G. Backes, A. Jahoor, A. Sabbagh & S. Ceccarelli, 2003. QTLs for agronomic traits in the Mediterranean environment identified in recombinant inbred lines of the cross 'Arta' × *H. spontaneum* 41-1. Theor. Appl. Genet. 107: 1215–1225.

Bres-Patry, C., M. Lorieux, G. Clement, M. Bangratz & A. Ghesquiere, 2001. Heredity and genetic mapping of domestication-related traits in a temperate japonica weedy rice. Theor. Appl. Genet. 102: 118–126.

Burke, J.M., S. Tang, S.J. Knapp & L.H. Rieseberg, 2002. Genetic analysis of sunflower domestication. Genetics 161: 1257–1267.

Cai, H.W. & H. Morishima, 2002. QTL clusters reflect character associations in wild and cultivated rice. Theor. Appl. Genet. 104: 1217–1228.

Cowie, R.H. & J.S. Jones, 1998. Gene frequency changes in *Cepaea* snails on the Marlborough Downs over 25 years. Biol. J. Linn. Soc. 65: 233–255.

Darwin, C., 1899. The Origin of Species. H.M. Caldwell Co., New York, NY., USA.

Doebley, J., A. Stec & L. Hubbard, 1997. The evolution of apical dominance in maize. Nature 386: 485–488.

Doebley, J., A. Stec, J. Wendel & M. Edwards, 1990. Genetic and morphological analysis of a maize-teosinte F-2 population implications for the origin of maize. Proc. Natl. Acad. Sci. USA 87: 9888–9892.

Doganlar, S., A. Frary, M.C. Daunay, R.N. Lester & S.D. Tanksley, 2002. Conservation of gene function in the Solanaceae as revealed by comparative mapping of domestication traits in eggplant. Genetics 161: 1713–1726.

Dudley, J.W. and R.J. Lambert, 1992. Ninety generations of selection for oil and protein in maize. Maydica 37:1–7

Eyre-Walker, A., R.L. Gaut, H. Hilton, D.L. Feldman & B.S. Gaut, 1998. Investigation of the bottleneck leading to the domestication of maize. Proc. Natl. Acad. Sci. USA 95: 4441–4446.

Frary, A. & S. Doganlar, 2003. Comparative genetics of crop plant domestication and evolution. Turkish J. Agr. For. 27: 59–69.

Frary, A., S. Doganlar, M.C. Daunay & S.D. Tanksley, 2003. QTL analysis of morphological traits in eggplant and implications for conservation of gene function during evolution of solanaceous species. Theor. Appl. Genet. 107: 359–370.

Frary, A., T.C. Nesbitt, A. Frary, S. Grandillo, E. van der Knaap, B. Cong, J. Liu, J. Meller, R. Elber, K.B. Alpert & S.D. Tanksley, 2000. fw2.2: A quantitative trait locus key to the evolution of tomato fruit size. Science 289: 85–88.

Fu, H., Z. Zheng & H.K. Dooner, 2001. Recombination rates between adjacent genic and retrotransposon regions in maize vary by 2 orders of magnitude. Proc. Natl. Acad. Sci. USA 99: 1082–1087.

Fulton, T.M., B.T. Beck, D. Emmatty, Y. Eshed, J. Lopez, V. Petiard, J. Uhlig, D. Zamir & S.D. Tanksley, 1997. QTL analysis of an advanced backcross of *Lycopersicon peruvianum* to the cultivated tomato and comparisons with QTLs found in other wild species. Theor. Appl. Genet. 95: 881–894.

Gill, K.S., B.S. Gill, T.R. Endo & E.V. Boyko, 1996. Identification and high-density mapping of gene-rich regions in chromosome group 5 of wheat. Genetics 143: 1001–1012.

Gottlieb, L.D., 1984. Genetics and morphological evolution in plants. Am. Nat. 123: 681–709.

Grant, P.R. & B.R. Grant, 2002. Unpredictable evolution in a 30-year study of Darwin's finches. Science 296: 707–711.

Grant, V., 1981. Plant Speciation. Columbia University Press, New York, NY., USA.

Harlan, J.R., 1992. Crops and Man. American Society of Agronomy: Crop Science Society of America, Madison WI, USA.

Hashizume, T., I. Shimamoto & M. Hirai, 2003. Construction of a linkage map and QTL analysis of horticultural traits for watermelon [*Citrullus lanatus* (Thunb.) Mastsum and Nakai] using RAPD, RFLP and ISSR markers. Theor. Appl. Genet. 106: 779–785.

Hillman, G.C. & M.S. Davies, 1990. Domestication rates in wild-type wheats and barley under primitive cultivation. Biol. J. Linn. Soc. 39: 39–78.

Hilu, K.W., 1983. The role of single-gene mutations in the evolution of flowering plants. Evol. Biol. 16: 97–128.

Huang, X.Q., H. Coester, M.W. Ganal & M.S. Roeder, 2003. Advanced backcross QTL analysis for the identification of quantitative trait loci alleles from wild relatives of wheat (*Triticum aestivum* L.). Theor. Appl. Genet. 106: 1379–1389.

Johnson, W.C., L.E. Jackson, O. Ochoa, R. van Wijk, J. Peleman, D.A. St. Clair & R.W. Michelmore, 2000. Lettuce, a shallow-rooted crop, and *Lactuca serriola*, its wild progenitor, differ at QTL determining root architecture and deep soil water exploitation. Theor. Appl. Genet. 101: 1066–1073.

Koinange, E.M.K., S.P. Singh & P. Gepts, 1996. Genetic control of the domestication syndrome in common bean. Crop Sci. 36: 1037–1045.

Ladizinsky, G., 1985. Founder effect in crop-plant evolution. Econ. Bot. 39: 191–199.

Lande, R., 1983. The response to selection on major and minor mutations affecting a metrical trait. Heredity 50: 47–65.

Le Thierry D'Ennequin, M., B. Toupance, T. Robert, B. Godelle & P.H. Gouyon, 1999. Plant domestication: a model for studying the selection of linkage. J. Evol. Biol. 12: 1138–1147.

Lenormand, T. & S.P. Otto, 2000. The evolution of recombination in a heterogeneous environment. Genetics 156: 423–428.

Lester, R.N., 1989. Evolution under domestication involving disturbance of genic balance. Euphytica 44: 125–132.

Mauricio, R., 2001. Mapping quantitative trait loci in plants: uses and caveats for evolutionary biology. Nat. Rev. Genet. 2: 370–381.

Orr, H.A., 1998a. The population genetics of adaptation: the distribution of factors fixed during adaptive evolution. Evolution 52: 935–949.

Orr, H. A., 1998b. Testing natural selection vs. genetic drift in phenotypic evolution using quantitative trait locus data. Genetics 149: 2099–2104.

Orr, H.A., 2003. The distribution of fitness effects among beneficial mutations. Genetics 163: 1519–1526.

Orr, H.A. & A.J. Betancourt, 2001. Haldane's sieve and adaptation from the standing genetic variation. Genetics 157: 875–884.

Orr, H.A. & J.A. Coyne, 1992. The genetics of adaptation: a reassessment. Am. Nat. 140: 725–742.

Otto, S.P. & N.H. Barton, 1997. The evolution of recombination: removing the limits to natural selection. Genetics 147: 879–906.

Otto, S.P. & N.H. Barton, 2001. Selection for recombination in small populations. Evolution 55: 1921–1931.

Paterson, A.H., 2002. What has QTL mapping taught us about plant domestication? New Phytol. 154: 591–608.

Paterson, A.H., S. Damon, J.D. Hewitt, D. Zamir, H.D. Rabinowitch, S.E. Lincoln, E.S. Lander & S.D. Tanksley, 1991. Mendelian factors underlying quantitative traits in tomato – comparison across species, generations, and environments. Genetics 127: 181–197.

Paterson, A.H., Y.R. Lin, Z. Li, K.F. Schertz, J.F. Doebley, S.R.M. Pinson, S.C. Liu, J.W. Stansel & J.E. Irvine, 1995. Convergent domestication of cereal crops by independent mutations at corresponding genetic loci. Science 269: 1714–1718.

Peng, J., Y. Ronin, T. Fahima, M.S. Roder, Y. Li, E. Nevo & A. Korol, 2003. Domestication quantitative trait loci in *Triticum dicoccoides*, the progenitor of wheat. Proc. Natl. Acad. Sci. USA 100: 2489–2494.

Pernes, J., 1983. The genetics of the domestication of cereals. Recherche 14: 910–919.

Pillen, K., A. Zacharias & J. Léon, 2004. Comparative AB-QTL analysis in barley using a single exotic donor of *Hordeum vulgare* ssp. *spontaneum*. Theor. Appl. Genet. 108: 1591–1601.

Poncet, V., F. Lamy, K.M. Devos, M.D. Gale, A. Sarr & T. Robert, 2000. Genetic control of domestication traits in pearl millet (*Pennisetum glaucum* L., Poaceae). Theor. Appl. Genet. 100: 147–159.

Poncet, V., F. Lamy, J. Enjalbert, H. Joly, A. Sarr & T. Robert, 1998. Genetic analysis of the domestication syndrome in pearl millet (*Pennisetum glaucum* L., Poaceae): Inheritance of the major characters. Heredity 81: 648–658.

Rao, G.U., C.A. Ben, Y. Borovsky & I. Paran, 2003. Mapping of yield-related QTLs in pepper in an interspecific cross of *Capsicum annuum* and *C. frutescens*. Theor. Appl. Genet. 106: 1457–1466.

Rieseberg, L.H., A. Widmer, A.M. Arntz & J.M. Burke, 2002. Directional selection is the primary cause of phenotypic diversification. Proc. Natl. Acad. Sci. USA 99: 12242–12245.

Ross-Ibarra, J., 2004. The evolution of recombination under domestication: a test of two hypotheses. Am. Nat. 163: 105–112.

Takeuchi, Y., S.Y. Lin, T. Sasaki & M.Yano, 2003. Fine linkage mapping enables dissection of closely linked quantitative trait loci for seed dormancy and heading in rice. Theor. Appl. Genet. 107: 1174–1180.

Thomson, M.J., T.H. Tai, A.M. McClung, X-H. Lai, M.E. Hinga, K.B. Lobos, Y. Xu, C.P. Martinez & S.R. McCouch, 2003. Mapping quantitative trait loci for yield, yield components and morphological traits in an advanced backcross population between *Oryza rufipogon* and the *Oryza sativa* cultivar Jefferson. Theor. Appl. Genet. 107: 479–493.

Varona, L., L. Gomez-Raya, W.M. Rauw, A. Clop, C. Ovilo & J.L. Noguera, 2004. Derivation of a Bayes factor to distinguish between linked or pleiotropic quantitative trait loci. Genetics 166: 1025–1035.

Vavilov, N., 1922. The law of homologous series in variation. J. Genet. 12: 47–89.

Visser, M.E., A.J. van Noordwijk, J.M. Tinbergen & C.M. Lessells, 1998. Warmer springs lead to mistimed reproduction in great tits (*Parus major*). Proc. R. Soc. Lond., Ser. B: Biol. Sci. 265: 1867–1870.

Wang, R.L., A. Stec, J. Hey, L. Lukens & J. Doebley, 1999. The limits of selection during maize domestication. Nature 398: 236–239.

Wang, D., G.L. Graef, A.M. Procopiuk, & B.W. Diers, 2004. Identification of putative QTL that underlie yield in interspecific soybean backcross populations. Theor. Appl. Genet. 108: 458–467.

Ware, D & L. Stein, 2003. Comparison of genes among cereals. Curr. Opin. Plant Biol. 6: 121–127.

Westerbergh, A. & J. Doebley, 2002. Morphological traits defining species differences in wild relatives of maize are controlled by multiple quantitative trait loci. Evolution 56: 273–283.

Xiao, J.H., J.M. Li, S. Grandillo, S.N. Ahn, L.P. Yuan, S.D. Tanksley & S.R. McCouch, 1998. Identification of trait-improving quantitative trait loci alleles from a wild rice relative, *Oryza rufipogon*. Genetics 150: 899–909.

Xiong, L.X., K.D. Liu, X.K. Dai, C.G. Xu & Q.F. Zhang, 1999. Identification of genetic factors controlling domestication-related traits of rice using an F-2 population of a cross between *Oryza sativa* and *O. rufipogon*. Theor. Appl. Genet. 98: 243–251.

Genetica (2005) 123: 205–209 © Springer 2005

Can ecology help genomics: the genome as ecosystem?

Rodney Mauricio
Department of Genetics, Davison Life sciences complex, University of Georgia, Athens, GA 30602-7223, USA (Phone: (706)-542-1417; Fax: (706)-542-3910; E-mail: mauricio@uga.edu)

Received 30 January 2002 Accepted 1 January 2004

Key words: autoecology, ecology, emergent properties, epigenetics, epistasis, genomics, microarrays, natural history, systems biology

Abstract

Ecologists study the rules that govern processes influencing the distribution and abundance of organisms, particularly with respect to the interactions of organisms with their biotic and abiotic environments. Over the past decades, using a combination of sophisticated mathematical models and rigorous experiments, ecologists have made considerable progress in understanding the complex web of interactions that constitute an ecosystem. The field of genomics runs on a path parallel to ecology. Like ecology, genomicists seek to understand how each gene in the genome interacts with every other gene and how each gene interacts with multiple, environmental factors. Gene networks connect genes as complex as the 'webs' that connect the species in an ecosystem. In fact, genes exist in an ecosystem we call the genome. The genome as ecosystem is more than a metaphor – it serves as the conceptual foundation for an interdisciplinary approach to the study of complex systems characteristic of both genomics and ecology. Through the infusion of genomics into ecology and ecology into genomics both fields will gain fresh insight into the outstanding major questions of their disciplines.

Introduction

Genomics has been described as the ultimate integrative discipline, crossing the full spectrum of the biological sciences. Without doubt, genomics is a multidisciplinary pursuit, combining primarily molecular biology and computer science. The genomics era has also brought a renewed interest in systems biology, conceptually a broader multidisciplinary endeavor, and said to bring together biology, chemistry, computer science, engineering, mathematics, and physics (Ideker et al., 2001; Kitano, 2002; Hood & Galas, 2003). Absent in these lists of the 21st century's new biology is a mention of the field of ecology, the scientific study of the processes influencing the distribution and abundance of organisms, particularly with respect to the interactions of organisms with their biotic and abiotic environments.

This absence is surprising – surprising because both ecologists and genomicists ask similar questions, their respective disciplines have developed along similar intellectual trajectories and share basic epistemological approaches. In many ways, the genome and the ecosystem are parallel constructs and can be studied using similar approaches. The thesis of this paper is that including the field of ecology as part of the study of genomics will lead to advances in both disciplines.

A metaphor

Imagine the Serengeti plain of east Africa: grasses, shrubs, and trees extend over the landscape; giraffe, elephants, and antelope graze over the grasslands; lions, leopards, and hyena hunt and scavenge; vultures, flies, and fungi linger over carrion. Over the past millennium, natural historians have discovered and described these, and many other, individual species of plants, animals and microbes. Ecologists stepped in over a century ago to study what an individual species *does* in its environment, its 'autoecology'. In other words, we now know how a giraffe manages to live in the Serengeti. In the past century, through a combination of manipulative experiments and mathematical theory, ecologists have made great strides in understanding interactions between individual species (e.g., Wilbur, 1987; Morin, 1999). As a result, to a large degree, we now know how giraffes interact with trees, with other giraffes, with other herbivores, with predators, and even with dung beetles (Jankielsohn et al., 2000): a fairly complex network of interactions.

However, the challenge of ecology is not to understand only the giraffe's role in the Serengeti ecosystem: a complete ecological understanding of the Serengeti would require that we understand the rules regulating how *each* and *every* species in the ecosystem, from bacteria to lions, interacts with every other species and how each species interacts with multiple environmental factors. Needless to say, this is a complicated problem. It is made more complicated by the fact that complex systems are rarely the sum of their parts: emergent properties lead to nonlinearities. Considering the complexity of the problem, ecologists have made astonishing inroads into understanding the natural world, although some remain skeptical (e.g., O'Connor, 2000). Keep the metaphor of the giraffe in the Serengeti in mind as we consider how examination of another 'species' – the gene in its genomic ecosystem – may further accelerate breakthroughs in ecology and genomics.

The metaphor extended: the genome as ecosystem

Although the pace of intellectual development has been much more rapid in genomics, the parallels to the development of ecology are unmistakable. Like those legions of systematists identifying the individual species in the ecosystem, geneticists made a cottage industry of identifying single genes until the advent of whole-genome sequencing (and bench geneticists continue to make remarkable progress in carefully reconciling predicted genes with actual ones). In many ways, genomicists reintroduced natural history to biology, albeit a molecular natural history, eschewing hypothesis-driven research and proclaiming a new phase of 'discovery-based' inquiry (Ideker et al., 2001) with the argument that the field needed to accumulate the basic information upon which hypotheses could later be based.

Like ecologists in the Serengeti, the mainstay of many modern molecular geneticists is attempting to understand the function, the autoecology, of each gene. For many pathways, we know how genes interact with other genes, like we know how giraffes interact with other giraffes or other animals. Molecular geneticists have long understood how genes interact with the environment. Genes live in an ecosystem like animals live in their ecosystem, and although the tools used to study genes and giraffes are clearly different, the broad intellectual approaches to understanding genes and giraffes are not so different.

However, like ecology, the ultimate challenge of genomics is to understand how each gene in the genome interacts with every other gene (epistasis) and how each gene interacts with multiple, environmental factors. Gene networks are just as complex as the 'web' that connects all the species in an ecosystem (Tong et al., 2004). Again, understanding that degree of complexity is a complicated, multidimensional problem. What emergent properties will arise from the complexities of the genome? Will understanding the function of every gene ever allow us to predict complex phenotypes? How pervasive are epigenetic effects (e.g., Waddington, 1942)?

If we see the genome as an ecosystem where genes live, how much more progress will genomicists make in understanding that ecosystem than ecologists have made in understanding their ecosystems? Regardless of the answer to that question, ecology and genomics do have enough to offer one another that the two disciplines may reach their common goal with a healthy interchange of ideas.

What can ecology and genomics offer each other?

Certainly molecular geneticists have offered ecologists a myriad of tools to understand ecology and in many ways those tools have revolutionized ecology. However, what does ecology offer genomics? The most important thing ecology can offer genomics is experience in simply thinking about, and being trained in thinking about, complex interactions. Most often, this training is manifested in being able to design experiments that test for complex interactions with both the environment and other individuals or species (Hairston, 1989; Resetarits & Bernardo, 1998).

For example, both geneticists and ecologists use manipulative 'field' experiments. Molecular geneticists use knockout experiments (experimentally excluding genes from a pathway with, for example, targeted mutagenesis or RNAi) to understand how genes interact within the genome and ecologists often experimentally exclude a species from an ecosystem (e.g., with a fence or pesticide) in order to understand the role of that species in the ecosystem. Since ecologists often manipulate multiple species in a factorial fashion, statistical and experimental approaches have been developed that allow for the analysis and interpretation of these data. Most molecular geneticists have tested single mutant, double mutants, and even triple mutants, but it gets exceedingly difficult to examine the factorial effects of every possible combination of four or more independent mutations. Genomics allows the investigator the opportunity to examine the global effects of mutants, but the statistical interpretation of such experiments often clouds the results. The ecologists' experience in designing experiments with an eye towards managing complexity will be directly applicable to the analysis of complex genomic datasets.

For example, many microarray experiments suffer from simple but significant flaws in design that make the data difficult to interpret (Tilstone, 2003). Technical problems arise that could be addressed simply by borrowing concepts from ecology. For example, the slides used for microarrays can sag, causing an attenuation of signal for those spots in the middle. Engineers have worked to improve the physical properties of the slides and computer scientists have worked to account for the signal attenuation. However, ecologists must always account for heterogeneity in their field sites and use a variety of experimental techniques to do so (Cochran & Cox, 1992; Scheiner & Gurevitch, 2001). The simplest field technique, 'spatial blocking,' is easily applied to a microarray (although at a cost of throughput). Rather than apply 10,000 unique spots on a chip, one could spot four replicates of each oligonucleotide or mRNA in distinct spatial blocks on a chip. A simple analysis of variance could account for the variation due to physical heterogeneity on the slide, whatever the underlying cause.

Beyond providing guidance in experimental design, ecologists can contribute a nuanced approach to studying the interactions of genes with the environment that goes beyond simple microarray gene expression studies done in a few different environments. For example, an investigation of mutant phenotypes performed under realistic ecological conditions could be valuable in shedding light on the 'genetic uncertainty principle' where a reverse genetics approach has not yielded an informative mutant phenotype (Tautz, 2000). The failure of a gene knockout to produce a visible phenotype could be due to genetic redundancy, but it could also be masked by the permissive environments in which most mutants are screened (Gilliland et al., 1998; Meagher et al., 2000).

In addition to being an experimental science, ecology is also a highly mathematical discipline. While some cell and molecular biologists have employed complex mathematics in their work, there remains an enormous potential in the synergy between the kind of datasets genomicists generate and the mathematical approaches that ecologists have refined over the last century. Very simple mathematical models were derived early in the history of ecology to predict population growth (logistic equation) and to study interactions among species (Lotka-Volterra equation). Today, ecology has developed a firm mathematical foundation (May 1976; Dieckmann et al., 2000; May, 2001; Okubo & Levin, 2001; Cushing et al., 2002). Mathematics is an essential tool to understanding complex systems. Models are used to generate hypotheses that can be experimentally tested. For example, a model of a complex network can be generated, along with a predicted response to a perturbation. Perturbation experiments can be performed and the observed results compared with the model. Mathematics will be essential to guide the course of experimentation in genomics as the complexity of systems increases. When applied to genomics, these models will focus in detail on the specific molecular mechanisms of individual genes and proteins and their interactions. Further models could explicitly incorporate deterministic environmental parameters as well as environmental stochasticity.

This approach has been recently advocated by systems biologists who favor an applied mathematics and computational approach to biology (Hood & Galas, 2003). Further evidence of the common path taken by ecology and genomics lies in the recent establishment of systems biology as an intellectual discipline. Systems biology has an antecedent in systems ecology. Systems ecology is a branch of ecology that attempts to understand the structure and function of ecosystems by concentrating on energy inputs and outputs of the system (Odum, 1983; Patten & Jørgensen, 1995). Systems ecology was developed partly as a way to confront the complexity of systems. The system itself is a black box and the approach trades off the ability to understand the details of the components of the system for understanding the system as a whole. Whether systems biologists embrace a deep systems approach or if they simply apply mathematics to molecular biology at a global scale (Ideker et al., 2001), the path of modern biology will be paved with mathematics; and ecologists have been strolling that way for decades (May, 1976).

Ecologists clearly have something to offer to genomics, but genomics will continue to be critical to advances in ecology. Certainly, techniques created for genomics have found application in ecology. Craig Venter's attempt to use sequencing to identify every microbe in the Sargasso Sea is an example of the power of genomics to identify all the players in a complex ecosystem. And ecologists have started using some of the tools of genomics in their own work (Jackson et al., 2002). Nevertheless, genomics could have an even more profound intellectual contribution to ecology. As physics infused ecology in the 1970s, a focused interest on the ecology of the genome may give great insight into biological systems at higher levels of organization. For example, perhaps gene networks are, at some level, fundamentally different from food webs. The present research interest in genetic networks could have substantial application to ecologists' work on species interactions (e.g., Barkai & Leibler 1997; Bergman & Siegal 2003). Genetic systems, like ecological systems, seem to be more stable the more connected they are. Although this result makes some intuitive sense in a genetic system, it is unclear why it seems to be the case in ecological systems. For many questions, modeling the genome as an ecosystem will have direct applications to understanding any complex system, including ecosystems.

Final thoughts

In this paper, I have attempted to outline some of the common approaches that genomics and ecology have taken to addressing the outstanding questions in their disciplines. I see unmistakable similarities in these two seemingly disparate fields. It strikes me that both ecology and genomics have much to offer each other. And since genomics is still in many ways establishing its paradigms, now seems the appropriate time for each field to take full advantage of the others' strengths. Will the infusion of ecological ideas into genomics help to make more sense of genomes than we presently have of ecosystems? Will a new synthesis of ecology and genomics lead us into this new century of biology? I do not know. But if I were a beginning graduate student in genetics, I would look at the course offerings in math. If I were a beginning ecology or math graduate student, I would look over at what the geneticists were doing. And if I were hiring systems biologists, I would take a careful look at ecologists.

Acknowledgments

I greatly appreciate my friend (and molecular geneticist) Greg Copenhaver's indulgence over the years: his comments on earlier versions of the manuscript were welcome. He has always been a welcome sounding board. Amy Bouck also made thoughtful comments on an earlier draft of this manuscript.

References

Barkai, N. & S. Leibler, 1997. Robustness in simple biochemical networks. Nature 387: 913–917.

Bergman, A. & M.L. Siegal, 2003. Evolutionary capacitance as a general feature of complex gene networks. Nature 424: 549–552.

Cochran, W.G. & G.M. Cox, 1992. Experimental Designs. Wiley & Sons, New York, 611 pp.

Cushing, J.M.R. Costantino, B. Dennis, R. Desharnais, & S. Henson (eds.), 2002. Chaos in Ecology: Experimental Nonlinear Dynamics. Academic Press, New York. 225 pp.

Dieckmann, U., R. Law & J.A.J. Metz (eds.), 2000. The Geometry of Ecological Interactions: Simplifying Spatial

Complexity. Cambridge University Press, Cambridge. 580 pp.

Gilliland, L.U., E.C. McKinney, M.A. Asmussen & R.B. Meagher, 1998. Detection of deleterious genotypes in multigenerational studies. I. disruptions in individual *Arabidopsis* actin genes. Genetics 149: 717–725.

Hairston, N.G., Sr., 1989. Ecological Experiments: Purpose, Design and Execution. Cambridge University Press, Cambridge. 384 pp.

Hood, L. & D. Galas, 2003. The digital code of DNA. Nature 421: 444–448.

Ideker, T., T. Galitski & L. Hood, 2001. A new approach to decoding life: Systems biology. Annu. Rev. Genomics Hum. Genet. 2: 343–372.

Jackson, R.B., C.R. Linder, M. Lynch, M. Purugganan, S. Somerville & S.S. Thayer, 2002. Linking molecular insight and ecological research. Trends Ecol. Evol. 17: 409–414.

Jankielsohn, A., C.H. Scholtz & S.V. Louw. 2000. Effect of habitat transformation on dung beetle assemblages: a comparison between a South African nature reserve and neighboring farms. Environ. Entomol. 30: 474–483.

Kitano, H., 2002. Systems biology: a brief overview. Science 295: 1662–1664.

May, R.M., 1976. Theoretical Ecology: Principles and Applications. W. B. Saunders, New York. 317 pp.

May, R.M., 2001. Stability and Complexity in Model Ecosystems. Princeton University Press, Princeton, New Jersey. 292 pp.

Meagher, S., D.J. Penn & W.K. Potts, 2000. Male–male competition magnifies inbreeding depression in wild house mice. Proc. Natl. Acad. Sci. USA 97: 3324–3329.

Morin, P.J., 1999. Community Ecology. Blackwell Science, Malden, MA. 424 pp.

O'Connor, R.J., 2000. Why ecology lags behind biology. The Scientist 14(20): 35.

Odum, H.T., 1983. Systems Ecology: An Introduction. Wiley and Sons, New York, 644 pp.

Okubo, A. & S.A. Levin (eds.), 2001. Diffusion and Ecological Problems. Springer Verlag. New York. 488 pp.

Patten, B.C. & S.E. Jørgensen (eds.), 1995. Complex Ecology. The Part-Whole Relation in Ecosystems. Prentice Hall, Englewood Cliffs, New Jersey. 705 pp.

Resetarits, W.J. & J. Bernardo (eds.), 1998. Experimental Ecology: Issues and Perspectives. Oxford University Press, New York. 488 pp.

Scheiner, S.M. & J. Gurevitch (eds.), 2001. Design and Analysis of Ecological Experiments. Oxford University Press, New York. 415 pp.

Tautz, D., 2000. A genetic uncertainty problem. Trends Genet. 16: 475–477.

Tilstone, C., 2003. Vital statistics. Nature 424: 610–612.

Tong, A. H. Y. et al., 2004. Global mapping of the yeast genetic interaction network. Science 303: 808–813.

Waddington, C.H., 1942. The epigenotype. Endeavour 1: 18–20.

Wilbur, H.M., 1987. Regulation of structure in complex systems: experimental temporary pond communities. Ecology 68:1437–1452.